Lecture Notes in Computer Science 9406

Commenced Publication in 1973
Founding and Former Series Editors:
Gerhard Goos, Juris Hartmanis, and Jan van Leeuwen

More information about this series at http://www.springer.com/series/7410

MHR Khouzani · Emmanouil Panaousis
George Theodorakopoulos (Eds.)

Decision and
Game Theory for Security

6th International Conference, GameSec 2015
London, UK, November 4–5, 2015
Proceedings

Springer

Editors
MHR Khouzani
Queen Mary University of London
London
UK

George Theodorakopoulos
Cardiff University
Cardiff
UK

Emmanouil Panaousis
University of Brighton
Brighton
UK

ISSN 0302-9743   ISSN 1611-3349   (electronic)
Lecture Notes in Computer Science
ISBN 978-3-319-25593-4   ISBN 978-3-319-25594-1   (eBook)
DOI 10.1007/978-3-319-25594-1

Library of Congress Control Number: 2015951801

LNCS Sublibrary: SL4 – Security and Cryptology

Springer Cham Heidelberg New York Dordrecht London

Printed on acid-free paper

Springer International Publishing AG Switzerland is part of Springer Science+Business Media
(www.springer.com)

Preface

Computers and IT infrastructure play ever-increasing roles in our daily lives. The technological trend toward higher computational power and ubiquitous connectivity can also give rise to new risks and threats. To ensure economic growth and prosperity, nations, corporations, and individuals constantly need to reason about how to protect their sensitive assets.

Security is hard: it is a multifaceted problem that requires a careful appreciation of many complexities regarding the underlying computation and communication technologies and their interaction and interdependencies with other infrastructure and services. Besides these technical aspects, security provision also intrinsically depends on human behavior, economic concerns, and social factors. Indeed, the systems whose security is concerned are typically heterogeneous, large-scale, complex, dynamic, interactive, and decentralized in nature.

Game and decision theory has emerged as a valuable systematic framework with powerful analytical tools in dealing with the intricacies involved in making sound and sensible security decisions. For instance, game theory provides methodical approaches to account for interdependencies of security decisions, the role of hidden and asymmetric information, the perception of risks and costs in human behavior, the incentives/limitations of the attackers, and much more. Combined with our classic approach to computer and network security, and drawing from various fields such as economic, social, and behavioral sciences, game and decision theory is playing a fundamental role in the development of the pillars of the "science of security."

Since its inception in 2010, GameSec has annually attracted original research in both theoretical and practical aspects of decision making for security and privacy. The past editions of the conference took place in Berlin (2010), College Park (2011), Budapest (2012), FortWorth (2013), and Los Angeles (2014). This year (2015), it was hosted for the first time in the UK, in the heart of London.

We received 37 submissions this year from which, 16 full-length and five short papers we selected after a thorough review process by an international panel of scholars and researchers in this field. Each paper typically received three reviews assessing the relevance, novelty, original contribution, and technical soundness of the paper. The topics of accepted papers include applications of game theory in network security, economics of cybersecurity investment and risk management, learning and behavioral models for security and privacy, algorithm design for efficient computation, and investigation of trust and uncertainty, among others.

We would like to thank Springer for its continued support of the GameSec conference and for publishing the proceedings as part of their *Lecture Notes in Computer*

Series (LNCS) with special thanks to Anna Kramer. We anticipate that researchers in the area of decision making for cybersecurity and the larger community of computer and network security will benefit from this edition.

November 2015

MHR Khouzani
Emmanouil Panaousis
George Theodorakopoulos

Organization

Steering Board

Tansu Alpcan   The University of Melbourne, Australia
Nick Bambos   Stanford University, USA
John S. Baras   University of Maryland, USA
Tamer Başar   University of Illinois at Urbana-Champaign, USA
Anthony Ephremides   University of Maryland, USA
Jean-Pierre Hubaux   EPFL, Switzerland
Milind Tambe   University of Southern California, USA

2015 Organizers

General Chair

Emmanouil Panaousis   University of Brighton, UK

TPC Chair

George Theodorakopoulos   Cardiff University, UK

Publication Chair

MHR Khouzani   Queen Mary University of London, UK

Local Arrangements

Andrew Fielder   Imperial College London, UK

Publicity Chairs

Europe

Mauro Conti   University of Padua, Italy

USA

Aron Laszka   University of California Berkeley, USA

Asia-Pacific

Benjamin Rubinstein   University of Melbourne, Australia

Web Chair

Johannes Pohl   University of Applied Sciences Stralsund, Germany

Technical Program Committee

TPC Chair

George Theodorakopoulos   Cardiff University, UK

TPC Members

Habtamu Abie	Norsk Regnesentral - Norwegian Computing Center, Norway
Ross Anderson	University of Cambridge, UK
John Baras	University of Maryland, USA
Alvaro Cardenas	University of Texas at Dallas, USA
Carlos Cid	Royal Holloway, University of London, UK
Andrew Fielder	Imperial College London, UK
Julien Freudiger	Apple Inc., USA
Jens Grossklags	Penn State University, USA
Murat Kantarcioglu	University of Texas at Dallas, USA
MHR Khouzani	Queen Mary University of London, UK
Aron Laszka	University of California, Berkeley, USA
Yee Wei Law	University of South Australia, Australia
Xinxin Liu	University of Florida, USA
Pasquale Malacaria	Queen Mary University of London, UK
Mohammad Hossein Manshaei	Isfahan University of Technology, Iran
John Musacchio	University of California, Santa Cruz, USA
Mehrdad Nojoumian	Florida Atlantic University, USA
Andrew Odlyzko	University of Minnesota, USA
Emmanouil Panaousis	University of Brighton, UK
Johannes Pohl	University of Applied Sciences Stralsund, Germany
David Pym	University College London, UK
Reza Shokri	University Texas at Austin, USA
Carmela Troncoso	Gradiant, Spain
Athanasios Vasilakos	NTUA, Greece
Yevgeniy Vorobeychik	Vanderbilt University, USA
Nan Zhang	The George Washington University, USA
Quanyan Zhu	New York University, USA
Jun Zhuang	SUNY Buffalo, USA

Contents

Short Papers

Full Papers

A Game-Theoretic Approach to IP Address Randomization in Decoy-Based Cyber Defense

Andrew Clark[1](✉), Kun Sun[2], Linda Bushnell[3], and Radha Poovendran[3]

[1] Department of Electrical and Computer Engineering,
Worcester Polytechnic Institute, Worcester, MA 01609, USA
aclark@wpi.edu
[2] Department of Computer Science, College of William and Mary,
Williamsburg, VA 23187, USA
ksun@wm.edu
[3] Network Security Lab, Department of Electrical Engineering,
University of Washington, Seattle, WA 98195, USA
{lb2,rp3}@uw.edu

Abstract. Networks of decoy nodes protect cyber systems by distracting and misleading adversaries. Decoy defenses can be further enhanced by randomizing the space of node IP addresses, thus preventing an adversary from identifying and blacklisting decoy nodes over time. The decoy-based defense results in a time-varying interaction between the adversary, who attempts to identify and target real nodes, and the system, which deploys decoys and randomizes the address space in order to protect the identity of the real node. In this paper, we present a game-theoretic framework for modeling the strategic interaction between an external adversary and a network of decoy nodes. Our framework consists of two components. First, we model and study the interaction between the adversary and a single decoy node. We analyze the case where the adversary attempts to identify decoy nodes by examining the timing of node responses, as well as the case where the adversary identifies decoys via differences in protocol implementations between decoy and real nodes. Second, we formulate games with an adversary who attempts to find a real node in a network consisting of real and decoy nodes, where the time to detect whether a node is real or a decoy is derived from the equilibria of the games in first component. We derive the optimal policy of the system to randomize the IP address space in order to avoid detection of the real node, and prove that there is a unique threshold-based Stackelberg equilibrium for the game. Through simulation study, we find that the game between a single decoy and an adversary mounting timing-based attacks has a pure-strategy Nash equilibrium, while identification of decoy nodes via protocol implementation admits only mixed-strategy equilibria.

1 Introduction

Cyber systems are increasingly targeted by sophisticated attacks, which monitor the system over a period of time, identify vulnerabilities, and mount efficient and

This work was supported by ARO grant W911NF-12-1-0448.

MHR Khouzani et al. (Eds.): GameSec 2015, LNCS 9406, pp. 3–21, 2015.
DOI: 10.1007/978-3-319-25594-1_1

effective attacks that are tailored to those vulnerabilities. An emerging approach
to thwarting such attacks is through a *moving target defense*, which proactively
varies the system protocol, operating system, and software configurations over
time, thus rendering vulnerabilities observed by the adversary obsolete before
the attack takes place.

One class of moving target defense consists of networks of virtual nodes,
which are created and managed by the system and include both real nodes that
implement services such as web servers and databases, as well as decoy nodes
whose only purpose is to mislead the adversary [18]. If the real and decoy nodes
have valid IP addresses that are visible to an external adversary, then the adversary may mount attacks on decoy nodes instead of the real node, wasting the
resources of the adversary and providing information to the system regarding the
goals and capabilities of the adversary. In order to maximize the probability that
the adversary interacts with a decoy node instead of a real node, the decoy nodes
should outnumber the real nodes in the network. When the number of decoys
is large, however, the amount of memory and CPU time that can be allocated
to each decoy is constrained, thus limiting the performance and functionality of
each decoy.

While limiting the functionality of decoy nodes reduces their memory and
processing cost, it also enables the adversary to detect decoys by observing deviations of the timing and content of node responses from their expected values [16].
Once a decoy node has been detected, its IP address is added to the adversary's
blacklist and the decoy is not contacted again by the adversary. By querying
and blacklisting decoy nodes over a period of time, the adversary can eventually
eliminate all decoys from consideration and mount attacks on the real node. The
time required to blacklist the decoy nodes depends on the amount of time needed
to identify a node as real or a decoy, which is a function of the resources given
to each decoy.

The effectiveness of decoy-based defenses can be further improved by periodically randomizing the IP address space [3]. IP randomization renders any
blacklist obsolete, effectively forcing the adversary to re-scan all network nodes.
This randomization, however, will also terminate higher-layer protocols such
as TCP on the real nodes, which depend on a stable IP address and must be
reestablished at a cost of extra latency to valid users [1]. Randomization of the IP
address space should therefore be performed based on a trade-off between the
performance degradation of valid users and the security benefit of mitigating
attacks.

The security benefit of IP randomization and decoy-based defenses depends
on the behavior of the adversary. The ability of the decoy nodes to mislead the
adversary is determined by the adversary's strategy for detecting decoy nodes.
Similarly, frequent IP randomization increases the latency of real users and hence
is only warranted when the adversary scans a large number of nodes. Modeling
and design of address randomization in decoy-based defenses should therefore
incorporate the strategic interaction between an intelligent adversary and the
system defense. Currently, however, no such analytical approach exists.

In this paper, we present a game-theoretic framework for modeling and design of decoy-based moving target defenses with IP randomization. Our modeling framework has two components, namely, the interaction between a single virtual node (real or decoy) and an adversary attempting to determine whether the node is real or a decoy, as well as the interaction between an adversary and a network of virtual nodes. These two components are interrelated, since the equilibria of the interaction games between a single virtual node and an adversary determine the time required for an adversary to detect a decoy node, and hence the rate at which an adversary can scan the network and identify real nodes. We make the following specific contributions:

- We develop game-theoretic models for two mechanisms used by adversaries to detect decoy nodes. In the timing-based mechanism, the adversary exploits the increased response times of resource-limited decoy nodes to detect decoys. In the fingerprinting-based mechanism, the adversary initiates a communication protocol with a node and, based on the responses, determines whether the node has fully implemented the protocol, or is a decoy with a partial implementation of the protocol.
- In the case of timing-based detection of a single decoy, we formulate a two-player game between an adversary who chooses the number of probe messages to send and a system that chooses the response time of the decoy subject to resource constraints. The utility of the system is equal to the total time spent by the adversary to query the network. We develop an efficient iterative procedure that converges to a mixed-strategy Nash equilibrium of the game.
- We present a game-theoretic model of decoy detection via protocol fingerprinting, in which we introduce protocol finite state machines as a modeling methodology for decoy detection. Under our approach, the system decides which states to implement, while the adversary attempts to drive the protocol to a state that has not been implemented in order to detect the decoy. We introduce algorithms for computing Nash equilibria of this interaction, which determine the optimal number of high- and low-interaction decoy nodes to be deployed.
- At the network level, we formulate a two-player Stackelberg game, in which the system (leader) chooses an IP address randomization policy, and the adversary (follower) chooses a rate at which to scan nodes after observing the randomization policy. We prove that the unique Stackelberg equilibrium of the game is achieved when both players follow threshold-based strategies. For the attacker, the trade-off is between the cost of scanning and the benefit of identifying and attacking the real node.
- We investigate the performance of the system under our framework through simulation study. For the timing-based game, we find that a pure strategy Nash equilibrium exists in all considered cases. For the fingerprinting game, we compute a mixed-strategy equilibrium, implying that at equilibrium the system should contain both high-interaction nodes that implement the full protocol and low-interaction nodes that only implement a subset of protocol states.

The paper is organized as follows. We discuss related work in Sect. 2. The system and adversary models are presented in Sect. 3. Our game-theoretic formulation for the interaction between the adversary and a single decoy node is given in Sect. 4. The interaction between an adversary scanning the decoy network and the system deciding when to randomize is considered in Sect. 5. Simulation results are contained in Sect. 6. Section 7 concludes the paper.

2   Related Work

Moving target defense is currently an active area of research aimed at preventing adversaries from gathering system information and launching attacks against specific vulnerabilities [13]. Moving target defense mechanisms in the literature include software diversity [9] and memory address layout randomization [10]. These approaches are distinct from decoy generation and IP address randomization and hence are orthogonal from our line of work.

Decoy networks are typically created using network virtualization packages such as honeyd [17]. Empirical studies on detection of decoys have focused on protocol fingerprinting, by identifying differences between the protocols simulated by decoys and the actual protocol specifications, including differences in IP fragmentation and implementation of TCP [11,22]. Decoy nodes can also be detected due to their longer response times, caused by lack of memory, CPU, and bandwidth resources [16]. The existing studies on decoy networks, however, have focused on empirical evaluation of specific vulnerabilities of widely-used decoy systems, rather than a broader analytical framework for design of dynamic decoy networks.

IP address space randomization has been proposed as a defense against scanning worms [1,3]. In [21], a framework for deciding when to randomize the IP address space in the presence of hitlist worms, based on a given estimate of whether the system is in a secure or insecure state, was proposed. A decision-theoretic approach to IP randomization in decoy networks was recently presented in [8], but this approach was concerned with the optimal system response to a given adversary strategy rather than the interaction between an intelligent adversary and the system. Furthermore, the work of [8] only considered timing-based attacks on decoy networks, and did not consider fingerprinting attacks.

Game-theoretic techniques have been used to model and mitigate a variety of network security threats [2]. A dynamic game-theoretic approach to designing a moving target defense configuration to maximize the uncertainty of the adversary was proposed in [26]. The method of [26], however, does not consider the timing of changes in the attack surface, and hence is complementary to our approach. The FlipIt game was formulated in [24] to model the timing of host takeover attacks; the FlipIt game does not, however, consider the presence of decoy resources.

In [6], platform randomization was formulated as a game, in which the goal of the system is to maximize the time until the platform is compromised by choosing a probability distribution over the space of available platforms. A game-theoretic approach to stochastic routing, in which packets are proactively allocated among multiple paths to minimize predictability, was proposed in [4]. In [12], game-theoretic methods for spatiotemporal address space randomization were introduced. While these approaches consider metrics such as time to compromise the system that are intuitively similar to our approach, the formulations are fundamentally different and hence the resulting algorithms are not directly applicable to our problem. To the best of our knowledge, game-theoretic approaches for decoy-based moving-target defenses are not present in the existing literature.

3   Model and Preliminaries

In this section, we present the models of the virtual network and the adversary.

3.1   Virtual Network Model

We consider a network consisting of n virtual nodes, including one real node and $(n-1)$ decoy nodes. Let $\pi = \left(1 - \frac{1}{n}\right)$ denote the fraction of nodes that are decoys. Decoy and real nodes have valid IP addresses that are chosen at random from a space of $M \gg n$ addresses, and hence decoy and real nodes cannot be distinguished based on the IP address. The assumption $M \gg n$ ensures that there is sufficient entropy in the IP address space for randomization to be effective. Decoy nodes are further classified as either high-interaction decoys, which implement the full operating system including application-layer services such as HTTP and FTP servers and SQL databases, and low-interaction decoys, which implement only partial versions of network and transport layer protocols such as IP, TCP, UDP, and ICMP [18].

Decoy nodes respond to messages from nodes outside the network. The decoy responses are determined by a configuration assigned to each decoy. Each possible configuration represents a different device (e.g., printer, PC, or server) and operating system that can be simulated by the decoy. Decoy nodes in the same network may have different configurations. Due to limited computation resources assigned to them, decoys will have longer communication delays than real nodes. The additional delay depends on the system CPU time and memory allocated to the decoy. Decoy node configurations can be randomized using software obfuscation techniques [15].

Based on models of service-oriented networks such as web servers, we assume that real nodes receive connection requests from valid users according to an M/G/1 queuing model [5]. Under this model, the service time of each incoming user is identically distributed and independent of both the service times of the other users and the number of users currently in the queue.

Since valid users have knowledge of the IP address of the real node, connections to decoy nodes are assumed to originate from errors or adversarial scanning. Decoy nodes will respond to suspicious, possibly adversarial queries in order to distract the adversary and delay the adversary from identifying and targeting the real node.

The virtual network is managed by a hypervisor, which creates, configures, and removes decoy nodes [7]. The hypervisor is assumed to be trusted and immune to compromise by the adversary. In addition to managing the decoy nodes, the hypervisor also assigns IP addresses to the nodes. In particular, the hypervisor can assign a new, uniformly random IP address to each node at any time. By choosing the new IP addresses to be independent of the previous IP addresses, the hypervisor prevents the adversary from targeting a node over a period of time based on its IP address. All IP addresses are assumed to be randomized simultaneously; generalizations to randomization policies that only update a subset of IP addresses at each time step are a direction for future work. Any communication sessions between valid users and the real node will be terminated when randomization occurs. Upon termination, the server sends the updated IP address to each authorized client. Each valid user must then reconnect to the real node, incurring an additional latency that depends on the connection migration protocol [23].

3.2   Adversary Model

We consider an external adversary with knowledge of the IP address space. The goal of the adversary is to determine the IP address of the real node in order to mount further targeted attacks. The adversary is assumed to know the set of possible IP addresses, if necessary by compromising firewalls or proxies, and attempts to identify the real node by sending query messages to IP addresses within this space. Based on the response characteristics, the adversary can evaluate whether a node is real or a decoy based on either timing analysis or protocol fingerprinting, as described below.

In timing-based blacklisting of nodes, an adversary exploits the response timing differences between real nodes and decoys. Since the decoy nodes have fewer CPU and memory resources than the real node, their response times will be longer. This longer delay can be used for detection. We assume that the adversary knows the response time distribution of a typical real node, which can be compared with response times of possible decoys for detection.

Protocol fingerprinting exploits the fact that the decoy nodes do not actually implement an operating system, but instead simulate an operating system using a prespecified configuration. As a result, differences between the decoys' behavior and the ideal behavior of the operating system allow the adversary to identify the decoy. Typical fingerprints include protocol versions, such as the sequence and acknowledgment numbers in TCP packets, the TCP options that are enabled, and the maximum segment size [25].

4   Modeling Interaction with Single Decoy

In this section, we provide a game-theoretic formulation for the interaction between the adversary and a single decoy node. We present a game-theoretic formulation for two attack types. First, we consider an adversary who attempts to identify decoy nodes through timing analysis. We then model detection based on fingerprinting techniques.

4.1   Timing-Based Decoy Detection Game

In timing-based detection, the adversary sends a sequence of probe packets (such as ICMP echo messages) and observes the delays of the responses from the node [16]. Let Z_k denote the delay of the response to the k-th probe packet. Based on the response times, the adversary decides whether the node is real or a decoy.

We let H_1 denote the event that the response is from a real node and H_0 denote the event that the response is from a decoy. The response times are assumed to be independent and exponentially distributed [16] with mean $\mu_1 = 1/\lambda_1$ for real nodes and $\mu_0 = 1/\lambda_0$ for decoys, where λ_1 and λ_0 represent the response rates of the real and decoy nodes, respectively. Note that the exponential response time is for a single query, while the M/G/1 assumption of Sect. 3.1 concerns the total length of a session between a valid user and the real node. The number of queries made by the adversary is denoted Q.

The adversary's utility function consists of three components, namely, the amount of time spent querying the node, the probability of falsely identifying a decoy as the real node (false positive), and the probability of falsely identifying the real node as a decoy (false negative). We let P_{FP} and P_{FN} denote the probabilities of false positive and false negative, respectively. The expected time spent querying is equal to $(\pi\mu_0 + (1 - \pi)\mu_1)Q$, where π denotes the fraction of decoy nodes.

The action space of the adversary consists of the number of times Q that the virtual node is queried, so that $Q \in \mathbb{Z}_{\geq 0}$. We assume that the adversary makes the same number of queries Q to each node, corresponding to a pre-designed, non-adaptive scanning strategy that does not consider feedback from past interactions. The system's action space consists of the mean of the decoy response time $\mu_0 \in [0, \infty)$.

The payoff of the adversary is equal to the total time required to scan the entire network. The expected utility of the adversary is given by

$$U_A(Q, \mu_0) = -(\pi\mu_0 + (1 - \pi)\mu_1)Q$$
$$-\pi c_{FP} P_{FP}(Q, \mu_0) - (1 - \pi)c_{FN} P_{FN}(Q, \mu_0), \quad (1)$$

where c_{FP} and c_{FN} denote the delays arising from false positive and false negative, respectively. The first term of (1) is the expected time to query a node. The second term is the additional time spent querying decoy nodes after a false positive occurs, which causes the adversary to attempt additional, time-intensive attacks on the decoys. The third term is the additional time spent querying decoy nodes after a false negative, when an adversary mistakes a real node for a decoy and scanning the rest of the network.

The cost of a given response rate is the additional delay experienced by the real nodes. Assuming that requests to the real node occur at rate θ and the network has a total capacity of c with variance σ^2, which is determined by the bandwidth, CPU, and memory constraints of the physical device, this delay is equal to $g(\mu_0) = \frac{\sigma^2\theta}{2(1-\theta/(c-1/\mu_0))} + \frac{1}{c-1/\mu_0}$, based on the assumption that the real node is an M/G/1 system [20, Chap. 8.5] (the M/G/1 assumption follows from the assumption of a single real node; generalization to M/G/m networks with m real nodes is a direction of future work). The payoff of the system is equal to

$$U_S(Q, \mu_0) = (\mu_0\pi + (1-\pi)\mu_1)Q + \pi c_{FP}P_{FP}(Q, \mu_0)$$
$$+ (1-\pi)c_{FN}P_{FN}(Q, \mu_0) - g(\mu_0). \quad (2)$$

The utility of the system is the total time spent by the adversary scanning the network, which increase the security of the real node.

In what follows, we introduce an algorithm for computing the Nash equilibrium of the timing-based interaction game. We first introduce a two-player zero-sum game with equivalent Nash equilibrium strategies. We then prove concavity of the utility functions of each player, implying that a unique equilibrium exists that can be computed using fictitious play.

Proposition 1. *Define the utility function*

$$\tilde{U}_A(Q, \mu_0) = -\pi\mu_0 Q - (1-\pi)\mu_1 Q - \pi c_{FP}P_{FP}(Q, \mu_0)$$
$$- (1-\pi)c_{FN}P_{FN}(Q, \mu_0) + g(\mu_0). \quad (3)$$

Then a pair of strategies (Q^*, μ_0^*) is a Nash equilibrium for the two-player game between a player 1 with utility function $\tilde{U}_A$ and a player 2 with utility function U_S if and only if it is the Nash equilibrium of a two-player game where player 1 has utility function U_A and player 2 has utility function U_S.

Proof. Let (Q^*, μ_0^*) be a Nash equilibrium for the game with utility functions $\tilde{U}_A$, U_S. The fact that μ_0^* is a best response to Q^* for the game with utility functions U_A and U_S follows trivially from the fact that U_S is the system's utility function in both cases. If Q^* satisfies $\tilde{U}_A(Q^*, \mu_0^*) \geq \tilde{U}_A(Q, \mu_0^*)$ for all $Q > 0$, then

$$\tilde{U}_A(Q^*, \mu_0^*) + g(\mu_0^*) \geq \tilde{U}_A(Q, \mu_0^*) + g(\mu_0^*),$$

and hence $U_A(Q^*, \mu_0^*) \geq U_A(Q, \mu_0^*)$, since $U_A(Q, \mu_0) = \tilde{U}_A(Q, \mu_0) + g(\mu_0)$ for all (Q, μ_0). Thus Q^* is the best response to μ_0^* under utility function U_A. The proof of the converse is similar.

By Proposition 1, it suffices to find a Nash equilibrium of the equivalent zero-sum game with adversary and system utilities $\tilde{U}_A$ and U_S, respectively. As a first step, we prove two lemmas regarding the structure of $\tilde{U}_A$ and U_S.

Lemma 1. *Let $\epsilon > 0$. Then there exists $\hat{Q}$ and a convex function $\hat{f} : \mathbb{R} \to \mathbb{R}$ such that $|\hat{f}(Q) - \tilde{U}_A(Q, \mu_0)| < \epsilon$ for all $Q > \hat{Q}$.*

Proof. Define $f(Q) = -(\pi\mu_0 + (1-\pi)\mu_1)Q - c_{FP}P_{FP}(Q, \mu_0) - c_{FN}P_{FN}(Q, \mu_0) + g(\mu_0)$. The first two terms are linear in Q and hence convex, while the last term does not depend on Q. In computing the probability of false positive, we first observe that the maximum-likelihood decision rule for the adversary is to decide that the node is real if $\mu_1 c_{FP} P_1(Z_1, \ldots, Z_Q) > \mu_0 c_{FN} P_0(Z_1, \ldots, Z_Q)$ and that the node is a decoy otherwise. Under the exponential assumption, this is equivalent to

$$Q \log \frac{\lambda_1}{\lambda_0} - (\lambda_1 - \lambda_0) \sum_{j=1}^{Q} Z_j > \log \frac{\mu_0 c_{FN}}{\mu_1 c_{FP}}.$$

Hence the probability of false positive is equal to

$$P_{FP}(Q) = Pr \left(Q \log \frac{\lambda_1}{\lambda_0} - (\lambda_1 - \lambda_0) \sum_{j=1}^{Q} Z_j > \log \frac{\mu_0 c_{FN}}{\mu_1 c_{FP}} \Big| H_0 \right).$$

Rearranging terms yields

$$P_{FP}(Q) = Pr \left(\overline{Z} < \frac{\log \lambda_1 - \log \lambda_0}{\lambda_1 - \lambda_0} - \frac{\log \frac{\mu_0 c_{FN}}{\mu_1 c_{FP}}}{Q(\lambda_1 - \lambda_0)} \Big| H_0 \right),$$

where $\overline{Z} = \frac{1}{Q} \sum_{j=1}^{Q} Z_j$.

By the Central Limit Theorem, $\overline{Z}$ can be approximated by an $N(\mu_0, \mu_0^2/Q)$-Gaussian random variable for Q sufficiently large. Letting $x = \frac{\log \lambda_1 - \log \lambda_0}{\lambda_1 - \lambda_0}$, the probability of false positive is equal to $Pr(X < \sqrt{Q}(x\lambda_0 - 1))$ where X is an $N(0,1)$-Gaussian random variable, so that

$$P_{FP} = \frac{1}{\sqrt{2\pi}} \int_{-\infty}^{\sqrt{Q}(x\lambda_0 - 1)} \exp\left(-\frac{x^2}{2} \right) dx.$$

Differentiating with respect to Q yields

$$\frac{x\lambda_0 - 1}{\sqrt{2\pi}} \frac{1}{2\sqrt{Q}} \exp\left(-\frac{Q(x\lambda_0 - 1)^2}{2} \right),$$

which is increasing in Q since $x\lambda_0 < 1$. Hence the probability of false positive can be approximated by a convex function for Q sufficiently large. The derivation for the probability of false negative is similar.

Approximate concavity of U_A implies that the best response of the adversary can be computed by enumerating the values of $U_A(Q, \mu_0)$ for $Q < \hat{Q}$, and using convex optimization to find the optimal value when $Q \geq \hat{Q}$.

The following lemma establishes concavity of the system utility function U_S as a function of μ_0 for a given T. The concavity of U_S enables efficient computation of the Nash equilibrium.

Lemma 2. *The function U_S is concave as a function of μ_0.*

Proof. It suffices to show that each term of U_S in Eq. (2) is concave. The first term of U_S is linear in μ_0 and therefore concave. The second derivative test implies that $g(\mu_0)$ is convex as a function of μ_0, and hence $-g(\mu_0)$ is concave. By the analysis of Lemma 1, in proving the concavity of the false positive probability, it is enough to show that $Pr\left(X < \frac{x\sqrt{Q}}{\mu_0} - \sqrt{T}\right)$ is concave as a function of μ_0. The derivative of $\frac{x}{\mu_0}$ with respect to μ_0 is equal to

$$\frac{\frac{1}{\mu_0}\left(\frac{\mu_0}{\mu_1} - 1\right) - (\log\mu_0 - \log\mu_1)\left(\frac{1}{\mu_1}\right)}{\left(\frac{\mu_0}{\mu_1} - 1\right)^2},$$

which is decreasing in μ_0. Hence the derivative of the false positive probability is equal to

$$\frac{\frac{1}{\mu_0}\left(\frac{\mu_0}{\mu_1} - 1\right) - (\log\mu_0 - \log\mu_1)\left(\frac{1}{\mu_1}\right)}{\left(\frac{\mu_0}{\mu_1} - 1\right)^2}\exp\left(-\frac{\left(\frac{x\sqrt{Q}}{\mu_0} - \sqrt{Q}\right)^2}{2}\right),$$

which is monotonically decreasing in μ_0 and hence concave.

Fictitious play can be used to find the Nash equilibrium of the interaction between the adversary and the network. The algorithm to do so proceeds in iterations. At each iteration m, there are probability distributions p_A^m and p_S^m defined by the prior interactions between the system and adversary. The system chooses μ_0 in order to maximize $\mathbf{E}_{p_A}(U_S(\mu_0)) = \sum_Q p_A^m(Q)U_S(Q,\mu_0)$, while the adversary chooses Q to maximize $\mathbf{E}_{p_S^m}(U_A(Q)) = \int_0^\infty p_S^m(\mu_0)U_A(Q,\mu_0)\,d\mu_0$. The strategies of the system and adversary at each iteration can be computed efficiently due to the concavity of U_S and the approximate convexity of U_A. Convergence is implied by the following proposition.

Proposition 2. *The fictitious play procedure converges to a mixed-strategy Nash equilibrium.*

Proof. Since the utility functions satisfy $\tilde{U}_A(Q,\mu_0)+U_S(Q,\mu_0) = 0$, the iterative procedure implies converge to a mixed-strategy Nash equilibrium [19, pg. 297]. Furthermore, by Proposition 1, the mixed-strategy equilibrium is also an NE for the game with utility functions U_A and U_S.

4.2 Fingerprinting-Based Decoy Detection Game

Operating system fingerprinting techniques aim to differentiate between real and decoy nodes by exploiting differences between the simulated protocols of the decoy and the true protocol specifications. In order to quantify the strategies of the adversary and the system, we model the protocol to be simulated (e.g., TCP) as a finite state machine $\mathcal{F}$, defined by a set of states S, a set of inputs I, and a set of outputs O. The transition function $\delta : I \times S \to S$ determines the next state of the system as a function of the input and current state, while the output is determined by a function $f : I \times S \to O$. We write $\mathcal{F} = (S, I, O, \delta, f)$.

The real and decoy protocols are defined by finite state machines $\mathcal{F}_R = (S_R, I_R, O_R, \delta_R, f_R)$ and $\mathcal{F}_D = (S_D, I_D, O_D, \delta_D, f_D)$. The goal of the decoy protocol is to emulate the real system while minimizing the number of states required. Under this model, the adversary chooses a state $s \in S_R$ and attempts to determine whether that state is implemented correctly in the decoy, i.e., whether the output o corresponding to an input i satisfies $o = f_R(s, i)$. In order to reach state s, the adversary must send a sequence of d_s inputs, where d_s denotes the minimum number of state transitions required to reach the state s from the initial state s_0.

The system's action space is defined by the set of states S_D, while the adversary's action space is the set s that the adversary attempts to reach. The choice of s will determine the sequence of messages sent by the adversary. The adversary's utility function is therefore given by

$$U_A(s, S_D) = -d_s - c_{FP}P_{FP}(s, S_D) - c_{FN}P_{FN}(s, S_D).$$

We note that the real node implements the state s correctly for all $s \in S_R$, and hence the probability of false negative is zero. Furthermore, we assume that the decoy returns the correct output at state s with probability 1 if $s \in S_D$ and returns the correct output with probability 0 otherwise. Hence the adversary's utility function is

$$U_A(s, S_D) = -d_s - \mathbf{1}(s \in S_D)c_{FP}, \tag{4}$$

where $\mathbf{1}(\cdot)$ denotes the indicator function.

For the system, the utility function is equal to the total time spent by the adversary querying a decoy node, minus the memory cost of the decoys. This utility is equal to

$$U_S(s, S_D) = d_s + \mathbf{1}(s \in S_D)c_{FP} - c_D(S_D), \tag{5}$$

where $c_D(S_D)$ is the cost of implementing a set of states. In order to avoid state space explosion for the system, we restrict the defender to strategies that implement all states within k steps of the initial state, where $k \in \{0, \ldots, |S_D|\}$. Intuitively, a strategy that implements a state $s \in S_D$ but does not implement a state $s' \in S_D$ with $d_{s'} < d_s$ may be suboptimal, because the protocol may reach state s before state s', thus enabling the adversary to identify the decoy in fewer steps.

A fictitious play algorithm for computing a mixed-strategy equilibrium is as follows. Probability distributions π_A^m and π_S^m, which represent the empirical frequency of each strategy of the adversary and system up to iteration m, are maintained. At the m-th iteration, the strategies $k^* = \arg\max \mathbf{E}_{\pi_A^m}(k)$ and $s^* = \arg\max \{\mathbf{E}_{\pi_S^m}(s)\}$ are computed and the corresponding entries of π_A^{m+1}

and π_S^{m+1} are incremented. Since there is an equivalent zero-sum game with adversary utility function $\tilde{U}_A(s) = d_s + \mathbf{1}(s \in S_D)c_{FP} - c_D(S_D)$, the empirical frequencies of each player converge to the mixed strategy equilibrium [19].

5   Characterization of Optimal IP Address Randomization Strategy by Network

In this section, we present a game-theoretic formulation for the interaction between the virtual network, which decides when to randomize the IP address space, and the adversary, which decides the scanning strategy. The optimal randomization policy of the network and the probability of detecting the real node at equilibrium are derived.

5.1   Game Formulation

We consider a game in which the adversary chooses a scanning strategy, determined by the number of simultaneous connections α. The parameter α is bounded above by α_{max}, which is chosen by the hypervisor to limit the total number of connections and hence avoid overutilization of the system CPU. The adversary incurs a cost ω for maintaining each connection with a node. The number of nodes scanned by the adversary per unit time, denoted Δ, is given by $\Delta = \frac{\alpha}{\tau}$, where τ is the time required to scan each node. The parameter τ depends on the detection method employed by the adversary, and is equal to the Nash equilibrium detection time of Sect. 4.1 if timing-based detection is used or the Nash equilibrium detection time of Sect. 4.2 if fingerprint-based detection is used.

At each time t, the system decides whether to randomize the IP address space; we let $t = 0$ denote the time when the previous randomization took place. Let R denote the time when randomization occurs. The system incurs two costs of randomization, namely, the probability that the adversary detects the real node and the number of connections that are terminated due to randomization. Since the real and decoy nodes cannot be distinguished based on IP addresses alone, the probability of detection at time t is equal to the fraction of nodes that are scanned up to time t, $\frac{\Delta t}{n}$.

The cost resulting from terminating connections is equal to the delay β resulting from migrating each connection to the real node's new IP address; TCP migration mechanisms typically have cost that is linear in the number of connections [23]. The cost of breaking real connections is therefore equal to $\beta Y(t)$, where $Y(t)$ is equal to the number of connections to the real node, so that the utility function of the system is given by $U_S(\alpha, R) = -\mathbf{E}\left(\frac{\alpha}{\tau n}R + \beta Y(R)\right)$.

For the adversary, the utility is equal to the detection probability, minus the cost of maintaining each connection, for a utility function of $U_A(\alpha, R) = \mathbf{E}\left(\frac{\alpha}{\tau n}R - \omega\alpha\right)$. The resulting game has Stackelberg structure, since the system first chooses the randomization policy, and the adversary then chooses a scanning rate based on the randomization policy.

5.2   Optimal Strategy of the System

The information set of the system is equal to the current number of valid sessions $Y(t)$ and the fraction of decoy nodes scanned by the adversary $D(t)$ at time t. The goal of the system is to choose a randomization time R in order to minimize its cost function, which can be expressed as the optimization problem

$$\underset{R}{\text{minimize}} \ \mathbf{E}(D(R) + \beta Y(R)) \tag{6}$$

where R is a random variable. The randomization policy can be viewed as a mapping from the information space $(Y(t), D(t))$ at time t to a $\{0, 1\}$ variable, with 1 corresponding to randomizing at time t and 0 corresponding to not randomizing at time t. Define L_t to be the number of decoy nodes that have been scanned during the time interval $[0, t]$.

The number of active sessions $Y(t)$ follows an M/G/1 queuing model with known arrival rate ζ and average service time $1/\phi$. We let $1/\phi_t$ denote the expected time for the next session with the real node to terminate, given that a time t has elapsed since the last termination. In what follows, we assume that ϕ_t is monotonically increasing in t; this is consistent with the M/M/1 and M/D/1 queuing models. The following theorem, which generalizes [8, Theorem 1] from an M/M/1 to an M/G/1 queuing model, describes the optimal strategy of the system.

Theorem 1. *The optimal policy of the system is to randomize immediately at time t if and only if $L_t = n$, $Y(t) = 0$, or $\frac{\Delta}{n}\phi + \beta\zeta\phi - \beta > 0$, and to wait otherwise.*

Proof. In an optimal stopping problem of the form (6), the optimal policy is to randomize at a time t satisfying

$$D(t) + \beta Y(t) = \sup \{\mathbf{E}(D(t') + \beta Y(t')|D(t), Y(t)) : t' \geq t\}.$$

If $L_t = n$, then the address space must be randomized to avoid detection of the real node. If $Y(t) = 0$, then it is optimal to randomize since $D(t)$ is increasing as a function of t.

Suppose that $L_t < n$ and $Y(t) > 0$. Let $\xi_1, \xi_2, \ldots$ denote the times when connections terminate. We prove by induction that, for each l, $t' \in [\xi_{l-1}, \xi_l]$ implies that $\mathbf{E}(D(t') + \beta Y(t')|Y(t)) > D(t) + \beta Y(t)$. First, consider $l = 1$, with $\xi_0 = t$. Then if $t' \in [\xi_0, \xi_1)$, $D(t') + \beta Y(t') > D(t) + \beta Y(t)$, since D is nondecreasing in time and no connections have terminated since time t. At time ξ_1, we have that

$$\mathbf{E}(D(\xi_1) + \beta Y(\xi_1)|Y(t)) = \frac{\Delta}{n}\mathbf{E}(\xi_1) \tag{7}$$

$$+ \beta(Y(t) + \zeta\mathbf{E}(\xi_1) - 1) \tag{8}$$

$$= \left(\frac{\Delta}{n} + \beta\zeta\right)\phi + \beta Y(t) - \beta \tag{9}$$

and so $\mathbf{E}(D(\xi_1) + \beta Y(\xi_1)|Y(t)) < D(t) + \beta Y(t)$ iff $\frac{\Delta}{n}\phi + \beta\zeta\phi - \beta > 0$.

Now, suppose that the result holds up to $(l-1)$. By a similar argument, $\mathbf{E}(D(\xi_{l-1}) + \beta Y(\xi_{l-1})|Y(t)) < \mathbf{E}(D(t') + \beta Y(t')|Y(t))$ for all $t' \in [\xi_{l-1}, \xi_l)$. The condition

$$\mathbf{E}(D(\xi_{l-1}) + \beta Y(\xi_{l-1})|Y(t)) < \mathbf{E}(D(\xi_l) + \beta Y(\xi_l)|Y(t))$$

holds iff $\frac{\Delta}{n}\phi + \beta\zeta\phi - \beta > 0$.

This result implies that a threshold-based policy is optimal for randomization over a broad class of real node dynamics.

5.3   Optimal Strategy of the Adversary

The optimal scanning rate is the solution to

$$\begin{aligned} \text{maximize} \quad & \mathbf{E}(D(R) - \omega\alpha) \\ \text{s.t.} \quad & \alpha \in [0, \alpha_{max}] \end{aligned} \tag{10}$$

which is a trade-off between the probability of identifying the real node and the adversary's cost of bandwidth.

The scanning rate is assumed to be constant and chosen based on the randomization policy of the system.

Since the scanning process is random, the detection probability at the time of randomization, $D(R)$, is equal to the fraction of the network scanned at time R, $\frac{\alpha}{\tau n}R$. Based on Theorem 1, the detection probability is given as

$$D(R) = \begin{cases} \frac{\alpha}{\tau n}T_0, & \left(\frac{\alpha}{\tau n} + \beta\zeta\right)\phi < \beta \\ 0, & \text{else} \end{cases} \tag{11}$$

where T_0 is the time for the number of connections to go to 0. Hence the value of α that maximizes $D(R)$ is $\alpha = \beta\tau n - \beta\zeta$. The overall utility of the adversary is equal to $\beta(\tau n - \zeta)(\tau n)\mathbf{E}(T_0) - \omega(\beta\tau n - \beta\zeta)$.

Proposition 3. *Let $\alpha^* = \min\{\alpha_{max}, \beta\tau n\left(\frac{1}{\phi} - \frac{1}{\zeta}\right)\}$. Then the unique Stackelberg equilibrium of the network interaction game is for the adversary to choose α based on*

$$\alpha = \begin{cases} \alpha^*, & \mathbf{E}(T_0) - \omega\tau n > 0 \\ 0, & else \end{cases} \tag{12}$$

Proof. The proof follows from Theorem 1 and the fact that the adversary's utility is negative unless the condition $\mathbf{E}(T_0) - \omega\tau n$ holds.

Proposition 3 indicates that the adversary follows a threshold decision rule, in which the adversary scans the system at the rate α^* if the expected time before randomization, T_0, exceeds the expected time to scan the entire network, τn. The adversary can determine the optimal scanning rate over a period of time by initially scanning at a low rate and incrementally increasing the rate until randomization occurs, signifying that the threshold scanning rate α^* has been found.

6 Simulation Study

A numerical study was performed using Matlab, consisting of three components. First, we studied the timing-based detection game of Sect. 4.1. Second, we considered the fingerprinting-based detection game of Sect. 4.2. Third, we analyzed the network-level interaction of Sect. 5.

For the timing-based detection game, we considered a network of 100 nodes, with 1 real node and 99 decoy nodes. The real nodes were assumed to have mean response time of 1, while the response time of the decoys varied in the range $[1, 1.25]$. The parameter α, representing the amount of real traffic, was set equal to 0, while the capacity c of the virtual network was equal to 1. The trade-off parameter γ took values from 1 to 5, while the number of queries by the adversary ranged from $T = 1$ to $T = 50$.

We observed that the timing-based detection game converged to a pure-strategy Nash equilibrium in each simulated case. Figure 1(a) shows the mean response time of the decoy nodes as a function of the trade-off parameter, γ. As the cost of delays to the real nodes increases, the response time of the decoys increases as well. For lower values of γ, it is optimal for the real and decoy nodes to have the same response time.

For detection via system fingerprinting, we considered a state machine of diameter 4, consistent with the simplified TCP state machine of [14], implying that there are 5 possible strategies in the game of Sect. 4.2. We considered a cost of 0.2 for the system and adversary, so that the normalized cost of implementing the entire state machine was equal to 1. Figure 1(b) shows a histogram representing the mixed strategy of the system. The mixed strategy indicates that roughly half of the decoy nodes should implement only the first level of states in the state diagram, while the remaining half should implement the entire state machine, for this particular choice of the parameter values. This suggests an optimal allocation of half high-interaction and half low-interaction decoys, leading to a resource-expensive strategy.

In studying the network-level interaction between the system and adversary, we considered a network of $n = 100$ virtual nodes with detection time $\tau = 5$ based on the previous simulation results. The trade-off parameter $\beta = 0.1$. The real node was assumed to serve users according to an M/M/1 process with arrival rate $\zeta = 0.4$ and service rate $\phi = 2$. The cost of each connection to the adversary was set at $\omega = 2$. Figure 1(c) shows the probability of detection for the adversary as a function of the number of simultaneous connections initiated by the adversary. The probability of detection increases linearly until the threshold is reached; beyond the threshold, the system randomizes as soon as the scanning begins and the probability of detection is 0. Furthermore, as the rate of connection requests to the real node, quantified by the parameter ζ, increases, the cost of randomization for the real node increases, leading to longer waiting times between randomization and higher probability of detection.

As shown in Fig. 1(d), the number of dropped connections due to randomization is zero when ζ is small, since the optimal strategy for the system is to wait until all connections terminate. As ζ approaches the capacity of the real node,

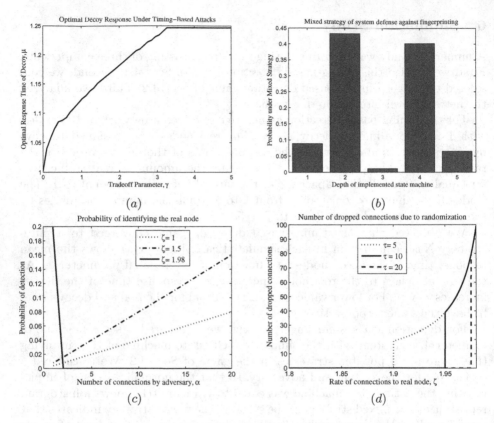

Fig. 1. Numerical results based on our proposed game-theoretic framework. (a) The timing-based detection game of Sect. 4.1 converged to a pure-strategy equilibrium in all experimental studies. The pure strategy of the system is shown as a function of the trade-off parameter, γ. A larger value of γ results in a slower response rate due to increased delay to the real nodes. (b) Histogram of the mixed strategy of the system for the fingerprinting game of Sect. 4.2 using the TCP state machine. The optimal strategy is to implement only the initial states of the protocol and the entire protocol with roughly equal probability. (c) Detection probability as a function of the number of simultaneous connections by the adversary. The detection probability increases before dropping to zero when the randomization threshold is reached. (d) Number of dropped connections when the number of adversary connections $\alpha = 5$. The number of dropped connections is initially zero, as the adversary scanning rate is below threshold, and then increases as the rate of connection to the real node approaches the capacity of the real node.

the number of dropped connections increases. The effectiveness of the decoy, described by the time τ required to detect the decoy, enables the system to operate for larger values of ζ (i.e., higher activity by the real nodes) without dropping connections.

7   Conclusion

We studied the problem of IP randomization in decoy-based moving target defense by formulating a game-theoretic framework. We considered two aspects of the design of decoy networks. First, we presented an analytical approach to modeling detection of nodes via timing-based analysis and protocol finger-printing and identified decoy design strategies as equilibria of two-player games. For the fingerprinting attack, our approach was based on a finite state machine model of the protocol being fingerprinted, in which the adversary attempts to identify states of the protocol that the system has not implemented. Second, we formulated the interaction between an adversary scanning a virtual network and the hypervisor determining when to randomize the IP address space as a two-player Stackelberg game between the system and adversary. We proved that there exists a unique Stackelberg equilibrium to the interaction game in which the system randomizes only if the scanning rate crosses a specific threshold. Simulation study results showed that the timing-based game consistently has a pure-strategy Nash equilibrium with value that depends on the trade-off between detection probability and cost, while the fingerprinting game has a mixed strat-egy equilibrium, suggesting that networks should consist of a mixture of high- and low-interaction decoys.

While our current approach incorporates the equilibria of the single-node interaction games as parameters in the network-level game, a direction of future work will be to compute joint strategies at both the individual node and network level simultaneously. An additional direction of future work will be to investigate dynamic game structures, in which the utilities of the players, as well as parameters such as the number of nodes and the system resource constraints, change over time. We will also investigate "soft blacklisting" techniques, in which an adversary adaptively increases the delays when responding to requests from suspected adversaries, at both the real and decoy nodes. Finally, modeling the ability of decoys to gather information on the goals and capabilities of the adversary is a direction of future work.

References

1. Abu Rajab, M., Monrose, F., Terzis, A.: On the impact of dynamic addressing on malware propagation. In: Proceedings of the 4th ACM Workshop on Recurring Malcode, pp. 51–56 (2006)
2. Alpcan, T., Başar, T.: Network Security: A Decision and Game-Theoretic App-roach. Cambridge University Press, Cambridge (2010)
3. Antonatos, S., Akritidis, P., Markatos, E.P., Anagnostakis, K.G.: Defending against hitlist worms using network address space randomization. Comput. Netw. **51**(12), 3471–3490 (2007)
4. Bohacek, S., Hespanha, J., Lee, J., Lim, C., Obraczka, K.: Game theoretic sto-chastic routing for fault tolerance and security in computer networks. IEEE Trans. Parallel Distrib. Syst. **18**(9), 1227–1240 (2007)

5. Cao, J., Andersson, M., Nyberg, C., Kihl, M.: Web server performance modeling using an M/G/1/K PS queue. In: 10th IEEE International Conference on Telecommunications (ICT), pp. 1501–1506 (2003)
6. Carter, K.M., Riordan, J.F., Okhravi, H.: A game theoretic approach to strategy determination for dynamic platform defenses. In: Proceedings of the First ACM Workshop on Moving Target Defense, pp. 21–30 (2014)
7. Chisnall, D.: The Definitive Guide to the Xen Hypervisor. Prentice Hall, Englewood (2007)
8. Clark, A., Sun, K., Poovendran, R.: Effectiveness of IP address randomization in decoy-based moving target defense. In: Proceedings of the 52nd IEEE Conference on Decision and Control (CDC), pp. 678–685 (2013)
9. Franz, M.: E unibus pluram: massive-scale software diversity as a defense mechanism. In: Proceedings of the 2010 Workshop on New Security Paradigms, pp. 7–16 (2010)
10. Giuffrida, C., Kuijsten, A., Tanenbaum, A.S.: Enhanced operating system security through efficient and fine-grained address space randomization. In: USENIX Security Symposium (2012)
11. Holz, T., Raynal, F.: Detecting honeypots and other suspicious environments. In: IEEE Information Assurance and Security Workshop (IAW), pp. 29–36 (2005)
12. Jafarian, J.H.H., Al-Shaer, E., Duan, Q.: Spatio-temporal address mutation for proactive cyber agility against sophisticated attackers. In: Proceedings of the First ACM Workshop on Moving Target Defense, pp. 69–78 (2014)
13. Jajodia, S., Ghosh, A.K., Subrahmanian, V., Swarup, V., Wang, C., Wang, X.S.: Moving Target Defense II. Springer, New York (2013)
14. Kurose, J., Ross, K.: Computer Networking. Pearson Education, New Delhi (2012)
15. Larsen, P., Homescu, A., Brunthaler, S., Franz, M.: Sok: automated software diversity. In: IEEE Symposium on Security and Privacy, pp. 276–291 (2014)
16. Mukkamala, S., Yendrapalli, K., Basnet, R., Shankarapani, M., Sung, A.: Detection of virtual environments and low interaction honeypots. In: IEEE Information Assurance and Security Workshop (IAW), pp. 92–98 (2007)
17. Provos, N.: A virtual honeypot framework. In: Proceedings of the 13th USENIX Security Symposium, vol. 132 (2004)
18. Provos, N., Holz, T.: Virtual Honeypots: From Botnet Tracking to Intrusion Detection. Addison-Wesley Professional, Reading (2007)
19. Robinson, J.: An iterative method of solving a game. Ann. Math. **54**(2), 296–301 (1951)
20. Ross, S.M.: Introduction to Probability Models. Academic Press, Orlando (2009)
21. Rowe, J., Levitt, K., Demir, T., Erbacher, R.: Artificial diversity as maneuvers in a control-theoretic moving target defense. In: Moving Target Research Symposium (2012)
22. Shamsi, Z., Nandwani, A., Leonard, D., Loguinov, D.: Hershel: single-packet OS fingerprinting. In: ACM International Conference on Measurement and Modeling of Computer Systems, pp. 195–206 (2014)
23. Sultan, F., Srinivasan, K., Iyer, D., Iftode, L.: Migratory TCP: connection migration for service continuity in the internet. In: Proceedings of the 22nd IEEE International Conference on Distributed Computing Systems, pp. 469–470 (2002)

24. Van Dijk, M., Juels, A., Oprea, A., Rivest, R.L.: Flipit: the game of stealthy takeover. J. Cryptology **26**(4), 655–713 (2013)
25. Wolfgang, M.: Host discovery with NMAP (2002). http://moonpie.org/writings/discovery.pdf
26. Zhu, Q., Başar, T.: Game-theoretic approach to feedback-driven multi-stage moving target defense. In: Das, S.K., Nita-Rotaru, C., Kantarcioglu, M. (eds.) GameSec 2013. LNCS, vol. 8252, pp. 246–263. Springer, Heidelberg (2013)

Attack-Aware Cyber Insurance for Risk Sharing in Computer Networks

Yezekael Hayel[1,2]([✉]) and Quanyan Zhu[1]

[1] Polytechnic School of Engineering, New York University, Brooklyn, NY 11201, USA
{yezekael.hayel,quanyan.zhu}@nyu.edu
[2] LIA/CERI, University of Avignon, Avignon, France

Abstract. Cyber insurance has been recently shown to be a promising mechanism to mitigate losses from cyber incidents, including data breaches, business interruption, and network damage. A robust cyber insurance policy can reduce the number of successful cyber attacks by incentivizing the adoption of preventative measures and the implementation of best practices of the users. To achieve these goals, we first establish a cyber insurance model that takes into account the complex interactions between users, attackers and the insurer. A games-in-games framework nests a zero-sum game in a moral-hazard game problem to provide a holistic view of the cyber insurance and enable a systematic design of robust insurance policy. In addition, the proposed framework naturally captures a privacy-preserving mechanism through the information asymmetry between the insurer and the user in the model. We develop analytical results to characterize the optimal insurance policy and use network virus infection as a case study to demonstrate the risk-sharing mechanism in computer networks.

Keywords: Cyber insurance · Incomplete information game · Bilevel optimization problem · Moral hazards · Cyber attacks

1 Introduction

Cyber insurance is a promising solution that can be used to mitigate losses from a variety of cyber incidents, including data breaches, business interruption, and network damage. A robust cyber insurance policy could help reduce the number of successful cyber attacks by incentivizing the adoption of preventative measures in return for more coverage and the implementation of best practices by basing premiums on an insureds level of self-protection. Different from the traditional insurance paradigm, cyber insurance is used to reduce risk that is not created by nature but by intelligent attacks who deliberately inflict damage on the network. Another important feature of cyber insurance is the uncertainties related to the risk of the attack and the assessment of the damage. To address

---

Q. Zhu—The work was partially supported by the NSF (grant EFMA 1441140) and a grant from NYU Research Challenge Fund.

MHR Khouzani et al. (Eds.): GameSec 2015, LNCS 9406, pp. 22–34, 2015.
DOI: 10.1007/978-3-319-25594-1_2

these challenges, a robust cyber insurance framework is needed to design policies to induce desirable user behaviors and mitigate losses from known and unknown attacks.

In this paper, we propose a game-theoretic model that extends the insurance framework to cyber security, and captures the interactions between users, insurance company and attackers. The proposed game model is established based on a recent game-in-games concept [1] in which one game is nested in another game to provide an enriched game-theoretic model to capture complex interactions. In our framework, a zero-sum game is used to capture the conflicting goals between an attacker and a defender where the defender aims to protect the system for the worst-case attack. In addition, a moral-hazard type of leader-follower game with incomplete information is used to model the interactions between the insurer and the user. The user has a complete information of his action while the insurer cannot directly observe it but indirectly measures the loss as a consequence of his security strategy. The zero-sum game is nested in the incomplete information game to constitute a bilevel problem which provides a holistic framework for designing insurance policy by taking into account the cyber attack models and the rational behaviors of the users.

The proposed framework naturally captures a privacy-preserving mechanism through the information asymmetry between the insurer and the user in the model. The insurance policy designed by the insurer in the framework does not require constant monitoring of users' online activities, but instead, only on the measurement of risks. This mechanism prevents the insurer from acquiring knowledge of users' preferences and types so that the privacy of the users is protected. The major contributions of the paper are three-fold. They are summarized as follows:

(i) We propose a new game-theoretic framework that incorporates attack models, and user privacy.
(ii) We holistically capture the interactions between users, attackers, and the insurer to develop incentive mechanisms for users to adopt protection mechanisms to mitigate cyber risks.
(iii) The analysis of our framework provides a theoretic guideline for designing robust insurance policy to maintain a good network condition.

1.1 Related Works

The challenges of cyber security are not only technical issues but also economic and policy issues [2]. Recently, the use of cyber insurance to enhance the level of security in cyber-physical systems has been studied [3,4]. While these works deal with externality effects of cyber security in networks, few of them take into account in the model the cyber attack from a malicious adversary to distinguish from classical insurance models. In [5], the authors have considered direct and indirect losses, respectively due to cyber attacks and indirect infections from other nodes in the network. However, the cyber attacks are taken as random inputs rather than a strategic adversary. The moral hazard model in economics literature [6,7] deal with hidden actions from an agent, and aims to address the question: How does a principal design the agent's wage contract in order to maximize his effort? This framework is related to insurance markets, and has been used to model cyber insurance [8] as a solution for mitigate losses from cyber attacks. In addition, in [9], the authors have studied a security investment problem in a network with externality effect. Each node determines his security investment level and competes with a strategic attacker. Their model does not focus on the insurance policies and hidden-action framework. In this work, we enrich the moral-hazard type of economic frameworks by incorporating attack models, and provide a holistic viewpoint towards cyber insurance and a systematic approach to design insurance policies.

Other works in the literature such as the robust network framework presented in [10] deal with strategic attacker model over networks. However, the network effect is modeled as a simple influence graph, and the stimulus of the good behavior of the network users is based on a global information known to every player. In [11], the authors propose a generic framework to model cyber-insurance problem. Moreover, the authors compare existing models and explain how these models can fit into their unifying framework. Nevertheless, many aspects, like the attacker model and the network effect, have not been analyzed in depth. In [12], the authors propose a mechanism design approach to the security investment problem, and present a message exchange process through which users converge to an equilibrium where they make investments in security at a socially optimal level. This paper has not yet taken into account both the network effect (topology) and the cyber attacker strategy.

1.2   Organization of the Paper

The paper is organized as follows. In Sect. 2, we describe the general framework of cyber moral hazard by first introducing the players and the interactions between them, and second, by defining the influence graph that models the network effect. In Sect. 3, we analyze the framework for a class of problems with separable utility functions. In addition, we use a case study to demonstrate the analysis of an insurance policy for the case of virus infection over a large-scale computer networks. The paper is concluded in Sect. 4.

2   Game-Theoretic Model for Cyber Insurance

In this section, we introduce the cyber insurance model between a user i and an insurance company I (Player I). A user i invests or allocates $a_i \in [0,1]$ resources for his own protection to defense against attacks. When $a_i = 1$, the user employs maximum amount of resources, e.g., investment in firewalls, frequent change of passwords, and virus scan of attached files for defense. When $a_i = 0$, the user does not invest resources for protection, which corresponds to behaviors such as reckless response to phishing emails, minimum investment in cyber protection, or infrequent patching of operating systems. The protection level a_i can also be interpreted as the probability that user i invokes a protection scheme. User i can be attacked with probability $q_i \in [0, 1]$. The security level of user i, Z_i, depends on a_i and q_i. To capture the dependency, we let $Z_i = p_i(a_i, q_i)$, where $p_i : [0, 1]^2 \to \mathbb{R}_+$ is a continuous function that quantifies the security level of user i. An insurance company cannot observe the action of the user, i.e., the action a_i if user i. However, it can observe a measurable risk associated with the protection level of user i. We let a random variable X_i denote the risk of user i that can be observed by the insurance company, described by

$$X_i := \mathcal{G}_i(Z_i, \theta_i), \tag{1}$$

where θ_i is a random variable with probability density function g_i that captures the uncertainties in the measurement or system parameters. The risk X_i can be measured in dollars. For example, a data breach due to the compromise of a server can be a consequence of low security level at the user end [13]. The economic loss of the data breach can be represented as random variable X_i measured in dollars. The magnitude of the loss depends on the content and the significance of the data, and the extent of the breach. The variations in these parameters are captured by the random variable θ_i. The information structure of the model is depicted in Fig. 1.

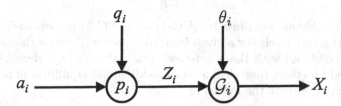

Fig. 1. Illustration of the information structure of the two-person cyber insurance system model: user i determines protection level a_i and an attacker chooses attack probability q_i. The security level Z_i is assessed using function p_i. The cyber risk X_i for user i is measured by the insurance company.

Note that the insurer cannot directly observe the actions of the attack and the user. Instead, he can measure an outcome as a result of the action pair. This type of framework falls into a class of moral hazard models proposed by Holmstrom in [6,7]. One important implication of the incomplete information of the insurer is on privacy. The user's decision a_i can often be related to personal habits and behaviors, which can be used to infer private information (e.g., online activities and personal preferences). This framework naturally captures a privacy-preserving mechanism in which the insurer is assumed to be uncertain about the user and his type. Depending on the choice of random variable θ_i, the level of uncertainties can vary, and hence θ_i can be used to determine the level of privacy of a user.

Player I measures the risk and pays the amount $s_i(X_i)$ for the losses, where $s_i : \mathbb{R}_+ \rightarrow \mathbb{R}_+$ is the payment function that reduces the risk of the user i if he is insured by Player I. Hence the effective loss to the user is denoted by $\xi_i = X_i - s_i(X_i)$, and hence user i aims to minimize a cost function U_i that depends on ξ_i, a_i and q_i given by $U_i(\xi_i, a_i, q_i)$, where $U_i : \mathbb{R}_+ \times [0,1]^2 \rightarrow \mathbb{R}_+$ is a continuous function monotonically increasing in ξ and q_i, and decreasing in a_i. The function captures the fact that a higher investment in the protection and careful usage of the network on the user side will lead to a lower cost, while a higher intensity of attack will lead to a higher cost. Therefore, given payment policy s_i, the interactions between an attacker and a defender can be captured by a zero-sum game in which the user minimizes U_i while the attacker maximizes it:

$$(\text{UG-1}) \quad \min_{a_i \in [0,1]} \max_{q_i \in [0,1]} \mathbb{E}[U_i(\xi_i, a_i, q_i)]. \tag{2}$$

Here, the expectation is taken with respect to the statistics of θ_i. The minimax problem can also be interpreted as a worst-case solution for a user who deploys best security strategies by anticipating the worst-case attack scenarios. From the attacker side, he aims to maximize the damage under the best-effort protection of the user, i.e.,

$$(\text{UG-2}) \quad \max_{q_i \in [0,1]} \min_{a_i \in [0,1]} \mathbb{E}[U_i(\xi_i, a_i, q_i)]. \tag{3}$$

The two problems described by (UG-1) and (UG-2) constitute a zero-sum game on at the user level. For a given insurance policy s_i, user i chooses protection level $a_i^* \in A_i(s_i)$ with the worst-case attack $q_i^* \in Q_i(s_i)$. Here, A_i and Q_i are set-valued functions that yield a set of saddle-point equilibria in response to s_i, i.e., a_i^* and q_i^* satisfy the following

$$\mathbb{E}[U_i(\xi_i, a_i^*, q_i)] \leq \mathbb{E}[U_i(\xi_i, a_i^*, q_i^*)] \leq \mathbb{E}[U_i(\xi_i, a_i, q_i^*)], \tag{4}$$

for all $a_i, q_i \in [0,1]$. In addition, in the case that $A_i(s_i)$, and $Q_i(s_i)$ are singleton sets, the zero-sum game admits a unique saddlepoint equilibrium strategy pair (a_i^*, q_i^*) for every s_i. We will use a shorthand notation val to denote the value of the zero-sum game, i.e.,

$$\mathbb{E}[U_i(\xi_i, a_i^*, q_i^*)] = \text{val}[\mathbb{E}[U_i(\xi_i, a_i, q_i)]], \tag{5}$$

and arg val to denote the strategy pairs that achieve the game value, i.e.,

$$(a_i^*, q_i^*) \in \arg\, \text{val}[\mathbb{E}[U_i(\xi_i, a_i, q_i)]]. \tag{6}$$

The outcome of the zero-sum game will influence the decision of the insurance company in choosing payment rules. The goal of the insurance company is twofold. One is to minimize the payment to the user, and the other is to reduce the risk of the user. These two objectives well aligned if the payment policy s_i is an increasing function in X_i, and we choose cost function $V(s_i(X_i))$, where $V : \mathbb{R}_+ \rightarrow \mathbb{R}_+$ is a continuous and increasing function. Therefore, with these assumptions, Player I aims to find an optimal policy among a class of admissible policies $\mathcal{S}_i$ to solve the following problem:

$$\text{(IP)} \quad \min_{s_i \in \mathcal{S}_i} \quad \mathbb{E}[V(s_i(X_i))]$$

$$\text{s.t.} \quad \text{Saddle-Point (6).}$$

This problem is a bilevel problem in which the insurance company can be viewed as the leader who announces his insurance policy, while the user behaves as a follower who reacts to the insurer. This relationship is depicted in Fig. 2. One important feature of the game here is that the insurer cannot directly observe the action a_i of the follower, but its state X_i. This class of problem differs from the classical complete information Stackelberg games and the signaling games where the leader (or the sender) has the complete information whereas the follower (or the receiver) has incomplete information. In this case the leader (the insurance company) has incomplete information while the follower (the user) has complete information. The game structure illustrated in Fig. 2 has a games-in-games structure. A zero-sum game between a user and a defender is nested in a bilevel game problem between a user and the insurer.

It is also important to note that user i pays Player I a subscription fee $T \in \mathbb{R}_{++}$ to be insured. The incentive for user i to buy insurance is when the average cost at equilibrium under the insurance is lower the cost incurred without insurance. Therefore, user i participates in the insurance program when

$$\mathbb{E}[U_i(\xi_i, a_i^*, q_i^*)] \geq T. \tag{7}$$

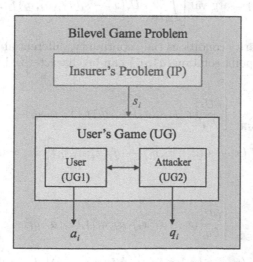

Fig. 2. The bilevel structure of the two-person cyber insurance game. The problem has a games-in-games structure. The user and the attacker interact through a zero-sum game while the insurer and the user interact in a bilevel game in which the user has complete information but the leader does not.

It can bee seen that the insurance policy plays an important role in the participation decision of the user. If the amount of payment from the insurer is low,
then the user tends not to be insured. On the other hand, if the payment is high,
then the risk for the insurer will be high and the user may behave recklessly in
the cyber space, as have been shown in Peltzman's effect [14].

3   Analysis of the Cyber Insurance Model

The formal framework introduced in Sect. 2 provides the basis for analysis and
design of cyber insurance to reduce risks for the Internet users. One challenge in
the analysis of the model comes from the information asymmetry between the
user and the insurer, and the information structure illustrated in Fig. 1. Since the
cost functions in (UG-1), (UG-2), and (IP) are expressed explicitly as a function
of X_i, the optimization problems can be simplified by taking expectations with
respect to the sufficient statistics of X_i. Let f_i be the probability density function
of X_i. Clearly, f_i is a transformation from the density function g_i (associated with
the random variable θ_i) under the mapping $\mathcal{G}_i$. In addition, f_i also depends on the
action pair (a_i, q_i) through the variable Z_i. Therefore, we can write $f_i(x_i; a_i, q_i)$
to capture the parametrization of the density function. To this end, the insurer's
bilevel problem (IP) can be rewritten as follows:

$$(\text{IP}')\quad \min_{s_i \in \mathcal{S}_i} \int_{x_i \in \mathbb{R}_+} V(s_i(x_i)) f_i(x_i, a_i^*, q_i^*) dx_i$$

$$\text{s.t.}\quad (a_i^*, q_i^*) = \arg \text{val} \left[\int_{x_i \in \mathbb{R}_+} U_i(x_i - s_i(x_i), a_i, q_i) f_i(x_i, a_i, q_i) dx_i \right].$$

Under the regularity conditions (i.e., continuity, differentiability and measur-
ability), the saddle-point solution (a_i^*, q_i^*) can be characterized by the first-order
conditions:

$$\int_{x_i \in \mathbb{R}_+} \left[\frac{\partial U_i}{\partial a_i}(x_i - s_i(x_i), a_i, q_i) f_i(x_i; a_i, q_i) \right.$$
$$\left. + U_i(x_i - s_i(x_i), a_i, q_i) \frac{\partial f_i}{\partial a_i} f_i(x_i; a_i, q_i) \right] dx_i = 0, \tag{8}$$

and

$$\int_{x_i \in \mathbb{R}_+} \left[\frac{\partial U_i}{\partial q_i}(x_i - s_i(x_i), a_i, q_i) f_i(x_i; a_i, q_i) \right.$$
$$\left. + U_i(x_i - s_i(x_i), a_i, q_i) \frac{\partial f_i}{\partial q_i} f_i(x_i; a_i, q_i) \right] dx_i = 0, \tag{9}$$

In addition, with the assumption that f_i and U_i are both strictly convex in
a_i and strictly concave in q_i, the zero-sum game for a given s_i admits a unique

saddle-point equilibrium [15]. Using Lagrangian methods from vector-space optimization [16], we can form a Lagrangian function with multipliers $\lambda_i, \mu_i \mathbb{R}_+$ as follows:

$$\mathcal{L}(s_i, \mu_i, a_i, q_i; \lambda_i, \mu_i) = \int_{x_i \in \mathbb{R}_+} V(s_i(x_i)) f_i(x_i, a_i, q_i) dx_i +$$

$$\lambda_i \int_{x_i \in \mathbb{R}_+} \left[\frac{\partial U_i}{\partial a_i}(x_i - s_i(x_i), a_i, q_i) f_i(x_i; a_i, q_i) \right.$$

$$\left. + U_i(x_i - s_i(x_i), a_i, q_i) \frac{\partial f_i}{\partial a_i} f_i(x_i; a_i, q_i) \right] dx_i +$$

$$\mu_i \int_{x_i \in \mathbb{R}_+} \left[\frac{\partial U_i}{\partial q_i}(x_i - s_i(x_i), a_i, q_i) f_i(x_i; a_i, q_i) \right.$$

$$\left. + U_i(x_i - s_i(x_i), a_i, q_i) \frac{\partial f_i}{\partial q_i} f_i(x_i; a_i, q_i) \right] dx_i.$$

The insurer's bilevel problem can thus be rewritten as a one-level optimization problem with Lagrange function $\mathcal{L}$:

$$\text{(IP')} \quad \max_{\lambda_i, \mu_i} \quad \min_{s_i \in \mathcal{S}_i, a_i \in [0,1], q_i \in [0,1]} \quad \mathcal{L}(s_i, \mu_i, a_i, q_i; \lambda_i, \mu_i).$$

Generally speaking, this Lagrangian is not simple to study but, as we see in the next section, several assumptions of the utility functions will help us to obtain the characterization of the optimal payment policies for the insurer.

3.1   Separable Utilities

One main assumption about player utility function is that it is separable into his variables, i.e.:

$$\forall i \in \{1, \ldots, N\}, \quad U_i(\xi_i, a_i, q_i) = H_i(\xi_i) + c_i(a_i, q_i).$$

In fact, the protection investment a_i induces a direct cost $c_i(a_i, q_i)$ on user i. This cost function is strictly increasing in a_i. Moreover, each player is typically risk-averse, and H_i is assumed to be increasing and concave. We give general results considering this particular case of separable utilities.

Following the first-order conditions (8) for user i, we obtain

$$\int_{x_i \in \mathbb{R}_+} \left[H_i(x_i - s_i(x_i)) \frac{\partial f_i}{\partial a_i}(x_i; a_i, q_i) + \frac{\partial c_i}{\partial a_i}(a_i, q_i) f_i(x_i; a_i, q_i) \right] dx_i = 0.$$

As we have $\frac{\partial c_i}{\partial a_i}(a_i, q_i) > 0$, the last equality is equivalent to:

$$\frac{H_i(x_i - s_i(x_i))}{-\frac{\partial c_i}{\partial a_i}(a_i, q_i)} \frac{\partial f_i}{\partial a_i}(x_i; a_i, q_i) = f_i(x_i; a_i, q_i)$$

Similarly, following (9), we obtain

$$\int_{x_i \in \mathbb{R}_+} \left[H_i(x_i - s_i(x_i)) \frac{\partial f_i}{\partial q_i}(x_i; a_i, q_i) + \frac{\partial c_i}{\partial q_i}(a_i, q_i) f_i(x_i; a_i, q_i) \right] dx_i = 0,$$

and arrive at

$$\frac{H_i(x_i - s_i(x_i))}{-\frac{\partial c_i}{\partial q_i}(a_i, q_i)} \frac{\partial f_i}{\partial q_i}(x_i; a_i, q_i) = f_i(x_i; a_i, q_i).$$

Therefore, we arrive at the following proposition:

Proposition 1. *The saddle-point strategy pair (a_i, q_i) satisfies the following relationship for every $x_i \in \mathbb{R}_+$:*

$$\frac{\frac{\partial f_i(x_i; a_i, q_i)}{\partial a_i}}{\frac{\partial f_i(x_i; a_i, q_i)}{\partial q_i}} = \frac{\frac{\partial c_i(a_i, q_i)}{\partial a_i}}{\frac{\partial c_i(a_i, q_i)}{\partial q_i}} \tag{10}$$

It can be seen that the saddle-point strategy pair depends on the state x_i. For different risk, the user will invest accordingly to protect his computer system.

3.2   Case Study: Cyber Insurance Under Infection Dynamics

We consider a possible virus or worm that propagates into a network. Each computer can be infected by this worm and we assume that if a node is infected, it induces a time window in which the node is vulnerable to serious cyber-attacks. The propagation dynamics follow a Susceptible-Infected-Susceptible (SIS) type infection dynamics [17] such that the time duration a node is infected follows an exponential distribution with parameter γ that depends on a and q. Note that we remove index i for the convenience of notations. Indeed, when a computer is infected, it is vulnerable to serious cyber-attacks. These can cause an outbreak of the machine and of the network globally. We thus assume that the parameter γ is increasing in a (resp. decreasing in q) meaning that more protection (resp. more attacks) reduces (resp. increases) the remaining time the node/computer is infected. Then, the action of the node decreases his risk whereas the action of the attacker increases the risk. We make also the following assumptions:

- The cost function is convex, i.e., the user is absolute risk-averse: $\forall \xi$, $H(\xi) = e^{r\xi}$;
- The cost function $c(a, q) = a - q$ is bi-linear;
- X follows an exponential distribution with parameter $\gamma(a, q)$, i.e., $X \sim \exp(\gamma(a, q))$. This random variable may represent the time duration a node is infected under an SIS epidemic process.
- The insurance policy is assumed to be linear in X, i.e., sX, where $s \in [0, 1]$. Hence the residual risk to the user is $\xi = (1 - s)X$.

Without loss of generality, we denote γ as a single constant when the notation does not lead to confusion. We thus have the following density function for the outcome:

$$\forall x \in \mathbb{R}_+, \quad f(x|a,q) = \gamma(a,q)e^{-\gamma(a,q)x}.$$

Then, we obtain

$$\forall x \in \mathbb{R}_+, \quad f_a(x|a,q) = \gamma_a e^{-\gamma x}(1 - \gamma x),$$

where by abuse of notation we denote $\gamma := \gamma(a,q)$ and $\gamma_a := \frac{\partial \gamma}{\partial a}(a,q)$. The average amount of damage is $\mathbb{E}(X) = \frac{1}{\gamma(a,q)} := \frac{q}{a}$. The expected utility of the node is given by:

$$\mathbb{E}U(X,a,q) = \int_0^\infty [H(x - sx) + c(a,q)]\, f(x|a,q)dx,$$

$$= c(a,q) + \int_0^\infty H(x(1-s))f(x|a,q)dx,$$

$$= c(a,q) + \frac{a}{q}\int_0^\infty e^{rx(1-s)-x\frac{a}{q}}dx,$$

$$= a - q + \frac{a}{q}\int_0^\infty e^{x[r(1-s)-\frac{a}{q}]}dx,$$

We assume that $a > qr(1-s)$ then:

$$\mathbb{E}U(X,a,q) = a - q + \frac{a}{a - qr(1-s)}.$$

We can observe that the optimal protection level of the node depends in a non-linear fashion of the cyber-attack level. For a given action of the attacker q and a contract s, the best action $a^*(s,q)$ for the node protection level is:

$$a^*(s,q) = \arg\min_a \mathbb{E}U(X,a,q) = q(1-s)r + \sqrt{q(1-s)r}.$$

Given the best protection, we can obtain the saddle point solution:

$$a^* = q^* = \frac{r(1-s)}{(1 - r(1-s))^2}.$$

If a player does not subscribe to cyber insurance, i.e., $s = 0$, then his best action becomes

$$a^*(0) = qr + \sqrt{qr}.$$

Hence, its expected cost is:

$$\mathbb{E}U^0 = \frac{a^*(0)}{a^*(0) - qr} + a^*(0) = qr + 2\sqrt{qr} + 1 = (1 + \sqrt{qr})^2.$$

If the player decides to be insured, then $s > 0$, i.e., part of his damage is covered and he has to pay a flat rate T for the participation. Then, his best action becomes $a^*(s)$ that depends on his coverage level s, and his expected cost is:

$$EU^s = \frac{a^*(s)}{a^*(s) - qr(1-s)} + a^*(s) + T = qr(1-s) + 2\sqrt{qr(1-s)} + 1 + T,$$

$$= (1 + \sqrt{qr(1-s)})^2 + T.$$

Proposition 2. *If the cyber insurance is too expensive, i.e.* $T \geq T_{\max} := qr + 2\sqrt{qr}$, *then the player will not subscribe to the cyber insurance independent of the coverage level* s.

Sketch of Proof. This proposition comes from the equivalence of $EU^1 \geq EU^0$ with $T \geq qr + 2\sqrt{qr}$. In this case, independent of the coverage level s, we have $EU^s \geq EU^0$, which implies that the node will not choose to pay the cyber insurance for any coverage level s.

Proposition 3. *For the subscription fee* $T < qr + 2\sqrt{qr}$, *there exists a minimum coverage* $s^0(T)$ *such that, for any coverage level* $s \in [s^0(T), 1]$, *the player will subscribe to the cyber-insurance. This minimum coverage is equal to:*

$$s^0(T) = 1 - \left(\frac{\sqrt{(1 + \sqrt{qr})^2 - T} - 1}{\sqrt{qr}} \right)^2.$$

Sketch of Proof. The function EU^s is strictly decreasing in s and $\lim_{s \to 0} EU^s > EU^0$. If $T < qr + 2\sqrt{qr}$, then $EU^1 < EU^0$. Hence, for a given $T < qr + 2\sqrt{qr}$, there exists a unique $s^0(T)$ such that $EU^{s^0(T)} = EU^0$. Moreover, for any $s \in [s^0(T), 1]$, we have $EU^s < EU^0$, then the player will subscribe to cyber insurance. By comparing the expressions of the expected utility functions, we obtain the following solution:

$$s^0(T) = 1 - \left(\frac{\sqrt{(1 + \sqrt{qr})^2 - T} - 1}{\sqrt{qr}} \right)^2.$$

We observe in Fig. 3 that for a same price T, for the node to subscribe to insurance, the level of cyber attack has to be sufficiently high. If we consider a competition framework in which the cyber insurer cannot change its price T, then for a fixed price, a higher cyber attack level leads to less minimum coverage accepted by the node. This shows that cyber attack plays an important role in insurance policy as it increases the willingness of the users to be insured.

The loss probability is defined as the probability that the damage covered by the insurance exceeds the price paid by the subscriber. We then define this loss of profit by $L(T) := \mathbb{P}(s^0(T)X(s^0(T)) > T)$, and obtain the following expression of the loss as:

$$L(T) = \exp\left(-\frac{q(1 - s^0(T))r + \sqrt{q(1 - s^0(T))r}}{qs^0(T)}T\right).$$

As we can see in Fig. 4, the loss is not monotone in the price, and a small price does not guarantee a profit (no loss) for the insurance company. One goal of the extended version of this work is to study the property of this loss depending on T.

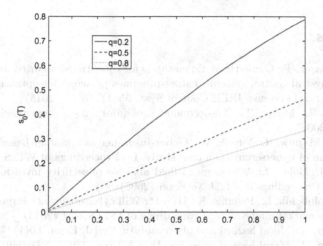

Fig. 3. Minimum coverage s_0 depending on the price T and cyber-attack level q with a risk-averse coefficient $r = 2$.

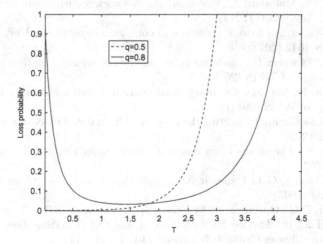

Fig. 4. Loss probability depending on the price T.

4   Conclusion

In this paper, we describe a game-theoretic framework for studying cyber insurance. We have taken into account complex interactions between users, insurer and attackers. The framework incorporates attack models, and naturally provides privacy-preserving mechanisms through the information asymmetry between the players. This work provides a first step towards a holistic understanding of cyber insurance and the design of optimal insurance policies. We would extend this framework to capture network effects, and address the algorithmic and design issues in cyber insurance.

References

1. Zhu, Q., Başar, T.: Game-theoretic methods for robustness, security, and resilience of cyberphysical control systems: games-in-games principle for optimal cross-layer resilient control systems. IEEE Control Syst. **35**(1), 46–65 (2015)
2. Anderson, R., Moore, T.: The economics of information security. Science **314**, 610–613 (2006)
3. Kesan, J., Majuca, R., Yurcik, W.: Cyber-insurance as a market-based solution to the problem of cybersercurity: a case study. In: Proceedings of WEIS (2005)
4. Lelarge, M., Bolot, J.: A local mean field analysis of security investments in networks. In: Proceedings of ACM NetEcon (2008)
5. Pal, R., Golubchik, L., Psounis, K., Hui, P.: Will cyber-insurance improve network security? a market analysis. In: Proceedings of INFOCOM (2014)
6. Hölmstrom, B.: Moral hazard and observability. Bell J. Econ. **10**(1), 74–91 (1979)
7. Holmstrom, B.: Moral hazard in teams. Bell J. Econ. **13**(2), 324–340 (1982)
8. Lelarge, M., Bolot, J.: Cyber insurance as an incentive for internet security. In: Proceedings of WEIS (2008)
9. Acemoglu, D., Malekian, A., Ozdaglar, A.: Network security and contagion. Perform. Eval. Rev. **42**(3) (2014)
10. Goyal, S., Vigier, A.: Attack, defense and contagion in networks. Rev. Econ. Stud. **81**(4), 1518–1542 (2014)
11. Böhme, R., Schwartz, G.: Modeling cyber-insurance: towards a unifying framework. In: Proceedings of WEIS (2010)
12. Naghizadeh, P., Liu, M.: Voluntary participation in cyber-insurance markets. In: Proceedings of WEIS (2014)
13. Raymond Law Group: Protecting the individual from data breach. In: The National LawReview (2014)
14. Peltzmann, S.: The effects of automobile safety regulation. J. Polit. Econ. **83**(4), 677–726 (1975)
15. Başar, T., Olsder, G.J.: Dynamic Noncooperative Game Theory, vol. 200. SIAM, Philadelphia (1995)
16. Luenberger, D.G.: Optimization by Vector Space Methods. Wiley, New York (1997)
17. Bailey, N.T.J., et al.: The Mathematical Theory of Infectious Diseases and its Applications. Charles Griffin & Company Ltd., London (1975). 5a Crendon Street, High Wycombe, Bucks HP13 6LE

Beware the Soothsayer: From Attack Prediction Accuracy to Predictive Reliability in Security Games

Benjamin Ford[✉], Thanh Nguyen, Milind Tambe, Nicole Sintov,
and Francesco Delle Fave

University of Southern California, Los Angeles, CA, USA
{benjamif,thanhhng,tambe,sintov,dellefav}@usc.edu

Abstract. Interdicting the flow of illegal goods (such as drugs and ivory) is a major security concern for many countries. The massive scale of these networks, however, forces defenders to make judicious use of their limited resources. While existing solutions model this problem as a Network Security Game (NSG), they do not consider humans' bounded rationality. Previous human behavior modeling works in Security Games, however, make use of large training datasets that are unrealistic in real-world situations; the ability to effectively test many models is constrained by the time-consuming and complex nature of field deployments. In addition, there is an implicit assumption in these works that a model's prediction accuracy strongly correlates with the performance of its corresponding defender strategy (referred to as predictive reliability). If the assumption of predictive reliability does not hold, then this could lead to substantial losses for the defender. In the following paper, we (1) first demonstrate that predictive reliability is indeed strong for previous Stackelberg Security Game experiments. We also run our own set of human subject experiments in such a way that models are restricted to learning on dataset sizes representative of real-world constraints. In the analysis on that data, we demonstrate that (2) predictive reliability is extremely weak for NSGs. Following that discovery, however, we identify (3) key factors that influence predictive reliability results: the training set's exposed attack surface and graph structure.

1 Introduction

By mathematically optimizing and randomizing the allocation of defender resources, Security Games provide a useful tool that has been successfully applied to protect various infrastructures such as ports, airports, and metro lines [16]. Network Security Games (NSGs), a type of Security Game, can be applied to interdict the flow of goods in smuggling networks (e.g., illegal drugs, ivory) or defend road networks from terrorist attacks (e.g., truck bombs). In comparison to previous work in Security Games [15], however, the number of possible actions for both attacker and defender grow exponentially for NSGs; novel scaling techniques have been developed to address this challenge by Jain et al. [10] for perfectly rational attackers.

© Springer International Publishing Switzerland 2015
MHR Khouzani et al. (Eds.): GameSec 2015, LNCS 9406, pp. 35–56, 2015.
DOI: 10.1007/978-3-319-25594-1_3

While early work in Security Games relied on the assumption of perfect adversary rationality, more recent work has shifted away towards modeling adversary bounded rationality [1,5,11,14]. In the effort to model human decision making, many human behavior models are being developed. As more Security Game applications are being deployed and used by security agencies [7,15], it becomes increasingly important to validate these models against real-world data to better ensure that these and future applications don't cause substantial losses (e.g., loss of property, life) for the defender. In efforts to generate real-world data, previous work [7,15] has demonstrated that field experiments are time-consuming and complex to organize for all parties involved; the amount of field experiments that can be feasibly conducted is grossly limited. Thus, in real-world situations, we will have limited field data.

By analyzing the prediction accuracy of many models on an existing large dataset of human subject experiments, previous works [1,5] empirically analyze which models most closely resemble human decision making for Stackelberg (SSG) and Opportunistic Security Games. While these works demonstrate the superiority of some models in terms of prediction accuracy and fitting performance, they do not address the larger, implicit question of how the models' corresponding strategies would perform when played against human subjects (i.e., average defender expected utility). We do not know how well the prediction accuracy of a model will correlate with its actual performance if we were to generate a defender strategy that was based on such a model; informally defined, **predictive reliability** refers to the percentage of strong correlations between a model's prediction accuracy and the model's actual performance. It is also unknown whether the prediction accuracy analysis approach will be suitable, especially for NSGs, in situations where we have limited field data from which to learn the models. As previously discussed, the amount of field experiments that can be conducted (and thus the amount of training data available for learning) is limited; it is important to know whether the model with superior prediction accuracy will actually result in higher defender gains than a model with worse prediction accuracy (especially when training data is limited). This raises the following question for NSG research: "Without the ability to collect very large amounts of data for training different bounded rationality models and without the ability to conduct very large amounts of tests to compare the performance of these models in action, how do we ensure high predictive reliability and choose the most promising models?"

We first lay the groundwork for determining whether our proposed construct of predictive reliability is valid in SSGs. As such, we first (i) conduct an empirical evaluation of predictive reliability in SSGs in situations where there is a large amount of training data. We then (ii) evaluate predictive reliability for NSGs. In this study, we use NSG human subject data from the lab and train our models on enough data such that prediction accuracies converge[1]. Following

---

[1] In other words, to simulate real-world scenarios, we do not assume the presence of very large amounts of data, but nonetheless, there is a sufficient amount of NSG data included in our study to at least see a stable prediction made by our different behavior models.

this primary analysis, we then examine the various factors that may influence predictive reliability. We propose a metric called Exposed Attack Surface (EAS) which is related to the degree of choice available to the attacker for a given training set. We then (iii) examine the effects of EAS on predictive reliability, and (iv) investigate which graph features influence predictive reliability.

Our primary analysis shows that (i) predictive reliability is strong for an SSG dataset where there is sufficient training data, (ii) even though there is sufficient training data (at least to see our models' prediction accuracies converge), predictive reliability is poor for NSGs. In our analysis to discover which factors have the most influence on predictive reliability, we find that (iii) a training set with a higher EAS score results in better predictive reliability than a training set with a lower EAS score. Note that this finding is independent of the training set's size (both training sets are of the same size). While it won't always be possible to obtain training data with a large exposed attack surface, if we do have it, we can be more confident in the predictive reliability of our models. In addition, we find that (iv) there is a strong correlation between poor predictive reliability and whether a graph has both a low to moderate number of intermediate nodes and a low to moderate number of outgoing edges from source nodes.

2   Background: Network Security Games

This paper will address zero-sum Network Security Games (NSGs). For a table of notations used in this paper, see Table 1. In NSGs, there is a network (shown in Fig. 1) which is a graph g containing a set of nodes/vertices V (the dots/circles in the figure) and a set of edges E (the arrows in the figure, labelled 1–6). In the network, there is a set of target nodes, denoted by $T \subset V$. While the defender

Table 1. Notations used in this paper

$g(V,E)$	General directed graph
J	Set of paths in graph g
k	Number of defender resources
X	Set of defender allocations, $X = \{X_1, X_2, ..., X_n\}$
X_i	i^{th} defender allocation $X_i = \{X_{ie}\}\ \forall e,\ X_{ie} \in \{0,1\}$
A	Set of attacker paths, $A = \{A_1, A_2, ..., A_m\}$
A_j	j^{th} attacker path $A_j = \{A_{je}\}\ \forall e,\ A_{je} \in \{0,1\}$
t_j	Target t in the graph g such that the attacker takes path j to attack t
$\mathcal{T}(t_j)$	The reward obtained for a successful attack on target t by taking path j s.t. $A_j \cap X_i = \emptyset$ where A_j is the attacker's selected path to attack target t and X_i is the selected defender allocation
x	Defender's mixed strategy over X
x_i	Probability of choosing defender pure strategy X_i
$EU_d(x)$	Defender's expected utility from playing x
z_{ij}	Function that refers to whether a defender allocation X_i intersects with an attacker path A_j. If there is an intersection, returns 1. Else, 0

attempts to allocate her limited resources to protect these target nodes, the attacker can observe the defender's patrolling strategy and then attack one of the target nodes based on that observation.

Attacker Strategies. The attacker can start at a source node $s \in S$ (where $S \subset V$ is the set of all source nodes in the network) and chooses a sequence of nodes and edges leading to a single target node $t \in T$. The attacker's decision corresponds to a single path $j \in J$ and is referred to as the attacker's path choice $A_j \in A$ where A is the set of all possible paths that the attacker can choose.

Defender Strategies. The defender can allocate her k resources to any subset of edges in the graph; each allocation is referred to as a pure strategy for the defender, denoted by X_i. There are $\binom{|E|}{k}$ defender pure strategies in total, and we denote this set of pure strategies by X. Then, a defender's *mixed* strategy is defined as a probability distribution over all pure strategies of the defender, denoted by $x = \{x_i\}_{i=1}^{N}$, where x_i is the probability that the defender will follow the pure strategy X_i and $\sum_i x_i = 1$.

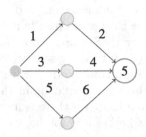

Fig. 1. Example graph

Defender and Attacker Utilities. An attack is successful if the attacker's path choice does not contain any edges in common with the defender's allocation ($X_i \cap A_j = \emptyset$), and the attacker will receive a reward $\mathcal{T}(t_j)$ while the defender receives a penalty of $-\mathcal{T}(t_j)$. Here, t_j is the target node on the path A_j. Conversely, if the attack is unsuccessful (i.e., the attacker's path intersected with the defender's allocation), both attacker and defender receive a payoff of 0.

Finally, the defender's expected utility of executing a mixed strategy x given an attacker path A_j can be computed as shown in Eq. 1 where the term $p_j(x)$ (defined in Eq. 2) refers to the probability that the adversary will be caught when choosing path A_j to attack target node t_j. In zero-sum games, the attacker's expected utility for choosing path A_j is equal to the opposite of the defender's expected utility, i.e., $EU_a(x, A_j) = -EU_d(x, A_j)$.

$$EU_d(x, A_j) = -\mathcal{T}(t_j) \cdot (1 - p_j(x)) \tag{1}$$

In Eq. 2, z_{ij} is an integer which indicates if the defender's pure strategy X_i intersects with the attacker path A_j ($z_{ij} = 1$) or not ($z_{ij} = 0$).

$$p_j(x) = \sum_{X_i \in X} z_{ij} x_i \tag{2}$$

3   Related Work

Human bounded rationality has received considerable attention in Security Game research [1,5,11,14]. The goal of these works was to accurately model human decision making such that it could be harnessed to generate defender strategies that lead to higher expected utilities for the defender. For the developed models and corresponding defender mixed strategies, some of these works conducted human subject experiments to validate the quality of their models [1,11,14]. Often in this research, different models' prediction accuracies are tested against human subjects, and the one that is most accurate is then used to generate defender strategies against human subjects [11,14]. However, these works do not evaluate whether or not the other models' prediction accuracies correlated with their actual performance (i.e., predictive reliability). In other words, prediction accuracy is used as a proxy for the defender's actual performance, but it has not been well established that this is a reasonable proxy to use. In order to evaluate predictive reliability for SSGs, we obtained the human subject experiment data from Nguyen et al. [14] and evaluated predictive reliability on this data between the Quantal Response (QR) and Subjective Utility Quantal Response (SUQR) models.

As yet another type of Security Game, NSG research covers a wide variety of applications and domains. NSGs have been applied to curbing the illegal smuggling of nuclear material [13], protecting maritime assets such as ports and ferries [15], studying ways to minimize road network disruptions [2], deterring fare evasion in public transit systems [4], and the assignment of checkpoints to urban road networks [9,17]. Although our NSG models most closely resemble the model used by Jain et al. [9,10], the primary difference is that we are not limited to modeling perfectly rational attackers.

In most NSG research, there is a basic assumption that the attacker is perfectly rational, but as demonstrated in work in Behavioral Game Theory by Camerer et al., humans do not behave with perfect rationality [3]. Gutfraind et al. [8] address one type of boundedly rational adversary, an unreactive Markovian evader, in their work. Even though the evader (i.e., attacker) is unreactive to the defender's actions, the relaxation of the rational adversary assumption still results in an NP-hard problem. Positing that humans will rely on heuristics due to the complex nature of solving an NSG, Yang et al. [18] address bounded rationality in a non-zero sum NSG setting by modeling the adversary's stochastic decision making with the Quantal Response (QR) model and various heuristic based variants of the QR model. While they demonstrated that attacker behavior is better captured with human behavior models, their work is limited to using one defender resource in generating defender strategies and only focused on much smaller networks. In order to adequately defend larger networks, like those modeled in previous work by Jain et al. [10] and the ones presented in this work, multiple defender resources are required. For the behavior models we present, multiple defender resources are supported in a zero-sum setting.

4   Adversary Behavioral Models

We now present an overview of all the adversary behavioral models which are studied in this paper.

4.1   The Perfectly Rational Model

In NSG literature, the adversary is often assumed to be perfectly rational and will always maximize his expected utility. In other words, the adversary will choose the optimal attack path that gives him the highest expected utility, i.e., $A_{opt} = \text{argmax}_{A_j} EU_a(x, A_j)$.

4.2   The Quantal Response Model

The Quantal Response (QR) model for NSGs was first introduced by Yang et al. [18]. However, their formulation only works under the assumption that there is one defender resource available, and as a result, we present a revised version of the QR model for a zero-sum NSG with multiple defender resources. In short, QR predicts the probability that the adversary will choose a path A_j, which is presented as the following:

$$q_j(\lambda|x) = \frac{e^{\lambda EU_j^a(x)}}{\sum_{A_k \in A} e^{\lambda EU_k^a(x)}} \tag{3}$$

where λ is the parameter that governs the adversary's rationality. For example, $\lambda = 0.0$ indicates that the adversary chooses each path uniformly randomly. On the other hand, $\lambda = \infty$ means that the adversary is perfectly rational. Intuitively, there is a higher probability that the adversary will follow a path with higher expected utility.

4.3   The Subjective Utility Quantal Response Model

Unlike QR, the Subjective Utility Quantal Response (SUQR) model [14] models the attacker's expected utility calculation as a weighted sum of decision factors such as reward and path coverage. As demonstrated by Nguyen et al. [14] for SSGs and Abbasi et al. [1] for Opportunistic Security Games (OSGs), SUQR performs better than QR for attack prediction accuracy. As such, we present an NSG adaptation of SUQR as shown in Eq. 4. Specifically, SUQR predicts the probability that the adversary chooses a path A_j as the following:

$$q_j(\omega|x) = \frac{e^{\omega_1 p_j(x) + \omega_2 \mathcal{T}(t_j)}}{\sum_{A_k \in A} e^{\omega_1 p_k(x) + \omega_2 \mathcal{T}(t_k)}} \tag{4}$$

where (ω_1, ω_2) are parameters corresponding to an attacker's preferences (i.e., weights) on the game features: the probability of capture $p_j(x)$ and the reward for a successful attack $\mathcal{T}(t_j)$.

4.4   The SUQR Graph-Aware Model

The previous models, designed for traditional Stackelberg Games, do not account for the unique features of Network Security Games. As such, we present some NSG-specific features that can be incorporated into the existing SUQR model in the form of additional parameters. Each of these features is computed for each path $A_j \in A$.

Path length simply refers to the number of edges in a path A_j, and the corresponding weight is referred to as ω_3 in Eq. 5. This model will henceforth be referred to as GSUQR1 (i.e., Graph-SUQR w/ 1 parameter). Yang et al. [18] also made use of path length as one of the tested QR heuristics.

$$q_j(\omega|x) = \frac{e^{\omega_1 p_j(x)+\omega_2 T(t_j)+\omega_3|A_j|}}{\sum_{A_k \in A} e^{\omega_1 p_k(x)+\omega_2 T(t_k)+\omega_3|A_k|}} \tag{5}$$

We also compute the maximum total degree (weight ω_4) of a path. This is an aggregate measure (maximum) of the path's nodes' indegrees (i.e., number of edges coming into the node) + outdegrees (i.e., number of edges leaving the node). We refer to this measure as MTO. A low value for this corresponds to simple paths with little connections to other areas of the graph; a high value corresponds to a path with one or more nodes that are highly connected to other paths. The resultant q_j function is shown in Eq. 6, and this model is henceforth referred to as GSUQR2.

$$q_j(\omega|x) = \frac{e^{\omega_1 p_j(x)+\omega_2 T(t_j)+\omega_3|A_j|+\omega_4 MTO_j}}{\sum_{A_k \in A} e^{\omega_1 p_k(x)+\omega_2 T(t_k)+\omega_3|A_k|+\omega_4 MTO_k}} \tag{6}$$

5   Defender Strategy Generation

In this section, we present the approach used to generate defender strategies for the boundedly rational adversary models.[2] Because the strategy space for NSGs can grow exponentially large, we address this by adapting a piecewise linear approximation approach, PASAQ, first introduced by Yang et al. [19]. Note that while we only show the PASAQ formulation as generating defender strategies for the QR model, we also adapted it for the SUQR, GSUQR1, and GSUQR2 models as well. Whereas the original PASAQ algorithm worked for SSGs involving independent targets and coverages, this paper has adopted PASAQ for NSGs, where non-independent path coverage probabilities ($p_j(x)$) must be taken into account. PASAQ works by performing a binary search to solve a non-linear fractional objective function. Determining whether the current solution is feasible, however, is a non-convex problem, and this feasibility checking problem is expressed as an inequality in Eq. 7, where r is the current binary search solution,

---

[2] The algorithm to generate a Maximin strategy can be found in [10].

x^* is the optimal defender mixed strategy, and $EU_d(x)$, the defender's expected utility given an adversary following the QR model, is defined in Eq. 8.[3]

$$r \leq EU_d(x^*) \tag{7}$$

$$EU_d(x) = \frac{\sum_{A_j \in A} e^{\lambda EU_a(x,A_j)} EU_d(x, A_j)}{\sum_{A_j \in A} e^{\lambda EU_a(x,A_j)}} \tag{8}$$

After rewriting Eq. 7 as a minimization function and further expansion, we obtain two non-linear functions $f_(j)^{(1)}(p_j(x)) = e^{\lambda(1-p_j(x))T(t_j)}$ and $f_(j)^{(2)}(p_j(x)) = (1 - p_j(x))e^{\lambda(1-p_j(x))T(t_j)}$ which are to be approximated. To do so, we divide the range $p_j(x) \in [0,1]$ into S segments (with endpoints $[\frac{s-1}{S}, \frac{s}{S}, s = 1 \ldots S]$) and will henceforth refer to each segment that contains a portion of $p_j(x)$ as $\{p_{js}, s = 1 \ldots S\}$. For example, p_{j2} refers to the second segment of $p_j(x)$ which is located in the interval $[\frac{1}{S}$ and $\frac{2}{S}]$. Our piecewise approximation follows the same set of conditions from [19]: each $p_{js} \in [0, \frac{1}{S}] \forall s = 1 \ldots S$ and $p_j = \sum_{s=1}^{S} p_{js}$. In addition, any $p_{js} > 0$ only if $p_{js'} = \frac{1}{S}, \forall s' < s$; in other words, p_{js} can be non-zero only when all previous partitions are completely filled (i.e., $= \frac{1}{S}$). Enforcing these conditions ensures that each p_{js} is a valid partition of $p_j(x)$. Following the definition from [19], the piecewise linear functions are represented using $\{p_{js}\}$. The S+1 segment end points of $f_j^{(1)}(p_j(x))$ can be represented as $\{(\frac{s}{S}, f_j^{(1)}(\frac{s}{S})),$ s=0...S} and the slopes of each segment as $\{\gamma_{js}, s=1 \ldots S\}$. Starting from $f_j^{(1)}(0)$, we denote the piecewise linear approximation of $f_j^{(1)}(p_j(x))$ as $L_j^{(1)}(p_j(x))$:

$$L_j^1(p_j(x)) = f_j^{(1)}(0) + \sum_{s=1}^{S} \gamma_{js} p_{js}$$
$$= e^{\lambda T(t_j)} + \sum_{s=1}^{S} \gamma_{js} p_{js} \tag{9}$$

The approximation of function $f_j^{(2)}(p_j(x))$ is performed similarly (slopes denoted as $\{\mu_{js}, s=1 \ldots S\}$) and yields $L_j^{(2)}(p_j(x))$.

$$L_j^2(p_j(x)) = e^{\lambda T(t_j)} + \sum_{s=1}^{S} \mu_{js} p_{js} \tag{10}$$

---

[3] Details on the binary search algorithm can be found in Yang et al.'s original PASAQ formulation [19].

Given the definition of these two piecewise linear approximations, the following system of equations details the solution feasibility checking function (invoked during the binary search):

$$\min_{x,b} \sum_{A_j \in A} (e^{\lambda T(t_j)} + \sum_{s=1}^{S} \gamma_{js} p_{js}) r \tag{11}$$

$$+ \sum_{A_j \in A} T(t_j)(e^{\lambda T(t_j)} + \sum_{s=1}^{S} \mu_{js} p_{js}) \tag{12}$$

$$s.t \sum_{X_i \in X} x_i \leq 1 \tag{13}$$

$$p_j(x) = \sum_{s=1}^{S} p_{js} \tag{14}$$

$$p_j(x) = \sum_{X_i \in X} z_{ij} x_i \tag{15}$$

$$b_{js} \frac{1}{S} \leq p_{js}, \forall j, s = 1 \ldots S - 1 \tag{16}$$

$$p_{j(s+1)} \leq b_{js}, \forall j, s = 1 \ldots S - 1 \tag{17}$$

$$0 \leq p_{js} \leq \frac{1}{S}, \forall j, s = 1 \ldots S \tag{18}$$

$$b_{js} \in \{0,1\}, \forall j, s = 1 \ldots S - 1 \tag{19}$$

$$z_{ij} \in \{0,1\}, \forall i, j \tag{20}$$

where b_{js} is an auxiliary integer variable that is equal to 0 only if $p_{js} < \frac{1}{S}$ (Eq. 16). Equation 17 enforces that $p_{j(s+1)}$ is positive only if $b_{js} = 1$. In other words, b_{js} indicates whether or not $p_{js} = \frac{1}{S}$ and thus enforces our previously described conditions on the piecewise linear approximation (ensuring each p_{js} is a valid partition). As demonstrated in [19], given a small enough binary search threshold ϵ and sufficiently large number of segments S, PASAQ is arbitrarily close to the optimal solution.

6   Human Subject Experiments

6.1   Experimental Overview

In order to test the effectiveness of these algorithms against human adversaries, we ran a series of experiments on Amazon Mechanical Turk (AMT). Even though we run these (effectively speaking) laboratory experiments, our goal is to collect this data in such a way as to simulate field conditions where there is limited data.[4]

---

[4] For a more detailed discussion of human subject experiment design considerations, such as steps taken to reduce sources of bias, please see the online appendix at: http://teamcore.usc.edu/people/benjamin/Ford15_GameSecAppendix.pdf.

Each participant was presented with a set of fifteen graphs in which they navigated a path from a source node to a destination node through using a series of intermediate nodes. Participants that successfully attacked a destination (without getting caught on an edge) received the corresponding reward; participants that got caught on an edge received zero points for that round. At the end of the experiment, participants received $1.50 plus the number of points they received (in cents) during the experiment. To avoid learning effects and other sources of bias, we took the following steps: randomized the order in which graphs were presented to participants, withheld success feedback until the end of the experiment, only allowed participants to participate in the experiment once, and finally, we divided participants into separate subject pools such that each participant only played against a single defender strategy and played on each of the fifteen graphs exactly once. Due to the inevitability of some participants playing randomly (thus confounding any behavioral analysis we may conduct), we included a set of validation rounds such that if participants chose a path that was covered by the defender 100 % of the time, we would drop their data from the analysis.

6.2   Experiment Data Composition

Participants and Dataset Sizes. In our experiments, all eligible AMT participants satisfied a set of requirements. They must have participated in more than 1000 prior AMT experiments with an approval rate of $\geq 95\,\%$, and we required that all participants were first-time players in this set of experiments. Out of 551 participants, 157 failed to complete all graphs or did not pass both validation rounds. The remainder, 394, successfully completed all rounds and passed both validation rounds, and we used only their data in the following data analyses.

Graph Design and Generation. To ensure our findings were not limited to a single set of homogeneous graphs, we generated three sets of random geometric graphs. Eppstein et al. demonstrated that geometric graphs were a suitable analogue to real-world road networks due to road networks' non-planar connectivity properties [6]. Each set was assigned a predefined neighborhood radius (r), corresponding to the maximum distance between two nodes for an edge to exist, and a predefined number of intermediate nodes (v_i). Set 1, a set of sparse random geometric graphs, had $r = 0.2$, $v_i = 10$, and was required to have at least 15 edges. Set 2, a set of densely connected graphs, had $r = 0.6$ and $v_i = 4$. Set 3, a set of intermediately connected graphs, had $r = 0.4$ and $v_i = 7$. In addition, all sets were generated with a set of common constraints; each graph was constrained to have no more than 30 edges, exactly two source nodes, and exactly three destination nodes (with reward values 3, 5, and 8).

For each set, we generated 100 unique random geometric graphs. For each graph, we first randomly placed the nodes in a 2-D region (a unit square), and edges were drawn between nodes that were, at most, a 2-norm distance r away from each other. During post-processing, invalid connections, such as edges connecting source nodes to other source nodes, were removed. After the set was generated, we computed a Maximin, QR, and SUQR strategy for each graph and computed a distance score. This distance score measured the 1-norm distance between the probability distributions (i.e., the mixed strategies) for two sets of strategies: QR and SUQR, and Maximin and SUQR; graphs with distinctly different defender strategies (in terms of the coverage probabilities on paths) would receive a high distance score. The five graphs with the highest distance scores were kept for the final set.

Model Parameter Learning. The full experiment set consists of eight subject pools. For the purposes of learning the model parameters for the human behavior models, however, we divided the experiment set into three separate experiment sets. The first experiment set consists solely of the Maximin subject pool (no model learning required). The latter two experiment sets are defined by the training dataset used to train the models (e.g., the experiment data from the Maximin subject pool). As was done in previous work on applying human behavior models to Security Games [1,11,14,18], we use Maximum Likelihood Estimation (MLE) to learn the parameter values (i.e., weights) for each behavior model. Because training data may be limited in the real-world, we limit the scope of each training dataset to contain data from only one subject pool. Unlike previous work in NSGs by Yang et al. [18], where one set of weights was learned across all graphs (i.e., an aggregate weight), we found that the log-likelihood was highest when weights were learned individually for each graph.

Experiment Set Composition. As mentioned previously, the experiments are divided into three separate experiment sets. Each combination of coverage strategy × graph set was assigned to their own subject pool. Prior to running these experiments, however, we had no training data on which to learn weights for the behavior models. Thus, the first experiment set, experiment set 1, only contains a coverage strategy generated by the Maximin algorithm.

Experiment set 2 contains coverage strategies generated by the corresponding PASAQ algorithms for the QR (Eq. 3), SUQR (Eq. 4), GSUQR1 (Eq. 5), and GSUQR2 (Eq. 6) models. For the models used to generate these strategies, we used the Maximin dataset as the training dataset to learn each model's weights. To help differentiate from the datasets in experiment set 3, we will refer to the datasets collected in experiment set 2 as QR-M, SUQR-M, GSUQR1-M, and GSUQR2-M.

Experiment set 3 also contains coverage strategies generated for the QR (Eq. 3), SUQR (Eq. 4), and GSUQR1 (Eq. 5) models. Instead of learning on Maximin data, however, we instead learn on GSUQR1-M data (from experiment set 2). As we will demonstrate later, learning from a non-Maximin dataset has a substantial positive impact on predictive reliability. As was done for experiment set 2, we will refer to the datasets collected in experiment set 3 as QR-S, SUQR-S, and GSUQR1-S.

6.3   Data Analysis Metrics

The following section discusses the various metrics used throughout our data analysis. First, we will introduce three metrics for computing model prediction accuracy (the degree to which a model correctly predicted attacker behavior). Next, we will introduce our proposed predictive reliability metric, which measures the degree to which models' predictions correspond to their actual performances. Finally, we introduce our last proposed metric, Exposed Attack Surface, which measures the number of unique path choices available to the attacker.

Model Prediction Accuracy. In previous empirical analyses [1,5] and in our own analysis, prediction accuracy measures are key to understanding the relative performance of behavior models; accuracy measures seek to answer the question "How well does this model predict human behavior?" Computed over all paths for each model × graph × coverage strategy combination, prediction accuracy quantifies the degree to which a model's predictions of attacker behavior were correct.

Regardless of a graph's size or coverage strategy, however, only a few paths have an actual probability of attack $(q_j) > 6\%$; most paths in most graphs are attacked with very low frequency. When looking at all paths in a graph, the average absolute prediction error (AAE) is 3%, regardless of the behavior model making the prediction. It appears that the error "outliers" are actually the primary values of interest. In other words, because there is no discriminatory power with the average, we instead analyze the maximum absolute prediction error (MAE) (Eq. 21) for each model, where $g \in G$ is a graph in the experiment set, ϕ is the behavior model (along with its weights) being evaluated, q_j is the behavior model ϕ's predicted attack proportion on path A_j given defender mixed strategy x, and $\hat{q}_j$ is the actual attack proportion on path A_j.

$$MAE(g, x, \phi) = \max_{A_j \in A} |q_j - \hat{q}_j| \tag{21}$$

As mentioned previously, only a few paths in a graph have some substantial probability of being attacked. Over all eight datasets, on average (across all graphs), 70% of all attacks occurred on only three paths (per graph). Thus, it is prudent to also analyze a model's prediction accuracy on these so-called "favored" paths.

Definition 1. *A path A_j is defined as a **favored path** A_{fj} if its actual probability of attack (q_j) is $\geq 10\%$.*

Similar to MAE but instead only over the favored paths $A_{fj} \subset A_j$ in a graph, we compute the maximum absolute error over favored paths (referred to as FMAE). Since this subset of paths does not suffer from excessive skewing, it is appropriate to also analyze the average absolute error (FAAE) over the set of favored paths A_{fj}.

Predictive Reliability. Now that we've introduced our prediction accuracy metrics, we turn our attention to the primary focus of our paper: predictive reliability - the degree to which models' prediction accuracies correspond with their corresponding strategies' performances in experiments. If predictive reliability is poor, then models chosen on the basis of having the best prediction accuracy may not perform the best when tested against actual humans; when field-deployment resources are limited, those resources should not be wasted on models that end up performing very poorly in the field!

After all human subject experiments have been conducted (we refer to the whole set of attack data as A_d), we can compute predictive reliability. Put simply, predictive reliability is the percentage of strong Pearson correlations. These correlations are computed separately for each combination of graph ($g \in G$), prediction accuracy metric (PAM), and testing dataset ($Te \in A_d$). For a given g, PAM, and Te, we compute the Pearson correlation over all models' (1) prediction accuracy on Te (using PAM), and (2) actual defender utility on the model's corresponding attack data (e.g., for model QR trained on Maximin, compute on the QR-M dataset). Note that if a model was trained on Te or if the model's corresponding attack data is Te, it is omitted from the Pearson correlation for that combination of g, PAM, and Te.

Definition 2. *Predictive reliability is defined as the percentage of correlations between actual utility values and prediction accuracies that are both (1) strong (magnitude > 0.70), and (2) in the desired direction (negative: as error decreases, actual utility increases). In other words, predictive reliability corresponds to the percentage of strong correlations (correlation < -0.70).*

Exposed Attack Surface. We now introduce our second proposed metric, Exposed Attack Surface (EAS). While early discussion of attack surface exposure was done by Manadhata et al. [12], more recently, Kar et al. [11] applied this concept to Repeated Stackelberg Security Games to improve the defender's utility against human subjects. EAS measures the number of unique attacker choices (i.e., paths) for a graph × strategy combination. To phrase this metric as a question, "Given a coverage strategy and graph, how many paths in the graph have a unique combination of path coverage and reward?" Referring to Fig. 2 as an example, there are three separate paths to target 5. While two of these paths have the same path coverage of {0.2, 0.2} (one attack surface), the other path has 0 path coverage (the second attack surface). Finally, the path to target 8 constitutes the last attack surface; the example figure's EAS score is 3. Although there are four paths in Fig. 2, two of these paths are equivalent to each other (i.e., same reward and coverage) and thus there are only three unique path choices (i.e., the EAS score) for the attacker.

Definition 3. *Exposed Attack Surface is defined as the number of unique combinations of reward $\mathcal{T}(t_j)$ and path coverage probability $p_j(x)$ over all paths A in a graph g.*

When computing this metric for a dataset $d_{\phi,G} \in D_{\Phi,G}$, we take the sum of EAS scores for each graph × coverage strategy (corresponding to a model ϕ) combination. To illustrate the simple (but important) intuition behind EAS, we present two extreme cases: (1) consider a training dataset that consists of a single graph × coverage strategy such that the graph's EAS score is one; all paths to

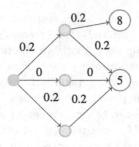

Fig. 2. Example graph 2

the single target have identical coverage (i.e., one unique path choice). When attempting to learn model parameters, it would be impossible to differentiate between attacker choices; obviously, this training set with a low EAS score is ill-suited for use in model learning. (2) In contrast, a training dataset with a high EAS score implies that there are many distinguishable attacker choices. Attacker choices over these many unique paths provide information about their preferences such that we can more effectively train a model; we hypothesize that a training dataset that contains more information about attacker preferences (i.e., one with high EAS) is superior to one that provides less information (i.e., low EAS).

7   Predictive Reliability Analysis

After defining predictive reliability in the previous section (Sect. 6.3), we now evaluate predictive reliability in previous work by Nguyen et al. [14] for SSGs, and then follow up with an evaluation of predictive reliability in our work for NSGs.

7.1   SSG Experiment

In this prior work on Stackelberg Security Games (SSGs), participants in human subject experiments were asked to play a game called "The Guards and Treasures". For one experiment, participants in each round (for 11 rounds total) picked one of 24 targets based on its defender coverage probability, reward and penalty to the attacker, and reward and penalty to the defender. For each of these rounds, five coverage strategies were generated: three corresponding to other defender strategy algorithms and two corresponding to the QR and SUQR human behavior models whose weights were learned from a prior dataset consisting of 330 data points. While the previous work demonstrated that SUQR's prediction accuracy was better than QR, and SUQR had the best corresponding strategy performance compared to other algorithms, it was an implicit assumption that the behavior model with the best prediction accuracy would also perform the best in human subject experiments. If predictive reliability was actually poor, then it could have been the case that QR and its strategy would have performed the best in experiments.

7.2   SSG Predictive Reliability

For the following analysis, we confirmed that predictive reliability was strong for this SSG experiment; prediction accuracy was reliably correlated with actual performance. In the dataset we obtained from Ngyuen et al. [14] (which contained human subject attack data), we computed the predictive reliability over the QR and SUQR models. Because there were only two models in this correlation, the correlation output was either -1 (i.e., supports good predictive reliability) or +1 (i.e., supports poor predictive reliability). This analysis was done across 11 different rounds and for each of the three non-QR/SUQR test datasets. In Table 2, we show the predictive reliability of the QR and SUQR models in this SSG dataset. When MAE was used as the error metric for each model, predictive reliability was 91 %. In other words, 91 % of correlations corresponded to prediction error being strongly inversely related to actual performance.

Table 2. Guards and treasures predictive reliability

	MAE	AAE
Predictive reliability	91 %	85 %

7.3   NSG Predictive Reliability

In the following predictive reliability evaluation analysis for NSGs, we demonstrate that while predictive reliability is strong for SSGs, it is weak for NSGs; in an NSG setting, model prediction accuracy does not consistently correspond to actual performance.

We computed the predictive reliability on the NSG dataset using the three different error metrics: Maximum Absolute Error (MAE), Favored Path Maximum Absolute Error (FMAE), and Favored Path Average Absolute Error (FAAE). Table 3 displays the predictive reliability analysis results. While the predictive reliability results for the SSG dataset were strong, it is surprising that predictive reliability is extremely poor for this NSG dataset. This result certainly serves as a cautionary note against relying solely on prediction accuracy (as in previous work [1,5]) to identify the best human behavior models; with weak predictive reliability, even the best model in terms of prediction accuracy may actually perform very poorly when its corresponding strategy is tested against human subjects (either in the lab or in field experiments).

Table 3. NSG predictive reliability

	MAE	FMAE	FAAE
Predictive reliability	23 %	24 %	22 %

7.4 Training Set Size

While the predictive reliability for NSGs is poor, an obvious question to ask is "Was there enough training data?" For any learning task, it is important to have sufficient training data. While we do not have nearly as much training data (33 data points) as the prior SSG experiments (330 data points), it is important to ensure that our training set size is sufficiently large for reliable training. In this analysis, we examine the effects of training set size on the Maximum Absolute Error (MAE) rates of each NSG model. While we expect MAE to be unstable when there is very little data in the training set, as we add more data to the training set, we expect the error rates to eventually stabilize. It is at this stabilization point (marked by a training set size) that we can conclude whether we have trained our models on enough data or not. For example, if the stabilization point is at 48 data points, it would indicate that our current training set size (33) is not large enough, and any poor predictive reliability (as was previously demonstrated to be the case) could easily be explained by this deficiency in training set size.

As such, the following analysis illustrates the MAE rates of all six NSG models as a function of changes in the size of the training set. In Figs. 3, 4, and 5, we show the results of this analysis on Graphs 7, 9, and 11 (respectively), where MAE is computed on the GSUQR2 testing set. Each line corresponds to a different model (e.g., QR-M refers to QR trained with Maximin data, SUQR-S refers to SUQR trained with GSUQR1 data), the Y-Axis displays the different MAE rates (higher is worse), and the X-Axis displays the change in training set size. While all the models appear to have different error rates and rates of

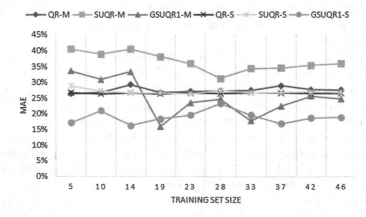

Fig. 3. MAE as a function of training set size (GSUQR2 testing set, graph 7)

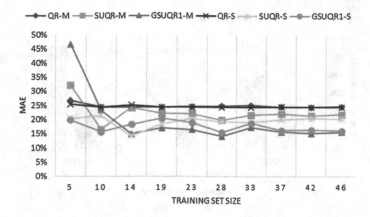

Fig. 4. MAE as a function of training set size (GSUQR2 testing set, graph 9)

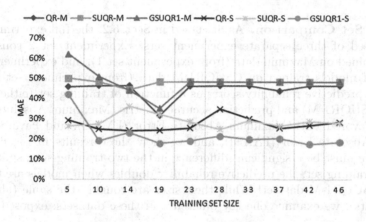

Fig. 5. MAE as a function of training set size (GSUQR2 testing set, graph 11)

convergence, most of the models appear to converge by the time 33 data points are introduced into the training set. Thus, we conclude that we have trained our models with a sufficient number of data points, and the poor predictive reliability results cannot be attributed to the size of the training set.

8 Predictive Reliability Factors

8.1 Training Set Feature: EAS

In the following analysis for our NSG dataset, we quantify the key difference in our experiment's two training sets: Exposed Attack Surface (EAS), and we demonstrate that having a higher EAS score can lead to substantial improvements in predictive reliability. Note that both training sets in this analysis are of the same size.

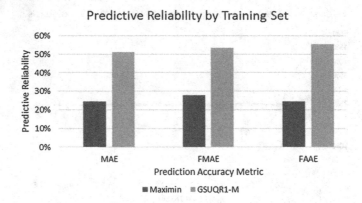

Fig. 6. Predictive reliability as a function of training set and error metric

Training Set Comparison. As discussed in Sect. 6.2, the full experiment set is comprised of three separate experiment sets. Experiment set 2 consists of models trained on Maximin data (from experiment set 1), and experiment set 3 consists of models trained on GSUQR1-M data (from experiment set 2). We computed predictive reliability scores as a function of training set (either Maximin or GSUQR1-M) and prediction accuracy metric (Maximum Absolute Error (MAE), Favored Path Maximum Absolute Error (FMAE), and Favored Path Average Absolute Error (FAAE)), and we show those results in Fig. 6. As is clear, there must be a significant difference in the two training sets; split solely on their training set, the predictive reliability doubles when models are trained on the GSUQR1-M dataset! While their sizes are roughly the same (about 47 participants), we examine one key difference in these datasets: exposed attack surface.

Exposed Attack Surface Analysis. Exposed Attack Surface (EAS), as defined in Sect. 6.3, refers to the number of unique combinations of reward $\mathcal{T}(t_j)$ and path coverage probability $p_j(x)$ over all paths A in a graph g. Since we are interested in computing this score for an entire dataset (consisting of 15 graphs $g \in G$), we compute the sum of EAS scores across all graphs. Table 4 shows the sum of each training dataset's EAS score. While the Maximin dataset had 50 unique Exposed Attack Surfaces, the GSUQR1-M dataset had 86 unique Exposed Attack Surfaces. This is not surprising, as a Maximin strategy's only goal is to conservatively minimize the attacker expected utility across all paths; for 11 out of 15 graphs in the Maximin dataset, the EAS score is equal to 3 (the minimum given three targets of different reward value). In contrast, an SUQR-based strategy seeks to actively predict which paths an attacker will choose (based on a linear combination of path coverage, reward, and potentially other factors), and as a result, the resultant defender coverage strategy is more varied (and thus only 3 out of 15 graphs have the minimum EAS score of 3).

Table 4. Training dataset comparison: sum of exposed attack surfaces

EAS-sum	Maximin	GSUQR1 M
	50	86

Based on this line of reasoning, we can view the EAS metric as a measure of dataset diversity. Since a diverse dataset would necessarily give more unique choices for attackers to make, we are able to obtain more information on which choices are favored or not favored by attackers. A higher EAS score could indicate that a dataset is better for training than another dataset; indeed, our current results strongly suggest that when there is a substantial difference in EAS-Sum scores, there will also be a substantial difference in predictive reliability. However, these results do not mean that a high EAS score will result in 100 % predictive reliability; if able to train on two datasets of equal size, it will likely improve predictive reliability to train on the dataset with the higher EAS score.

9 Graph Features and Their Impacts on Predictive Reliability

In addition to training set features, we also investigated the impacts that a graph's features may have on predictive reliability. For example, some graphs may be inherently more difficult to make predictions on than others, and it would be useful to characterize the factors that add to this complexity. Because this analysis is evaluating how a graph's features impact predictive reliability, the predictive reliability will be computed on a per graph basis. Figure 7 shows the predictive reliability scores for each graph, where each bin of three bars corresponds to a single graph, each bar corresponds to a prediction error metric, and the Y-axis corresponds to predictive reliability. As can be seen, the predictive reliability varies greatly as a function of the graph g. As such, it is logical to investigate what graph features could have led to such significant differences in predictive reliability.

We analyzed the correlation between a graph's features and the predictive reliability score for that graph. Initially, we tested many different features such as graph size (i.e., the number of paths in the graph), number of edges, number of intermediate nodes, average path length, and the average in-degree (incoming edges) and out-degree (outgoing edges) of source, destination, and intermediate nodes. What we found, however, is that none of these had a strong, direct correlation with predictive reliability. For example, the lack of a strong correlation between graph size and predictive reliability states: "A graph's size does not impact the ability to make reliable predictions".

Upon further investigation, we found one interesting relationship: there is a strong correlation (+0.72) between poor predictive reliability and graphs with both a low to moderate average out-degree for source nodes (<3) and a low to

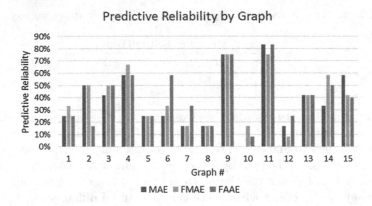

Fig. 7. Predictive reliability as a function of graph

moderate number of intermediate nodes (≤ 6). While we could not find a correlation among the other features' values and the average out-degree of source nodes, we did find a strong correlation between the number of intermediate nodes and the average in-degree of destination nodes (-0.75). Informally stated, as the number of intermediate nodes increases, the number of edges going into destination nodes decrease. This balance is perhaps due to the edge limit imposed during graph creation. Regardless, when there are less edges going into destination nodes (due to many intermediate nodes), it is likely easier for the defender to allocate resources which, in turn, reduces the number of good attack options for the attacker. If the attacker does not have many good attack options to choose from, they may act in a way that it is easier to predict by human behavior models.

10   Conclusion

Interdicting the flow of illegal goods (such as drugs and ivory) is a major security concern for many countries. However, the massive scale of these networks forces defenders to make judicious use of their limited resources. While existing solutions model this problem as a Network Security Game (NSG), they do not consider humans' bounded rationality. While existing techniques for modeling human behavior make use of large training datasets, this is unrealistic in real-world situations; the ability to effectively test many models is constrained by the time-consuming and complex nature of field deployments. In addition, there is an implicit assumption in these works that a model's prediction accuracy strongly correlates with the performance of its corresponding defender strategy (referred to as predictive reliability). If the assumption of predictive reliability does not hold, then this could lead to substantial losses for the defender. In this paper, we (1) first demonstrated that predictive reliability was strong for previous Stackelberg Security Game experiments. We also ran our own set of human subject experiments in such a way that models were restricted to learning on dataset sizes representative of real-world constraints. In the analysis on that data, we demonstrated that (2) predictive reliability was extremely weak for NSGs. Following that discovery, however, we identified (3) key factors that influenced predictive reliability results: exposed attack surface of the training data and graph structure.

Acknowledgments. This research was supported by MURI Grant W911NF-11-1-0332 and by CREATE under grant number 2010-ST-061-RE0001.

References

1. Abbasi, Y.D., Short, M., Sinha, A., Sintov, N., Zhang, C., Tambe, M.: Human adversaries in opportunistic crime security games: evaluating competing bounded rationality models. In: 3rd Conference on Advances in Cognitive Systems (2015)
2. Bell, M.G.H., Kanturska, U., Schmöcker, J.D., Fonzone, A.: Attacker-defender models and road network vulnerability. Philos. Trans. Roy. Soc. A: Math. Phys. Eng. Sci. **366**(1872), 1893–1906 (2008)
3. Camerer, C.: Behavioral Game Theory: Experiments in Strategic Interaction. Princeton University Press, Princeton (2003)
4. Correa, J.R., Harks, T., Kreuzen, V.J.C., Matuschke, J.: Fare evasion in transit networks. In: CoRR (2014)
5. Cui, J., John, R.S.: Empirical comparisons of descriptive multi-objective adversary models in stackelberg security games. In: Poovendran, R., Saad, W. (eds.) GameSec 2014. LNCS, vol. 8840, pp. 309–318. Springer, Heidelberg (2014)
6. Eppstein, D., Goodrich, M.T.: Studying (non-planar) road networks through an algorithmic lens. In: Proceedings of the 16th ACM SIGSPATIAL International Conference on Advances in Geographic Information Systems, p. 16. ACM (2008)
7. Fave, F.M.D., Jiang, A.X., Yin, Z., Zhang, C., Tambe, M., Kraus, S., Sullivan, J.: Game-theoretic security patrolling with dynamic execution uncertainty and a case study on a real transit system. J. Artif. Intell. Res. **50**, 321–367 (2014)
8. Gutfraind, A., Hagberg, A., Pan, F.: Optimal interdiction of unreactive markovian evaders. In: Hooker, J.N., van Hoeve, W.-J. (eds.) CPAIOR 2009. LNCS, vol. 5547, pp. 102–116. Springer, Heidelberg (2009)
9. Jain, M., Conitzer, V., Tambe, M.: Security scheduling for real-world networks. In: AAMAS (2013)
10. Jain, M., Korzhyk, D., Vanek, O., Conitzer, V., Pechoucek, M., Tambe, M.: A double oracle algorithm for zero-sum security games on graphs. In: AAMAS (2011)
11. Kar, D., Fang, F., Fave, F.D., Sintov, N., Tambe, M.: "A game of thrones": when human behavior models compete in repeated stackelberg security games. In: AAMAS (2015)
12. Manadhata, P., Wing, J.M.: Measuring a system's attack surface. Technical report, DTIC Document (2004)
13. Morton, D.P., Pan, F., Saeger, K.J.: Models for nuclear smuggling interdiction. IIE Trans. **39**(1), 3–14 (2007)
14. Nguyen, T.H., Yang, R., Azaria, A., Kraus, S., Tambe, M.: Analyzing the effectiveness of adversary modeling in security games. In: AAAI (2013)

15. Shieh, E., An, B., Yang, R., Tambe, M., Baldwin, C., DiRenzo, J., Maule, B., Meyer, G.: Protect: a deployed game theoretic system to protect the ports of the united states. In: AAMAS (2012)
16. Tambe, M.: Security and Game Theory: Algorithms, Deployed Systems. Lessons Learned. Cambridge University Press, New York (2011)
17. Tsai, J., Yin, Z., Kwak, J.y., Kempe, D., Kiekintveld, C., Tambe, M.: Urban security: Game-theoretic resource allocation in networked physical domains. In: AAAI (2010)
18. Yang, R., Fang, F., Jiang, A.X., Rajagopal, K., Tambe, M., Maheswaran, R.: Modeling human bounded rationality to improve defender strategies in network security games. In: HAIDM Workshop at AAMAS (2012)
19. Yang, R., Ordonez, F., Tambe, M.: Computing optimal strategy against quantal response in security games. In: AAMAS (2012)

Games of Timing for Security in Dynamic Environments

Benjamin Johnson[1], Aron Laszka[2]([✉]), and Jens Grossklags[3]

[1] CyLab, Carnegie Mellon University, Pittsburgh, USA
[2] Institute for Software Integrated Systems,
Vanderbilt University, Nashville, USA
laszka@berkeley.edu
[3] College of Information Sciences and Technology,
Pennsylvania State University, University Park, USA

Abstract. Increasing concern about insider threats, cyber-espionage, and other types of attacks which involve a high degree of stealthiness has renewed the desire to better understand the timing of actions to audit, clean, or otherwise mitigate such attacks. However, to the best of our knowledge, the modern literature on games shares a common limitation: the assumption that the cost and effectiveness of the players' actions are time-independent. In practice, however, the cost and success probability of attacks typically vary with time, and adversaries may only attack when an opportunity is present (e.g., when a vulnerability has been discovered).

In this paper, we propose and study a model which captures dynamic environments. More specifically, we study the problem faced by a defender who has deployed a new service or resource, which must be protected against cyber-attacks. We assume that adversaries discover vulnerabilities according to a given vulnerability-discovery process which is modeled as an arbitrary function of time. Attackers and defenders know that each found vulnerability has a basic lifetime, i.e., the likelihood that a vulnerability is still exploitable at a later date is subject to the efforts by ethical hackers who may rediscover the vulnerability and render it useless for attackers. At the same time, the defender may invest in mitigation efforts to lower the impact of an exploited vulnerability. Attackers therefore face the dilemma to either exploit a vulnerability immediately, or wait for the defender to let its guard down. The latter choice leaves the risk to come away empty-handed.

We develop two versions of our model, i.e., a continuous-time and a discrete-time model, and conduct an analytic and numeric analysis to take first steps towards actionable guidelines for sound security investments in dynamic contested environments.

Keywords: Security · Game Theory · Games of timing · Vulnerability discovery

© Springer International Publishing Switzerland 2015
MHR Khouzani et al. (Eds.): GameSec 2015, LNCS 9406, pp. 57–73, 2015.
DOI: 10.1007/978-3-319-25594-1_4

1   Introduction

Since at least the Cold War era there has been a considerable interest in the study of games of timing to understand *when* to act in security-relevant decision-making scenarios [1]. The recent rise of insider threats, cyber-espionage, and other types of attacks which involve a high degree of stealthiness has renewed the desire to better understand the timing of actions to audit, clean, or otherwise mitigate such attacks. However, to the best of our knowledge, the modern literature on games and decision-theoretic approaches (including the FlipIt model [3,31]) shares a common limitation: the assumption that the cost and effectiveness of the players' actions are time-independent. For example, in the FlipIt model and its derivatives (see section on related work), an adversary may make a move at any time for exactly the same fixed cost, and these moves always succeed.

In practice, the cost and success probability of attacks typically vary with time. Moreover, an adversary may only attack when an opportunity is present (e.g., when a vulnerability has been discovered). These observations motivate the development of games of timing which take into account the dynamic environment of contested computing resources. Defenders need to develop an optimal defensive strategy which considers the nature of vulnerability discovery by adversaries. At the same time, the attacker faces the decision-making dilemma on when to exploit an identified vulnerability.

For example, the black hat community knew already for a long time that Microsoft would stop supporting Windows XP in April 2014, which would significantly lower the defense and mitigation effort for this software product.[1] Security professionals conjectured that attackers would begin stockpiling vulnerabilities to exploit them more profitably. However, under what circumstances is such behavior optimal for the attacker, when there is a risk that the vulnerability is rediscovered by an internal security team or external ethical hackers before the planned time of exploitation [22,35]?

In this paper, we propose and study a model which captures dynamic environments. More specifically, we study the problem faced by a defender who has deployed a new service or resource, which must be protected against cyberattacks. We assume that adversaries discover vulnerabilities according to a given vulnerability-discovery process which is modeled as an arbitrary function of time. Attackers and defenders know that each found vulnerability has a basic lifetime, i.e., the likelihood that a vulnerability is still exploitable at a later date is subject to the efforts by ethical hackers who may rediscover the vulnerability and render it useless for attackers. At the same time, the defender may invest in mitigation efforts to lower the impact of an exploited vulnerability. Attackers therefore face the dilemma to either exploit a vulnerability immediately, or wait for the defender to let its guard down. The latter choice leaves the risk to come away empty-handed.

---

[1] In July 2011, Microsoft made the announcement that support for the operating system will end in 2014. Note that previously Microsoft already stopped the so-called full mainstream support for Windows XP in April 2009.

We develop two versions of our model, i.e., a continuous-time and a discrete-time model, to increase the applicability of our work. We provide fundamental constraints on the shape of equilibria for both models, and give necessary and sufficient conditions for the existence of non-waiting equilibria in terms of the shape of the vulnerability discovery function. We further provide numerical results to illustrate important properties of our findings.

The remainder of this paper is organized as follows. In Sect. 2, we summarize related theoretical and behavioral work on security games of timing. In Sect. 3, we introduce our game-theoretic model including players and the decision-making environment. In Sect. 4, we derive theoretical results for our model. In Sect. 5, we present numerical examples. Finally, in Sect. 6, we discuss our results and offer concluding remarks.

2 Related Work

2.1 Security Economics and Games of Timing

The economics of security decision-making is a rapidly expanding field covering theoretical, applied, and behavioral research. Theoretical work utilizes diverse game-theoretic and decision-theoretic approaches, and addresses abstract as well as applied scenarios. A central research question has been how to optimally determine security investments [7,11,25,32], e.g., by selecting from different canonical defense actions (i.e., protection, mitigation, risk-transfer) [12,19], and how such investments are influenced by the actions of strategic attackers [6,30]. Another frequently addressed aspect has been the consideration of interdependence of security decision-making and the propagation of risks [4,8,13,14]. Recent surveys summarize these research efforts in great detail [2,15,20].

An often overlooked but critical decision dimension for successfully securing resources is the consideration of *when* to act to successfully thwart attacks. Scholars have studied such time-related aspects of tactical security choices since the cold-war era by primarily focusing on zero-sum games called *games of timing* [1]. The theoretical contributions on some subclasses of these games have been surveyed by [27].

Recently, the question of the optimal timing of security decisions has again become a lively research topic with the development of the FlipIt game [3,31]. In the following, we discuss FlipIt as well as theoretical and behavioral follow-up research.

2.2 Theoretical Analyses of FlipIt

The FlipIt model identifies optimal timing-related security choices under targeted attacks [3,31]. In FlipIt, two players compete for a resource that generates a payoff to the current owner. Players can make costly moves (i.e., "flips") to take ownership of the resource, however, they have to make moves under incomplete information about the current state of possession. In the original FlipIt papers, equilibria and dominant strategies for simple cases of interaction are studied [3,31].

In follow-up research, Pham and Cid studied a version of FlipIt with periodic strategies with random phase. They also considered the impact of a move to check the state of the game (i.e., audit) [26].

Laszka et al. study games of timing with non-covert defender moves. They consider also non-instantaneous attacker moves, and different types of adversaries, e.g., targeting and non-targeting attackers [18]. A follow-up paper further generalizes the results of this line of research [17].

The previous papers considered FlipIt with one resource. This limitation has been addressed with the strategic analysis of the game with multiple contested resources [16]. Similarly, an extension of the game has been proposed with multiple defenders [24].

Feng et al. [5] and Hu et al. [10] study games with multiple layers in which in addition to external adversaries the actions of insiders (who may trade information to the attacker for a profit) need to be considered. Hu et al. [10] study the scenario in a dynamic game framework.

Zhang et al. [34] study the FlipIt game with resource constraints on both players.

Drawing on the setup of FlipIt, Wellman and Prakash develop a discrete-time model with multiple, ordered states in which attackers may compromise a server through cumulative acquisition of knowledge rather than in a one-shot takeover [33].

2.3   Behavioral Studies of FlipIt

Nochenson and Grosenklags describe and analyze two experiments which draw from the theoretical model of the FlipIt game [21]. They conduct a Mechanical Turk experiment with over 300 participants in which each participant is matched with a computerized opponent in several fast-paced rounds of the FlipIt game. Preliminary analysis of this experiment shows that participant performance improves over time (however, older participants improve less than younger ones). They also found significant performance differences with regards to gender and a measure of the desire for deep reasoning about a problem (i.e., need for cognition).

In follow-up work, Reitter et al. contrast two experiments where the feedback to the human decision maker in the decision-environment is varied between visual feedback with history, and temporal feedback without history. The authors study the human strategies and develop a model backed by a cognitive architecture, which described human heuristics that practically implement risk-taking preference in timing decisions [28].

Grosenklags and Reitter extend these preliminary works with an in-depth analysis of the experimental data of these previous studies [9]. In particular, they study the interaction effects between the psychometric measures including also the general propensity of risk taking with task experience and how those factors explain task performance.

The behavioral studies will help to develop theoretical models which take the imperfections of human decision-making into account. Likewise, theoretical studies of rational behavior serve as an important comparison baseline for experimentally generated human data or measurements from the field.

3   Model

Our model captures the motivational aspects of timing, as it pertains to the discovery, repair, and exploitation of software vulnerabilities. The salient features of our model may be enumerated as follows.

1. The life cycle of a software product is finite with a known end time $t = T$.
2. The rate of vulnerability discovery $V(t)$ is an arbitrary function of time, specified as an exogenous parameter. We make this modeling choice to maximize applicability for varieties of software products and services that may differ in quality, attention, and life cycle.[2]
3. The lifetime of a vulnerability decays at a fixed rate λ without action by either player. This choice is made to account for the fact that unknown vulnerabilities are often repaired by chance only, so that one might reasonably assume they die with some fixed probability in a unit of time.[3]
4. The defender's security investment $d(t)$ is a function of time, and serves to mitigate losses when a vulnerability is exploited.
5. The timing of vulnerability exploitation $a(t)$ is chosen by an attacker for optimal exploitation dependent on the defender's security investments.

To further extend the applicability of our model, we describe and analyze two distinct versions – one with continuous time, and one with discrete time. In the continuous version of the model, attackers and defenders choose strategies as continuous functions of time, and the payoffs are determined by integrating expected losses over the range of all time. In the discrete version, time is divided into a finite number of steps; attackers and defenders choose an action at each time step, and the payoffs are determined by summing the expected outcomes over all time periods. Both versions of the game adhere to the paradigms described above.

We begin by describing the game's players and their respective choices. We then proceed to describe the environment. Finally we discuss the consequences from a configuration of choices. Whenever applicable, we separate the specification and discussion according to either the continuous or the discrete model. For reference, a list of symbols used in this paper may be found in Table 1.

---

[2] A small number of studies investigate the social utility of vulnerability discovery. On the one hand, Rescorla studied the ICAT dataset of 1,675 vulnerabilities and found very weak or no evidence of vulnerability depletion. He thus suggested that the vulnerability discovery efforts might not provide much social benefit [29]. On the other hand, this conclusion is challenged by Ozment and Schechter, who showed that the pool of vulnerabilities in the foundational code of OpenBSD is being depleted [22,23]. Zhao et al. present evidence that the number of discovered vulnerabilities is declining for a majority of public company-specific vulnerability bounty programs on HackerOne [36].

[3] Unsurprisingly, statistical evidence is lacking regarding how often defenders and attackers discover the same vulnerabilities. However, empirical research by Ozment about the ethical hacker community found that vulnerability rediscovery is common in the OpenBSD vulnerability discovery history [22].

Table 1. List of Symbols

Symbol	Description
R	scaling factor between security costs and losses
λ	vulnerability repair rate
	Continuous-time Model
T	end time
$V(t)$	vulnerability discovery rate at time t
$d(t)$	defender's security investment at time t
$a(t)$	attacker's waiting time before exploiting a vulnerability discovered at time t
	Discrete-time Model
K	number of time periods
$V(k)$	expected number of vulnerabilities discovered in time period k
$d(k)$	defender's security investment in time period k
$a(k)$	attacker's waiting time before exploiting a vulnerability discovered in time period k

3.1   Players and Choices

Our game has two players, a defender and an attacker. The defender's objective is to mitigate damages from vulnerability exploitation through security investment, while the attacker's objective is to maximally exploit vulnerabilities as they are discovered. Neither the attacker nor the defender control the rate of vulnerability discovery $V(t)$, which is an exogenous function of time.

We may construe the defender's investments quite broadly, in ways other than monetary investments. For example, we may understand them as a measure of strictness in policy enforcement, which can be optimized to minimize usability loss.

On the attacker side, it is interesting to note that we would obtain the same results if we modeled the game as one containing several attackers, where each attacker randomly finds vulnerabilities according to a given rate, and then independently chooses the timing of their exploitation. However, for the sake of clear exposition, we frame the interaction as a two-player game with a single attacker.

Continuous-Time Model. In the continuous-time model over a time interval $[0, T]$, the defender chooses a continuous function $d(t) : [0, T] \to \mathbb{R}_{\geq 0}$ which specifies the level of her security investment at each time t. The attacker chooses a continuous function $a(t) : [0, T] \to \mathbb{R}_{\geq 0}$ which specifies how long to wait before exploiting a vulnerability discovered at time t.

Discrete-Time Model. In the discrete-time model with discrete time periods $0, 1, \ldots, K$, the defender chooses a function $d(k) : \{0, 1, \ldots, K\} \to \mathbb{R}_{\geq 0}$ specifying her security investment level at each distinct time period. The attacker chooses a function $a(k) : \{0, 1, \ldots, K\} \to \mathbb{Z}_{\geq 0}$ specifying how many discrete time steps to wait before launching an attack using a vulnerability discovered in the kth time period.

3.2 Environment

Here we construe the environment primarily as the security state of a software system over a finite period of time. More specifically, the rate of vulnerability discovery by attackers, $V(t)$, is a function of time, specified as an exogenous parameter. We anticipate that this modeling choice increases the applicability for different types of software products and services that may differ in quality, attention, and life cycle.

The fixing of vulnerabilities, on the other hand, follows a random process as defenders eventually rediscover vulnerabilities which have been found by the attacker. More specifically, we assume that the lifetime of a vulnerability follows an exponential distribution (parameterized by λ) without action by either player. The net effect of this eventual rediscovery is that an attacker who learns of a vulnerability at one time, cannot simply wait indefinitely for the defender's security investment to lapse.

Continuous-Time Model. In the continuous-time model, the vulnerability function has the form $V(t) : [0, T] \to \mathbb{R}_{\geq 0}$. The interpretation is that $V(t)$ gives the precise rate at which vulnerabilities are being discovered by the attacker for each moment of time. In terms of our analysis and computation, we will obtain the expected number of vulnerabilities discovered during any fixed time interval by integrating $V(t)$ with respect to t over that time interval.

The vulnerability repair process is determined by an exponential decay function of the form $e^{-\lambda \tau}$. This function determines the probability that a vulnerability still remains exploitable τ time after its discovery. The structured formulation guarantees that this exploit probability decays at a constant rate of λ. An approximate interpretation is that in each unit of time, a constant fraction of its exploit probability is lost.

Discrete-Time Model. In the discrete-time model, the vulnerability function has the form $V(k) : \{0, 1, \ldots, K\} \to \mathbb{R}_{\geq 0}$. Here, $V(k)$ gives directly the expected number of vulnerabilities discovered during the time period k. Computationally, we may obtain the expected number of vulnerabilities discovered over any sequence of time periods by summing $V(k)$ over those periods.

To capture the analogous fixed rate reduction phenomenon for vulnerability repair in the discrete-time model, we use a geometric distribution function of the form $(1 - \lambda)^\tau$, which gives us the probability that a vulnerability is not repaired in τ number of time periods after its discovery. The interpretation is that a λ fraction of a vulnerability's exploit potential is lost in each time period.

3.3   Consequences

Suppose that both defender and attacker have simultaneously chosen their strategies for defense d and wait times a, respectively. The consequences for the defender involve both the defense costs and the loss from vulnerability exploitation. We construe the defense function in terms of direct costs, while the amount of loss resulting from an attack is inversely proportional to the defense rate, scaled by a fixed constant R.

On the attacker's side, we are only concerned with the gain from maximally exploiting the vulnerabilities. Thus, the overall structure is that the defender's payoff is always negative, while the attacker's payoff is always positive. The sum of payoffs related to vulnerability exploitation is zero; but the game itself is not zero-sum, unless the defender abstains from any defensive investment (i.e., when $d \equiv 0$).

Continuous-Time Model. In the continuous-time model, the defender's objective is to minimize her total losses over the course of the time interval $[0, T]$. The defender's costs over this time interval may be easily computed as

$$\int_{t=0}^{T} d(t)\,dt,$$

while her losses depend in part on the waiting time of an attacker. If the attacker immediately exploits a vulnerability discovered at time t, the expected loss per unit time due to vulnerabilities discovered around time t may be expressed as

$$\frac{R}{d(t)}.$$

On the other hand, if the attacker instead waits for some time $a(t)$ before exploiting a vulnerability discovered at time t, then we must account for both the decay in vulnerability exploitability as well as adjust the timing relative to the defense investment. In this case, the expected loss per unit of time due to vulnerabilities discovered around time t will be given by

$$e^{-\lambda a(t)} \frac{R}{d(t + a(t))}.$$

Putting everything together along with the vulnerability discovery function, the defender's total payoff in the continuous-time model is given by

$$U_d = -\int_{t=0}^{T} \left(d(t) + V(t)e^{-\lambda a(t)} R \frac{1}{d(t + a(t))} \right) dt; \tag{1}$$

while the attacker's payoff is given by

$$U_a = \int_{t=0}^{T} V(t)e^{-\lambda a(t)} R \frac{1}{d(t + a(t))} dt. \tag{2}$$

Discrete-Time Model. In the discrete-time model, the defender's objective is to minimize her total losses over the course of the time stages $\{0, 1, \ldots, K\}$. The defender's costs are computed as a sum

$$\sum_{k=0}^{K} d(k),$$

while losses depend on the waiting time of an attacker. Suppose that an attacker waits for $a(k)$ time periods before exploiting a vulnerability discovered in time period k; then, the defender's losses due to vulnerabilities discovered in time step k will be given by

$$(1 - \lambda)^{a(k)} \frac{R}{d(k + a(k))}.$$

Assembling everything together, the payoff for the defender in the discrete-time model is given by

$$U_d = -\sum_{k=0}^{K} \left(d(k) + V(k)(1 - \lambda)^{a(k)} \frac{R}{d(k + a(k))} \right); \qquad (3)$$

while the payoff for the attacker is given by

$$U_a = \sum_{k=0}^{K} V(k)(1 - \lambda)^{a(k)} \frac{R}{d(k + a(k))}. \qquad (4)$$

4   Analysis

In this section, we analyze the model to find applicable consequences for the software vulnerability scenario. We will primarily focus on Nash equilibrium configurations, in which each player is responding optimally in the current context.

We begin by giving a result in the continuous-time model that constrains the attacker's strategy at the temporal boundaries.

Proposition 1. *If $V(0) > 0$, then every equilibrium in the continuous-time model satisfies $a(0) = 0$ and $a(T) = 0$. In words, the attacker should never wait to attack at either the beginning or the end of the game.*

Proof. Suppose $a(0) > 0$. Since there is no previous time at which the attacker may have discovered a vulnerability, the defender may safely choose $d(0) = 0$ as an optimal investment. However, if the attacker knew $d(0) = 0$, she would rather prefer not to wait, in order to cause maximum damage in case a vulnerability were found at that time. This contradiction shows $a(0) > 0$ cannot be an equilibrium if $V(0) > 0$.

The second part of the proposition is more trivially deduced since it would not benefit the attacker to wait longer because there is no time remaining at the end of the game. In fact, for this reason more generally, the attacker's strategy in equilibrium must satisfy the constraint $a(t) \leq T - t$. $\qquad \square$

Our second result constrains the attacker's strategy in any pure-strategy equilibrium. These conditions are considerably more restrictive than those in the continuous-time case. They tell us that if there is an ubiquitous risk of vulnerability discovery, then there can be no pure-strategy equilibrium in which the attacker uses any positive wait times.

Proposition 2. *If $V(k) > 0$ for each time period k, then for every pure-strategy equilibrium in the discrete-time game, we have $a(k) = 0$ for each $k = 0, 1, \ldots, K$. In words, if the attacker uses any positive wait time in the discrete-time game, then it must be part of a mixed strategy.*

Proof. We prove the result by induction on the number of time periods. When $k = 0$, the claimed result is perfectly analogous to the continuous-time model's result from the previous proposition. Obviously, there can be no previous vulnerability discovery. If the attacker waits to attack in round 0, then the defender can optimally save herself the trouble of making any security investment in round 0 (i.e., $d(0) = 0$). But if $V(0) > 0$, then this configuration is clearly not an optimal response configuration for the attacker.

But now that we know $a(0) = 0$, a very similar argument also holds for $k = 1$. We do not have any vulnerabilities from the one earlier round, because the attacker did not wait in round 0. If the attacker now waits in round 1, the defender may optimally choose not to invest in security protection in this round (i.e., $d(1) = 0$). But this configuration is not optimal for the attacker and so cannot be part of an equilibrium. The argument can now be iterated inductively for $k = 2, \ldots, K$.   □

The crux of these two results is that the attacker may only optimally wait to attack in a given time period if there is some attack probability arising from a previous time period. In the continuous case, this implies only that the attacker cannot wait at the beginning of the game, because continuously increasing the wait time from $t = 0$ can still lead to positive attack probability at every point in time. On the discrete side, however, this observation precludes having any simple optimal attack strategy in which the attacker waits at all.

The next two propositions give necessary and sufficient conditions for "never waiting" to be the attacker's strategy in an equilibrium. In both the continuous-time model and the discrete-time model, the conditions involve only a simple relation between the vulnerability discovery function V and the discovery rate λ.

Proposition 3. *In the continuous-time model, there exists an equilibrium in which the attacker never waits before attacking if the vulnerability function satisfies*

$$\frac{V(t + a)}{V(t)} \geq e^{-2\lambda a} \tag{5}$$

for every $t \in [0, T]$ and $a \in [0, T - t]$.

Proof. Suppose that the attacker never waits. Let us consider the defender's best response to this strategy. Simplifying Eq. (1), the defender's utility function becomes

$$-\int_0^T \left(d(t) + V(t)\frac{R}{d(t)} \right) dt.$$

This utility is maximized by choosing $d(t)$ at each time t to minimize the cost plus risk. Setting

$$\frac{d}{dx}\left(x + V(t)\frac{R}{x} \right) = 0$$

and solving for x, we obtain the optimal $d(t)$ as

$$d(t) = \sqrt{V(t)R}. \tag{6}$$

Now, the part of the equilibrium condition that says $a(t) = 0$ is the attacker's best response function implies that for every t and a, we have

$$\frac{V(t)R}{d(t)} \geq \frac{V(t)Re^{-\lambda a}}{d(t+a)}.$$

Incorporating the defender's strategy and simplifying, we obtain

$$\frac{d(t+a)}{d(t)} \geq e^{-\lambda a}$$

$$\frac{\sqrt{V(t+a)R}}{\sqrt{V(t)R}} \geq e^{-\lambda a}$$

$$\frac{V(t+a)}{V(t)} \geq e^{-2\lambda a}.$$

Now conversely, suppose that

$$\frac{V(t+a)}{V(t)} \geq e^{-2\lambda a}.$$

Let $d(t) = \sqrt{V(t)R}$ be the defender's investment strategy. Because the sequence of inequalities above is reversible, we have that $a(t)$ is a best response to $d(t)$; and we have already showed that $d(t)$ is a best response to $a(t)$. So there exists an equilibrium in which the attacker never waits.   □

The following proposition gives an analogous result for the discrete-time model.

Proposition 4. *In the discrete-time model, there is an equilibrium in which the attacker never waits before attacking if the vulnerability function satisfies*

$$\frac{V(k+a)}{V(k)} \geq (1-\lambda)^{2a} \tag{7}$$

for every $k \in \{0,\ldots,K-1\}$ and $a \in \{1,\ldots,K-k\}$.

Proof. Suppose that the attacker never waits. Let us consider the defender's best response to this strategy. Simplifying Equation(3), the defender's utility function becomes

$$-\sum_{k=0}^{K}\left(d(k) + V(k)\frac{R}{d(k)}\right).$$

This utility is maximized by choosing $d(k)$ at each step k to minimize the cost plus risk, giving

$$d(k) = \sqrt{V(k)R}. \tag{8}$$

To say that $a(k) = 0$ is the attacker's best response function now implies that for every k and a, we have

$$\frac{V(k)R}{d(k)} \geq \frac{V(k)R(1-\lambda)^a}{d(k+a)}.$$

Incorporating the defender's strategy and simplifying, we obtain

$$\frac{d(k+a)}{d(k)} \geq (1-\lambda)^a$$

$$\frac{\sqrt{V(k+a)R}}{\sqrt{V(k)R}} \geq (1-\lambda)^a$$

$$\frac{V(k+a)}{V(k)} \geq (1-\lambda)^{2a}.$$

The argument that the condition implies existence of an equilibrium is analogous to the continuous version. □

5   Numerical Examples

In this section, we present numerical examples to illustrate our model and our theoretical results, focusing on the vulnerability-discovery function and the defender's equilibrium strategy. For these numerical examples, we use the discrete-time version of our model.

First, in Figs. 1 and 2, we study two example vulnerability functions with the corresponding equilibrium defense strategies. In the first example (Fig. 1), the vulnerability discovery rate grows and decays exponentially. More formally, the vulnerability discovery rate in this example is given by the following formula:

$$V(k) = e^{-\frac{(k-33)^2}{200}}. \tag{9}$$

In the second example (Fig. 2), the vulnerability discovery rate grows and decays linearly (i.e., according to an affine function). In both cases, we let $R = 1$, $K = 100$, and $\lambda = 0.3$.

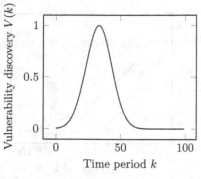
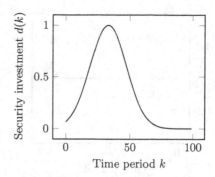

(a) Vulnerability discovery rate as a function of time

(b) Defender's equilibrium security investment as a function of time

Fig. 1. Example based on exponentially growing and decaying vulnerability discovery rate with the corresponding equilibrium defense strategy.

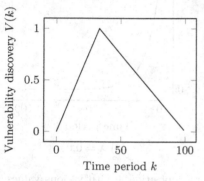
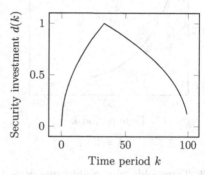

(a) Vulnerability discovery rate as a function of time

(b) Defender's equilibrium security investment as a function of time

Fig. 2. Example based on linearly growing and decaying vulnerability discovery rate with the corresponding equilibrium defense strategy.

We can see that, in both examples, the rise and fall of the defender's security investment is dampened compared to those of the vulnerability functions. However, the security investments are very far from being constant, which indicates that dynamic environments play an important role in determining equilibrium investments.

Second, in Fig. 3, we study the condition given by Proposition 4. Recall that Proposition 4 establishes a threshold on the maximum rate of decrease in vulnerability discovery such that the attacker never waiting is an equilibrium. In Fig. 3, for various values of λ, we plot vulnerability discovery functions that decrease with this maximum rate.

Firstly, in Fig. 3(a), we can see that if $\lambda = 0$, then the vulnerability discovery rate has to be constant in order for the attacker not waiting to be an equilibrium.

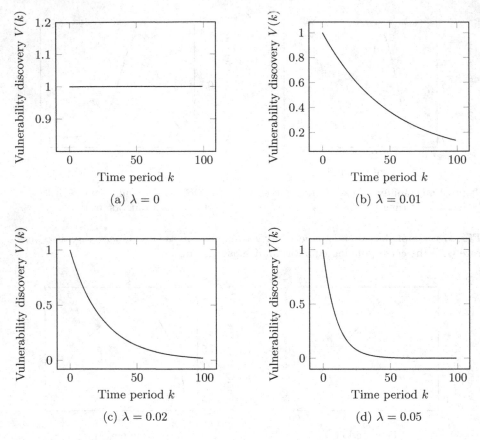

Fig. 3. Threshold vulnerability functions $V(k)$ for Proposition 4 with various values of λ.

The explanation for this corner case is that $\lambda = 0$ means that the attacker can stockpile vulnerabilities without taking any risk; hence, the attacker will wait only if security investments are constant over time, which implies that the vulnerability discovery rate must also be constant for the equilibrium to exist. Secondly, in Figs. 3(b), (c), and (d), we see that the higher the value of λ, the more steeply the vulnerability discovery rate may decrease. Again, the explanation for this is that higher values of λ mean higher risk for stockpiling vulnerabilities; hence, the higher λ is, the more steeply the discovery rate can decrease without the attacker opting to wait.

6   Conclusion

The recent rise of attacks involving a high degree of stealthiness has sparked considerable interest in games of timing for security. However, to the best of our knowledge, the previously proposed models in the recent literature share a common limitation: the assumption that the cost and effectiveness of the attackers' actions are time-independent. In this paper, we proposed and studied a model which captures dynamic environments, i.e., in which the attackers' actions depend on the availability of exploitable vulnerabilities. More specifically, we assumed that attackers discover vulnerabilities according to a given vulnerability-discovery process, which we modeled as an arbitrary function of time. Based on this assumption, we formulated a two-player game of timing between a defender, who tries to protect a service or resource through security investments, and an attacker, who can choose when to exploit a vulnerability. The most interesting novel feature of our model is the attacker's dilemma: whether to wait in hope of exploiting the vulnerability at a time when security is lower, but risking that the vulnerability is rediscovered and fixed in the meantime.

In our theoretical analysis, we primarily focused on characterizing equilibria in which the attacker does not stockpile vulnerabilities (i.e., never waits to exploit a vulnerability). The question of vulnerability stockpiling is interesting in many practical scenarios, most importantly in the case of software products that are widely used even after their end of official support. Our results relate the vulnerability discovery process to the rate of repairing vulnerabilities, and hence provide guidelines for finding vulnerability repair rates that will not lead to a vulnerability stockpiling equilibrium in practice. In our numerical examples, we considered multiple specific vulnerability functions, and studied the corresponding equilibrium strategies.

There are multiple directions for extending our current work. Firstly, we plan to provide a theoretical characterization of the game's equilibria in the case when the attacker does not stockpile vulnerabilities (i.e., when never waiting is not an equilibrium). Secondly, we plan to study and characterize the Stackelberg equilibria of our game. In our current work, we assume that the defender and the attacker choose their strategies at the same time, which captures scenarios with uninformed players. However, in [17], it was shown – for a different timing-game model – that a defender can substantially decrease its losses by publicly committing to a strategy and letting the attacker choose its strategy in response. We expect that a similar result holds for the model presented in this paper as well.

Acknowledgment. We thank the anonymous reviewers for their helpful comments. This work was supported in part by the National Science Foundation (CNS-1238959).

References

1. Blackwell, D.: The noisy duel, one bullet each, arbitrary accuracy. Technical report, The RAND Corporation, D-442 (1949)
2. Böhme, R., Schwartz, G.: Modeling cyber-insurance: towards a unifying framework. In: 9th Workshop on the Economics of Information Security (WEIS) (2010)
3. Bowers, K.D., van Dijk, M., Griffin, R., Juels, A., Oprea, A., Rivest, R.L., Triandopoulos, N.: Defending against the unknown enemy: applying FlipIt to system

security. In: Grossklags, J., Walrand, J. (eds.) GameSec 2012. LNCS, vol. 7638, pp. 248–263. Springer, Heidelberg (2012)

4. Chen, P., Kataria, G., Krishnan, R.: Correlated failures, diversification, and information security risk management. MIS Q. **35**(2), 397–422 (2011)

5. Feng, X., Zheng, Z., Hu, P., Cansever, D., Mohapatra, P.: Stealthy attacks meets insider threats: a three-player game model. Technical report

6. Fultz, N., Grossklags, J.: Blue versus red: towards a model of distributed security attacks. In: Dingledine, R., Golle, P. (eds.) FC 2009. LNCS, vol. 5628, pp. 167–183. Springer, Heidelberg (2009)

7. Gordon, L., Loeb, M.: The economics of information security investment. ACM Trans. Inf. Syst. Secur. **5**(4), 438–457 (2002)

8. Grossklags, J., Christin, N., Chuang, J.: Secure or insure?: a game-theoretic analysis of information security games. In: Proceedings of the 17th International World Wide Web Conference, pp. 209–218 (2008)

9. Grossklags, J., Reitter, D.: How task familiarity and cognitive predispositions impact behavior in a security game of timing. In: Proceedings of the 27th IEEE Computer Security Foundations Symposium (CSF), pp. 111–122 (2014)

10. Hu, P., Li, H., Fu, H., Cansever, D., Mohapatra, P.: Dynamic defense strategy against advanced persistent threat with insiders. In: Proceedings of the 34th IEEE International Conference on Computer Communications (INFOCOM) (2015)

11. Ioannidis, C., Pym, D., Williams, J.: Investments and trade-offs in the economics of information security. In: Dingledine, R., Golle, P. (eds.) FC 2009. LNCS, vol. 5628, pp. 148–166. Springer, Heidelberg (2009)

12. Johnson, B., Böhme, R., Grossklags, J.: Security games with market insurance. In: Baras, J.S., Katz, J., Altman, E. (eds.) GameSec 2011. LNCS, vol. 7037, pp. 117–130. Springer, Heidelberg (2011)

13. Johnson, B., Laszka, A., Grossklags, J.: The complexity of estimating systematic risk in networks. In: Proceedings of the 27th IEEE Computer Security Foundations Symposium (CSF), pp. 325–336 (2014)

14. Kunreuther, H., Heal, G.: Interdependent security. J. Risk Uncertain. **26**(2), 231–249 (2003)

15. Laszka, A., Felegyhazi, M., Buttyan, L.: A survey of interdependent information security games. ACM Comput. Surv. **47**(2), 23:1–23:38 (2014)

16. Laszka, A., Horvath, G., Felegyhazi, M., Buttyán, L.: FlipThem: modeling targeted attacks with flipit for multiple resources. In: Poovendran, R., Saad, W. (eds.) GameSec 2014. LNCS, vol. 8840, pp. 175–194. Springer, Heidelberg (2014)

17. Laszka, A., Johnson, B., Grossklags, J.: Mitigating covert compromises. In: Chen, Y., Immorlica, N. (eds.) WINE 2013. LNCS, vol. 8289, pp. 319–332. Springer, Heidelberg (2013)

18. Laszka, A., Johnson, B., Grossklags, J.: Mitigation of targeted and non-targeted covert attacks as a timing game. In: Das, S.K., Nita-Rotaru, C., Kantarcioglu, M. (eds.) GameSec 2013. LNCS, vol. 8252, pp. 175–191. Springer, Heidelberg (2013)

19. Lelarge, M., Bolot, J.: Economic incentives to increase security in the internet: the case for insurance. In: Proceedings of the 33rd IEEE International Conference on Computer Communications (INFOCOM), pp. 1494–1502 (2009)

20. Manshaei, M., Zhu, Q., Alpcan, T., Başar, T., Hubaux, J.: Game theory meets network security and privacy. ACM Comput. Surv. **45**(3), 25:1–25:39 (2013)

21. Nochenson, A., Grossklags, J.: A behavioral investigation of the FlipIt game. In: 12th Workshop on the Economics of Information Security (WEIS) (2013)

22. Ozment, A.: The likelihood of vulnerability rediscovery and the social utility of vulnerability hunting. In: Proceedings of the 4th Workshop on the Economics of Information Security (WEIS) (2005)
23. Ozment, A., Schechter, S.: Milk or wine: does software security improve with age? In: Proceedings of the 15th USENIX Security Symposium (2006)
24. Pal, R., Huang, X., Zhang, Y., Natarajan, S., Hui, P.: On security monitoring in sdns: a strategic outlook
25. Panaousis, E., Fielder, A., Malacaria, P., Hankin, C., Smeraldi, F.: Cybersecurity games and investments: a decision support approach. In: Poovendran, R., Saad, W. (eds.) GameSec 2014. LNCS, vol. 8840, pp. 266–286. Springer, Heidelberg (2014)
26. Pham, V., Cid, C.: Are we compromised? modelling security assessment games. In: Grossklags, J., Walrand, J. (eds.) GameSec 2012. LNCS, vol. 7638, pp. 234–247. Springer, Heidelberg (2012)
27. Radzik, T.: Results and problems in games of timing. In: Lecture Notes-Monograph Series. Statistics, Probability and Game Theory: Papers in Honor of David Blackwell, vol. 30, pp. 269–292 (1996)
28. Reitter, D., Grossklags, J., Nochenson, A.: Risk-seeking in a continuous game of timing. In: Proceedings of the 13th International Conference on Cognitive Modeling (ICCM), pp. 397–403 (2013)
29. Rescorla, E.: Is finding security holes a good idea? IEEE Secur. Priv. $3(1)$, 14–19 (2005)
30. Schechter, S.E., Smith, M.D.: How much security is enough to stop a thief? In: Wright, R.N. (ed.) FC 2003. LNCS, vol. 2742, pp. 122–137. Springer, Heidelberg (2003)
31. Van Dijk, M., Juels, A., Oprea, A., Rivest, R.: Flipit: the game of "stealthy takeover". J. Crypt. $26(4)$, 655–713 (2013)
32. Varian, H.: System reliability and free riding. In: Camp, J., Lewis, S. (eds.) Economics of Information Security, pp. 1–15. Kluwer Academic Publishers, Dordrecht (2004)
33. Wellman, M.P., Prakash, A.: Empirical game-theoretic analysis of an adaptive cyber-defense scenario (preliminary report). In: Poovendran, R., Saad, W. (eds.) GameSec 2014. LNCS, vol. 8840, pp. 43–58. Springer, Heidelberg (2014)
34. Zhang, M., Zheng, Z., Shroff, N.: Stealthy attacks and observable defenses: a game theoretic model under strict resource constraints. In: Proceedings of the IEEE Global Conference on Signal and Information Processing (GlobalSIP), pp. 813–817 (2014)
35. Zhao, M., Grossklags, J., Chen, K.: An exploratory study of white hat behaviors in a web vulnerability disclosure program. In: Proceedings of the ACM Workshop on Security Information Workers, pp. 51–58 (2014)
36. Zhao, M., Grossklags, J., Liu, P.: An empirical study of web vulnerability discovery ecosystems. In: Proceedings of the 22nd ACM Conference on Computer and Communications Security (CCS) (2015)

Threshold FlipThem: When the Winner Does Not Need to Take All

David Leslie[1], Chris Sherfield[2], and Nigel P. Smart[2]([✉])

[1] Department Mathematics and Statistics,
University of Lancaster, Lancaster, UK
d.leslie@lancaster.ac.uk
[2] Department of Computer Science, University of Bristol, Bristol, UK
c.sherfield@bristol.ac.uk, nigel@cs.bris.ac.uk

Abstract. We examine a FlipIt game in which there are multiple resources which a monolithic attacker is trying to compromise. This extension to FlipIt was considered in a paper in GameSec 2014, and was there called FlipThem. Our analysis of such a situation is focused on the situation where the attacker's goal is to compromise a threshold of the resources. We use our game theoretic model to enable a defender to choose the correct configuration of resources (number of resources and the threshold) so as to ensure that it makes no sense for a rational adversary to try to attack the system. This selection is made on the basis of the relative costs of the attacker and the defender.

1 Introduction

At its heart security is a game played between an attacker and a defender; thus it is not surprising that there have been many works which look at computer security from the point of view of game theory [1,9,12,15]. One particularly interesting example is the FlipIt game developed by van Dijk et al. [16]. In FlipIt the attacker and defender are competing to control a resource. Both players are given just a single button each. The attacker gets control of the resource by pressing her button, whilst the defender can regain control by pressing his button. Pressing the button has a cost for each player, and owning the resource has a gain.

In this work we examine the FlipIt game in the situation where the defender has multiple resources, and the attacker is trying to obtain control of as many of these resources as possible. This was partially considered before in the paper [7], who introduced a variant of FlipIt called FlipThem in which the defender has control of multiple resources. Instead of flipping the state of a single resource from good to bad, the attacker is trying to flip the states of multiple resources. In [7] the authors examine the simplest situations in which an attacker "wins" if he has control of all resources, and a defender "wins" if she has control of at least one resource. Thus using the terminology of secret sharing schemes the paper [7] considers only the *full threshold* situation.

In this paper we study non-full threshold cases. This is motivated by a number of potential application scenarios which we now outline:

© Springer International Publishing Switzerland 2015
MHR Khouzani et al. (Eds.): GameSec 2015, LNCS 9406, pp. 74–92, 2015.
DOI: 10.1007/978-3-319-25594-1_5

- Large web sites usually have multiple servers responding to user requests so as to maintain high availability and response times. An APT attack on a web site may try to knock out a proportion of the servers so as to reduce the owners quality of service below an acceptable level.
- Large networks contain multiple paths between different nodes; again to protect against attacks. An attacker will not usually be successful if he knocks out a single path, however knocking out all paths is overkill. There will be a proportion of the paths which will result in a degradation of the network connectivity which the attacker may want to achieve.
- In many computer systems multiple credentials are needed to access a main resource. Thus an attacker only needs to obtain enough credentials to compromise a main resource. Thus modelling attacks on credentials (e.g. passwords, certificates, etc.) should really examine the case of multiple credentials in the non-full threshold case.
- Multi-party Computation (MPC) has always used threshold adversaries; an external attacker trying to compromise a system protected with MPC technology will only be interested in obtaining a threshold break above the tolerance limit of the MPC system. In such a situation however one is interested in proactively secure MPC systems, since when modelled by FlipThem a defender may regain control of a compromised party.
- Related to the last point is that of fault tolerance. It is well known that Byzantine agreement is not possible if more than $n/3$ of the parties are compromised. Thus an adversary who simply wants to inject errors into a network protected by some Byzantine agreement protocol only needs to compromise more than $n/3$ of the servers.

Thus we examine variants of the FlipThem game of [7] in which an attacker is trying to obtain control of at least t of the resources. We call this the (n,t)-FlipThem game.

Our main results are to examine Nash equilibria in the case of stochastic models of play. These are models in which the players strategies are defined by some random process. The random process defines, for each player, the next time point at which it will make a play (with time being considered as continuous). In all of the models we consider, we think of an attacker's play as being to attack a single resource; in the case of a stealthy defender the machine to attack at a given point in time will be random, whereas in the case of a non-stealthy defender the attacker will always attack a non-compromised resource. For the defender we allow two possible moves; in the first type the defender gains control of *all* resources with a single play. This models the situation where a defender might reset and reinstall a whole cloud infrastructure in one go, or reset all credentials/passwords in a given move; we call this a *full reset*. In the second type of move the defender needs to select a single resource to reset. Just like in the case of the attacker, the defender can do this in two ways depending on whether the attacker is stealthy or not. We call this type of defender move a *single reset*. This paper introduces continuous time Markov chains as a method of finding the benefit functions and calculating Nash equilibria of the two player partial threshold multi-party FlipIt game, FlipThem. For full reset, it finds that the equilibria depend solely on the threshold of the resources and the costs of play, not the number of resources involved. As the cost for the attacker increases the necessary amount of servers (threshold) required for the defender to maximise his benefit decreases. For single reset, the analysis is harder by hand. However, using numerical methods, one can find analogous results.

1.1   Prior Work

The FlipIt game has attracted attention as it focuses on the situation where the attacker always gets in; building on the modern appreciation that perimeter defence on its own is no longer enough. For example the paper [2] examines the FlipIt game as applied to various different situations in computer security; for example password reset strategies, key management, cloud auditing and virtual machine refresh methodologies.

Despite its simplicity the FlipIt game is rather complex in terms of the possible different attacker and defender strategies, and can be modified in various ways. In the original FlipIt game both the attacker and the defender are 'stealthy' in the sense that neither knows if the other controls the resource before they execute a button press. In [13] the authors introduce a new mechanism where by a player can test who controls the resource. The idea being to model the situation whereby investigating whether a breach has occured is less costly than clearing up after a breach. Thus a 'peek'/'probe' at the resource state costs less than taking control of the resource. The paper [13] then moves onto discuss situations where a resource becomes hardened over time; meaning that every time a player moves on a resource he already controls, part of the move consists of making it harder for the opponent to regain control of the resource. An example would be a system administrator resetting the system to regain control and then patching the system so the attacker can not use the same method of infiltration.

One can think of the 'peek'/'probe' at the resource state from [13] as a way of removing the stealthiness from the FlipIt game. In [8] a different approach is proposed in which *defender* moves are not stealthy, i.e. an attacker knows if the defender controls the resource. This is introduced to model situations such as password resetting, in which an attacker knows when the password is reset (as he is no longer able to login), but the defender may not notice that their password is compromised. As well as this non-stealthy mode of operation the paper also introduces the idea of a defender trying to defend against multiple (independent) attackers.

The main prior work related to the current paper is that of Laszka et al. [7]. They consider the same situation as us of multiple resources being attacked by a single monolithic adversary. However, their work has a number of distinct differences. Firstly, and most importantly, they focus on the case where an attacker wins if he controls all resources, and the defender wins when he controls one resource. We on the other hand examine a general threshold structure. Secondly, the paper of Laszka et al. considers two types of strategies defined by periodic and non-arithmetic renewal processes[1]. The paper establishes some basis facts on these strategies, but does not consider constructing full benefit functions for either of these strategies and nor does it find analytic Nash equilibria for the strategies. This is due to the analytic difficulty in obtaining such formulae.

Given this (relatively) negative result the paper moves onto considering strategies arising from Markov processes. They develop a model for two resources, considering discrete time steps and set up a linear programming solution that becomes more complicated as the finite time horizon extends. We on the other hand are able to obtain simpler analytic formulae by considering a continous Markov process. This is because in [7] when constructing the Markov chain, they consider the state space to be the inter-arrival times of each resource with respect to the attacker.

In our paper we set up the state space to be the number of resources compromised at a specific (continuous) time. Thus moving from discrete to continuous time, and Markov to Stochastic processes simplifies the analysis somewhat. Without this simplification the paper [7] looks at two specific examples; trying to find the optimal strategy of the attacker given the strategy of the defender, and then the optimal flip rates that maximise the benefit at the defender side given that the attacker plays optimally. Finally they briefly mention how to find a Nash equilibrium, stating there is a simple iterative algorithm to find one but they state that algorithm will not converge for the majority of cases.

The paper [17] also considers a number of extensions of the FlipIt paper, and much like that of Laszka et al. comments on the difficulty of obtaining analytic solutions to the Nash equilibrium. Therefore, they adopt a simulation based method. The attackers probability of compromising increases progressively with probing, while the defender uses a moving-target technique to erase attacker progress. The paper extends the model to multiple resources and considers a time dependent 'reimage' initiated by the defender, much like our full reset play of the defender described above. In addition [17], much like our own work, sets up a situation of asymmetric stealth in that the attacker can always tell when the defender has moved however the defender does not know when the attacker has compromised the resource but finds this out when he has probes the resource.

Having multiple resources which an attacker needs to compromise also models the situation of a moving target defence and a number of game theoretic works are devoted to other aspects of moving target defence including [3, 18]. Since these works are not directly related to our own work we do not discuss them here.

2 The Multi-party FlipIt Model

Our basic multi-party FlipIt game, or FlipThem game, consists of a defender who is trying to protect against an attacker getting control of n different resources. It may help the reader to notice how at each point our game degenerates to the FlipIt game when $n = 1$.

---

[1] A renewal process is called non-arithmetic if there is no positive real number $d > 0$ such that the inter-arrival times are all the integer multiples of d.

At a given point in time the attacker will control a given threshold k of the resources. The attacker is deemed to be "in control", or have won, if k exceeds some value t. For example in a denial-of-service attack on a web site, the web-site may still be able to function even if $2/3$ of the servers are down, thus we will set $t = 2 \cdot n/3$. In the case of an attacker trying to disrupt a consensus building network protocol, i.e. an instantiation of the problem of Byzantine agreement, the value of t would be $n/3$. In the case of a multi-party computation protocol the threshold t would correspond to the underlying threshold tolerated by the MPC protocol; e.g. $t = n/3$, $t = n/2$ or $t = n$. Note, in the case of MPC protocols, the ability of the defender to reset all resources is a common defence against mobile adversaries, and is thus related to what is called pro-active security in the MPC community [11].

The variable D_B is the multiplicative factor of the defender's benefit (i.e. the benefit obtained per unit time), the same for the attacker's A_B. The values are potentially distinct, since the defender could gain more (or less) than the attacker for being in control of the system for an amount of time. The values D_c and A_c are respectively the defender and attacker's cost per action they perform. We set $d = \frac{D_c}{D_B}$ to be the ratio of the defender's cost and benefit. Similarly for the attacker, $a = \frac{A_c}{A_B}$. We then consider the ratio $\rho = \frac{a}{d} = \frac{A_c \cdot D_B}{A_B \cdot D_c}$. Much of our analysis will depend on whether ρ is large or small; which itself depends on the relative ratios of the benefit/costs of the attacker and defender. With each application scenario being different. A game where the costs are normalized in this way we shall call a "normalized game".

For each time period for which the attacker obtains control of t or more of the resources it obtains a given benefit, whereas for each time period that he does not have such control the defender obtains a benefit. In the normalized game we assume the attacker's benefit lies in $[0, 1]$ and is the proportion of time that he controls the resource; whilst the defenders benefit is the proportion of time in which they control the resource. Thus in the normalized game the benefits always sum to one.

In all games the utility for the attacker is their benefit minus their cost of playing (i.e. the cost of pushing the buttons), with the utility for the defender obtained in the same manner. Therefore, the game is non-zero sum. The attacker (resp. defenders) goal is to derive a strategy which maximises their respective utility.

In one basic normalised "Single Reset" game the defender has a set of n buttons; there is one button on each resource which when pressed will return that resource to the defenders control, or do nothing if the resource is already under the defenders control. Pressing the resource's button costs the defender a given value, which in the normalized game is the value d. In another normalised "Full Reset" game addition there is a "master button" which simultaneously returns all resources to the defenders control. Pressing the master button costs the defender a value which we shall denote by D_n, the value of which depends on n, the number of resources. The reason for having a master button is to capture the case when resetting the entire system in one go is simpler than resetting each resource individually. In particular we assume that $d \leq D_n$. To simplify our games we assume that the defender does not have access to the master button and the individual resource buttons in a single game. This property could be relaxed which would result in a much more complex analysis than that given here.

The attacker has a set of n buttons, one for each resource. When the attacker presses a resources button it will allow the adversary control of that resource, or again do nothing if the resource is already under the attackers control. The cost to the attacker of pressing one of its buttons is a in the normalized game.

As can be inferred from the above discussion we do not assume that the defender knows whether it controls a resource, nor do we assume that an attacker knows whether it controls a resource at a given time point. This situation is called the two-way stealthy situation, if we assume a defender is not stealthy (but the attacker is) we are said to be in a one-way stealthy situation.

Throughout the paper we model a number of games. We denote $\text{FlipThem}_\epsilon^{\mathcal{R}}(n, t, d, \rho)$ to be the game of partial threshold FlipThem. By abuse of notation we also think of $\text{FlipThem}_\epsilon^{\mathcal{R}}(n, t, d, \rho)$ as a function which returns all the rates of play strategy pairs for the defender and attacker that are Nash Equilibria where $\mathcal{R} \in \{\mathcal{F}, \mathcal{S}\}$. Here we denote by $\mathcal{F}$ the full reset game and $\mathcal{S}$ the single reset game, both to be described in detail in later sections. The variables n, t, d ρ and ϵ denote the number of resources, the threshold, the defender's cost of play, the ratio between the attacker's and defender's cost and the lowest rate of play in the defender's strategy space $(\epsilon, \infty]$ respectively. Having $\epsilon > 0$ recognises the fact that the defender will never actually set the reset rate to 0. It also ensures that the benefit functions are well defined for all valid attacker-defender strategy pairs. We will not treat the choice of our ϵ to be strategic, it will be a very small number, close to zero to represent that even when the attacker has given up (plays a rate of zero) the defender will not.

We also use a function $\text{Opt}_{N,\epsilon}^{\mathcal{R}}(d, \mathcal{T}, \rho)$ to answer the following question: Given the ratio ρ of costs of play between the attacker and defender and a limit N for the number of resources the defender can own, what is the best set up for the defender in order to maximise their benefit function? The function $\text{Opt}_{N,\epsilon}^{\mathcal{R}}(d, \mathcal{T}, \rho)$ plays the first game $\text{FlipThem}_\epsilon^{\mathcal{R}}(n, t, d, \rho)$ for all n and all t subject to some constraint space $\mathcal{T}^2$. The function $\text{Opt}_{N,\epsilon}^{\mathcal{R}}(d, \mathcal{T}, \rho)$ then finds the values of n and t which produce the greatest possible benefit for the defender.

3   Obtaining Nash Equilibria in Continuous Time for a Stochastic Process

In this section we analyse various different cases of our basic game $\text{FlipThem}_\epsilon^{\mathcal{R}}(n, t, d, \rho)$. To explain the basic analysis techniques in a simple example; we first examine the game $\text{FlipThem}_0^{\mathcal{F}}(n, n, d, \rho)$. In this game the defender can perform a full reset and the attacker is trying to compromise all n servers (i.e. the full threshold case). We also, again for initial simplicity and exposition purposes, assume that the defender could decide not to play, i.e. $\epsilon = 0$. A moments thought

---

[2] For example $t \leq n$, or $t \leq n/2$, or $n - t \geq B$ for some bound B.

will reveal in practice that such a strategy is not realistic. In the later sub-sections we remove these two simplifying assumptions and examine other cases. In particular in Sect. 3.3 when we consider defender performing single resets, the analysis becomes more complex.

3.1 Simple Example, $\text{FlipThem}_0^{\mathcal{F}}(n, n, d, \rho)$: Full Threshold, Full Reset

We first consider a simple example of our framework in which the time an attacker takes to successfully compromise an individual resource follows an exponential distribution with rate λ, and the defender performs a full reset, and thus regains control of all resources, at intervals with lengths given by an exponential distribution with rate μ. An alternative description is that individual resources are compromised on at the arrival times of a Poisson process with rate λ, and the state is reset at the arrival times of a Poisson process with rate μ.

In this context we think of the attacker as being stealthy, i.e. the defender does not know how many resources are compromised when he does a full reset. A moment's thought will also reveal that in this situation it makes no difference if the defender is stealthy or not; if the defender is not stealthy then the attacker will always pick an uncompromised resource to attack, whereas if the defender is stealthy then the attacker is more likely to compromise an uncompromised resource by picking one which he knows he controlled the longest time ago. Thus an attacker simply attacks each resource in turn, given some specific ordering.

We model the number of resources compromised by the attacker at time τ as a family of random variables $X = \{X(\tau) : \tau \geq 0\}$ in the finite space $S = \{0, \ldots, n\}$. Since both the defender and attacker follow memoryless strategies (with memoryless exponential random variables determining the times between changes of state) the process X is a continuous time Markov chain. Following the analysis of continuous time Markov chains in Grimmet et al. [5], for such a process there exists an $|S| \times |S|$ generator matrix G with entries $\{g_{ij} : i, j \in S\}$ such that

$$\Pr[X(\tau + h) = j \mid X(\tau) = i] = \begin{cases} 1 + g_{ii} \cdot h + o(h), & \text{if } j = i, \\ g_{ij} \cdot h + o(h), & \text{if } j \neq i. \end{cases}$$

The generator matrix G for continuous time Markov chains replaces the transition matrix P for discrete time Markov chains; entry g_{ij} for $i \neq j$ is the "rate" of transition from state i to state j. Summing equation (3.1) over j implies that $\sum_{j \in S} g_{ij} = 0$, so that $g_{ii} = -\sum_{j \neq i} g_{ij} \leq 0$. Basic theory [5] tells us that when the chain arrives in state i it remains there for an amount of time following a Exponential$(-g_{ii})$ distribution, then jumps to state $j \neq i$ with probability $-g_{ij}/g_{ii}$.

Considering our specific example with the defender using full reset, we can consider our model as a "birth-reset process" (by analogy with a "birth–death process") in which

$$\Pr[X(\tau + h) = j \mid X(\tau) = i] = \begin{cases} \lambda \cdot h + o(h), & \text{if } j = i+1, \\ \mu \cdot h + o(h), & \text{if } j = 0, \\ 1 - (\lambda + \mu) \cdot h + o(h), & \text{if } j = i, \\ o(h), & \text{otherwise.} \end{cases}$$

Thus, $g_{i0} = \mu$, $g_{i,i+1} = \lambda$, $g_{ii} = -(\mu + \lambda)$ and $g_{ij} = 0$ otherwise. From this the generator matrix can be constructed:

$$G = \begin{pmatrix} -\lambda & \lambda & 0 & 0 \cdots & 0 & 0 \\ \mu & -(\mu + \lambda) & \lambda & 0 \cdots & 0 & 0 \\ \mu & 0 & -(\mu + \lambda) & \lambda \cdots & 0 & 0 \\ \vdots & \vdots & \vdots & \vdots \ddots & \vdots & \vdots \\ \mu & 0 & 0 & 0 \cdots & -(\mu + \lambda) & \lambda \\ \mu & 0 & 0 & 0 \cdots & 0 & -\mu \end{pmatrix}.$$

Thus when the state is $i \in \{1, \ldots, n-1\}$ the system will jump to either state $i+1$ with probability $\lambda/(\lambda+\mu)$ (when the attacker compromises another resource before reset occurs) or to state 0 with probability $\mu/(\lambda+\mu)$ (when the reset occurs before another resource is compromised). Clearly the chain is never going to settle in one state; it will continue to randomly fluctuate between various states depending on the rates of play μ and λ. However further theory [5] indicates that the long run proportion of time the system spends in each state is given by the stationary distribution, a row vector $\pi = (\pi_0, \ldots, \pi_n)$ such that $\pi G = 0$ and $\sum_{i=0}^{n} \pi_i = 1$.

Using our specific generator matrix G it can be shown that

$$\pi = \left(\frac{\mu}{\mu + \lambda}, \frac{\mu \cdot \lambda}{(\mu + \lambda)^2}, \ldots, \frac{\mu \cdot \lambda^{n-1}}{(\mu + \lambda)^n}, \frac{\lambda^n}{(\mu + \lambda)^n} \right). \tag{1}$$

This tells us the proportion of time spent in each state. We therefore obtain the benefit functions of

$$\beta'_D(\mu, \lambda) = D_B \cdot (1 - \pi_n) - D_c \cdot \mu$$

and

$$\beta'_A(\mu, \lambda) = A_B \cdot \pi_n - A_c \cdot \lambda$$

where β'_D is the benefit function of the defender and β'_A is the benefit function of the attacker. We can then normalise β'_D and β'_A such that

$$\beta_D(\mu, \lambda) = \frac{\beta'_D}{D_B} = 1 - \pi_n - d \cdot \mu = 1 - \frac{\lambda^n}{(\mu + \lambda)^n} - d \cdot \mu$$

and

$$\beta_A(\mu, \lambda) = \frac{\beta'_A}{A_B} = \pi_n - a \cdot \lambda = \frac{\lambda^n}{(\mu + \lambda)^n} - a \cdot \lambda,$$

where β_D is the normalized benefit function of the defender and β_A is the normalized benefit function of the attacker.

Recall that in this model, when the defender plays he is resetting all resources at once. Therefore, the normalized cost of the defenders move d is likely to depend on n, the number of resources. We represent this by setting $d = D_n$.

Using the stationary distribution described above the benefit functions for the normalized game are

$$\beta_D(\mu, \lambda) = 1 - \frac{\lambda^n}{(\mu + \lambda)^n} - D_n \cdot \mu \quad \text{and} \quad \beta_A(\mu, \lambda) = \frac{\lambda^n}{(\mu + \lambda)^n} - a \cdot \lambda. \quad (2)$$

We are assuming that both players are rational, in that they are both interested in maximising their benefit functions, and will therefore choose a rate (λ or μ) to maximise their benefit given the behaviour of their opponent. A pair of rates at which each player is playing optimally against the other is called a Nash equilibrium [10]. At such a point neither player can increase their benefit by changing their rate; we are looking for pairs (λ^*, μ^*) such that

$$\beta_D(\mu^*, \lambda^*) = \max_{\mu \in R_+} \beta_D(\mu, \lambda^*) \quad \text{and} \quad \beta_A(\mu^*, \lambda^*) = \max_{\lambda \in R_+} \beta_A(\mu^*, \lambda).$$

Note that $\mu^* = \lambda^* = 0$ is an equilibrium of the game defined by Eq. (2). This is an artefact of assuming the existence of a unique distribution for all μ, λ, where as when $\lambda = \mu = 0$ the Markov chain never makes any transitions. In later sections we will bound μ below to remove this solution and for now we will search for non-trivial solutions.

Differentiating the defender's benefit function β_D with respect to μ and solving for μ gives at most one non-negative real solution, given by

$$\hat{\mu}(\lambda) = \sqrt[n+1]{\frac{n\lambda^n}{D_n}} - \lambda$$

If $\lambda < \frac{n}{D_n}$ then this is positive, and checking the second derivative confirms this corresponds to a maximum. If $\lambda \geq \frac{n}{D_n}$ then $\frac{\partial \beta_D}{\partial \mu} < 0$ for all $\mu \geq 0$ and so the optimal rate for the defender is $\mu = 0$. Hence the best response of the defender is given by

$$\hat{\mu}(\lambda) = \begin{cases} \sqrt[n+1]{\frac{n\lambda^n}{D_n}} - \lambda & \text{if} \quad \lambda < \frac{n}{D_n} \\ 0 & \text{if} \quad \lambda \geq \frac{n}{D_n}. \end{cases}$$

We now calculate

$$\frac{\partial \beta_A}{\partial \lambda} = \frac{n \cdot \mu \cdot \lambda^{n-1}}{(\mu + \lambda)^{n+1}} - a.$$

A closed form solution for λ which equates this to 0 is not easy to calculate directly. However, plugging in $\hat{\mu}(\lambda^*)$ we see that λ^* must be either 0 or satisfy

$$\frac{n \cdot \hat{\mu}(\lambda^*) \cdot (\lambda^*)^{n-1}}{(\hat{\mu}(\lambda^*) + \lambda^*)^{n+1}} - a = 0. \quad (3)$$

If it were the case that $\lambda^* \geq \frac{n}{D_n}$ then $\hat{\mu}(\lambda^*) = 0$ and there are no solutions to this equation. Note that this indicates that no equilibrium exists when the attacker's rate is too high — the intuition for this is if the attacker's rate is sufficiently high, the defender ceases to defend, and thus the attacker can do just as well by reducing their rate. Thus at any equilibrium we must have $\lambda^* < \frac{n}{D_n}$, and therefore $\mu^* = \hat{\mu}(\lambda^*) = \sqrt[n+1]{\frac{n(\lambda^*)^n}{D_n}}$. Plugging this back into Eq. (3) we see that either

$$\lambda^* = \frac{n \cdot D_n^n}{(D_n + a)^{n+1}}, \quad \mu^* = \hat{\mu}(\lambda^*) = \frac{n \cdot a \cdot D_n^{n-1}}{(D_n + a)^{n+1}}, \tag{4}$$

or $\mu^* = \lambda^* = 0$. The non-zero solution will only correspond to a Nash equilibrium if $\beta_A(\mu^*, \lambda^*) \geq \beta_A(\mu^*, 0) = 0$, since otherwise λ^* is not a best response against μ^*. Note that this is the case if

$$0 < \frac{(\lambda^*)^n}{(\mu^* + \lambda^*)^n} - a \cdot \lambda^* = \frac{(D_n)^n}{(D_n + a)^{n+1}} (D_n + a \cdot (1 - n))$$

i.e. if $a/D_N < 1/(n-1)$.

In the game $\text{FlipThem}_0^{\mathcal{F}}(n, n, D_n, \rho)$ we have defined ρ to be the ratio between the attacker and defender's costs, so that $\rho = a/D_n$. Therefore, the game $\text{FlipThem}_0^{\mathcal{F}}(n, n, D_n, \rho)$ returns the list $\{(0,0)\}$ for all $\rho > 1/(n-1)$. If $\rho < 1/(n-1)$ we have a further equilibrium (μ^*, λ^*) such that the game returns the list $\{(0,0), (\mu^*, \lambda^*)\}$ where

$$\mu^* = \frac{n \cdot \rho}{D_n \cdot (1+\rho)^{n+1}}, \quad \lambda^* = \frac{n}{D_n \cdot (1+\rho)^{n+1}} = \mu^*/\rho.$$

The attacker's cost per move is independent of n, which implies that the defender will be successful, assuming $\frac{D_n}{n-1}$ is a decreasing function of n, as long as n is large enough. Thus for the defender to always win we require the cost of a full reset to be a sublinear function of the number of resources.

In the case of resetting a cloud or web service this might be a reasonable assumption, but in the case of requiring n users to reset their passwords it is likely that the cost is a superlinear function as opposed to sublinear due to the social cost in needing to implement such a password policy.

3.2 $\text{FlipThem}_\epsilon^{\mathcal{F}}(n, t, d, \rho)$: (n,t)-Threshold, Full Reset

We now generalize the previous easy case to the threshold case $\text{FlipThem}_\epsilon^{\mathcal{F}}(n, t, d, \rho)$, i.e. we treat the number of servers which the attacker has to compromise as a parameter t, and in addition we bound the defenders strategy away from zero. Thus the defender not playing at all is not considered a valid strategy[3]. Much of the prior analysis carries through, since we are still assuming the defender performs a full reset on his turn. Thus the stationary

---

[3] Of course if the attacker decides not to play that is considered a good thing.

distribution is once more,

$$\pi = \left(\frac{\mu}{\mu + \lambda}, \ldots, \frac{\mu \cdot \lambda^{k-1}}{(\mu + \lambda)^k}, \ldots, \frac{\mu \cdot \lambda^{n-1}}{(\mu + \lambda)^n}, \frac{\lambda^n}{(\mu + \lambda)^n} \right).$$

The (normalized) benefit functions are now derived from the ratio of times which the attacker has compromised at least t resources, which simplifies due to the formula for geometric series:

$$\beta_D(\mu, \lambda) = 1 - \frac{\lambda^n}{(\mu + \lambda)^n} - \sum_{i=t}^{n-1} \frac{\mu \cdot \lambda^i}{(\mu + \lambda)^{i+1}} - D_n \cdot \mu$$

$$= 1 - \frac{\lambda^t}{(\mu + \lambda)^t} - D_n \cdot \mu.$$

Using the same analysis, the attacker's benefit is $\beta_A(\mu, \lambda) = \frac{\lambda^t}{(\mu+\lambda)^t} - a \cdot \lambda$. Note that these benefit functions are identical to those in the full threshold case of the previous section, but with n replaced by t. If we were still considering the lower bound for the defender's rate of play ϵ to be zero the conclusions would be as before, but with the modification that we use t instead of n. Since we are now considering the more realistic assumption that $\epsilon > 0$ the analysis gets slightly more involved, but remains similar to that above. In particular

$$\beta_D(\mu, \lambda) = 1 - \left(\frac{\lambda}{\lambda + \mu} \right)^t - D_n \cdot \mu, \quad \text{and} \quad \frac{\partial \beta_D}{\partial \mu} = \frac{t \cdot \lambda^t}{(\lambda + \mu)^{t+1}} - D_n.$$

This derivative is decreasing in μ, and 0 at $\lambda \cdot \left[\left(\frac{t}{\lambda \cdot D_n} \right)^{\frac{1}{t+1}} - 1 \right]$. It follows immediately that β_D is a unimodal function of μ, so that the maximising μ value in $[\epsilon, \infty)$ is given by

$$\hat{\mu}(\lambda) = \min \left\{ \epsilon, \lambda \cdot \left[\left(\frac{t}{\lambda \cdot D_n} \right)^{\frac{1}{t+1}} - 1 \right] \right\}. \tag{5}$$

As above, we have that

$$\beta_A(\mu, \lambda) = \left(\frac{\lambda}{\mu + \lambda} \right)^t - a \cdot \lambda \quad \text{and} \quad \frac{\partial \beta_A}{\partial \lambda} = \frac{t \cdot \mu \cdot \lambda^{t-1}}{(\lambda + \mu)^{t+1}} - a. \tag{6}$$

Thus for a particular value of μ the maximising λ must either be 0 or be a root of the derivative. However, explicitly solving for λ does not appear to be possible, but we note that

$$\frac{\partial^2 \beta_A}{\partial \lambda^2} = \frac{t \cdot \mu \cdot \lambda^{t-2}}{(\lambda + \mu)^{t+2}} \cdot [\mu \cdot (t - 1) - 2 \cdot \lambda]$$

so that the first derivative, $\frac{\partial \beta_A}{\partial \lambda}$, is increasing when $\lambda < \mu \cdot (t - 1)/2$ then decreasing. Since $\frac{\partial \beta_A}{\partial \lambda}$ is equal to $-a$ when $\lambda = 0$ and asymptotes to $-a$ as $\lambda \to \infty$

we have the derivative increasing from $-a$ to a maximum when $\lambda = \mu \cdot (t-1)/2$ then decreasing back to $-a$. The maximal value of $\frac{\partial \beta_A}{\partial \lambda}$ is given by

$$\frac{4 \cdot t \cdot (t-1)^{t-1}}{\mu \cdot (t+1)^{t+1}} - a, \tag{7}$$

which is positive only if μ is sufficiently small. As a function of λ, β_A therefore initially decreases (from 0), has a period of increase only if μ is sufficiently small, then decreases again. It follows that β_A has at most one non-zero maximum, which occurs in the region $(\mu \cdot (t-1)/2, \infty)$ once the derivative is decreasing, and this fixed point maximises $\beta_A(\mu, \lambda)$ on $\lambda \in [0, \infty)$ if and only if $\beta_A(\mu, \lambda) > 0$; otherwise the best response must be $\lambda = 0$. We use these insights to explore Nash equilibria directly. First consider the existence of a Nash equilibrium (μ^*, λ^*) with $\mu^* > \epsilon$. Note that if λ^* were equal to 0 then this would force $\mu^* = \epsilon$, so it must be the case that $\mu^* = \hat{\mu}(\lambda^*)$ and $\frac{\partial \beta_A}{\partial \lambda}(\mu^*, \lambda^*) = 0$. It follows from (5) and (6) that

$$a = \frac{t \cdot \mu^* \cdot \lambda^{*t-1}}{(\lambda^* + \mu^*)^{t+1}} = D_n \cdot \left[\left(\frac{t}{\lambda^* \cdot D_n} \right)^{\frac{1}{t+1}} - 1 \right]$$

and hence

$$\lambda^* = \frac{t}{D_n \cdot (1+\rho)^{t+1}}, \qquad \mu^* = \frac{t \cdot \rho}{D_n \cdot (1+\rho)^{t+1}}. \tag{8}$$

We have checked necessary conditions so far, but have still not verified that this λ^* does correspond to a maximum of β_A. As observed above, the necessary and sufficient condition is that

$$0 < \beta_A(\mu^*, \lambda^*) = \frac{1 + \rho - \rho \cdot t}{(1+\rho)^{t+1}}.$$

Thus an equilibrium of this form exists when

$$\rho < \frac{1}{t-1} \quad \text{and} \quad \mu^* = \frac{t \cdot \rho}{D_n \cdot (1+\rho)^{t+1}} > \epsilon.$$

Therefore, if the ratio ρ of the attacker's cost and defender's cost is less than $\frac{1}{t-1}$ then the game FlipThem$_\epsilon^{\mathcal{F}}(n, t, d, \rho)$ returns the list consisting of two pairs, the trivial equilibrium of no play (from the attacker, the defender plays at minimal rate ϵ) and an equilibrium at

$$\mu^* = \frac{t \cdot \rho}{D_n \cdot (1+\rho)^{t+1}}, \qquad \lambda^* = \frac{t}{D_n \cdot (1+\rho)^{t+1}} = \mu^*/\rho.$$

Note that if the maximal value of the derivative of β_A is non-positive then no stationary point of β_A exists, and so λ will be 0. By removing all local maxima of the attacker's payoff function we really would expect the attacker to just stop playing; i.e. this would be the perfect defenders strategy. From (7) we see that by taking

$$\epsilon \geq \frac{4 \cdot t \cdot (t-1)^{t-1}}{a \cdot (t+1)^{t+1}} \tag{9}$$

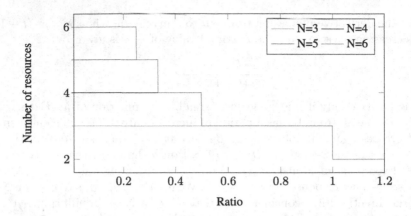

Fig. 1. Number of resources used by the defender to maximise his benefit given a specific ρ

we can ensure there is only the trivial equilibrium. Note that a simpler lower bound on ϵ, which trivially implies the one above, is to take $\epsilon \geq \frac{4}{a \cdot (t+1)}$. Note that choosing a sufficiently high ϵ in this way is very conservative. The rate of decrease of β_A is $-a$ at $\lambda = 0$ and as $\lambda \to \infty$, so by insisting there is no local maximum at all we ensure β_A stays well away from 0.

Picking ϵ to force out the attacker only makes sense if the defender's benefit is actually maximised. It might be the case that stopping the attacker completely is not economically viable. Therefore, in such a case ϵ should be chosen to be very small, close to zero and the other equilibria in Eq. (8) should be used; implying that μ^* is less than the right hand side of Eq. (9). Thus an expected amount of attacker success may be tolerated if completely eliminating such success comes at too much of a price. Recall our function $\text{Opt}_{N,\mathcal{T},\epsilon}^{\mathcal{F}}(d, \rho)$. If we fix $\epsilon = 0.01/d$ and set $\mathcal{T} = \{t \leq n\}$, and run this programmatically for ρ from 0 to 1, Fig. 1 shows the smallest $n \leq N$ that maximises the defenders benefit for various N. Recall that the attacker will not play if $\rho > 1/t - 1$, meaning that as ρ increases the level of threshold decreases and therefore the number of servers required decrease. The optimum defender's benefit occurring when $t = n$. This explains the step down in Fig. 1.

We end this section by examining the classic case of a threshold situation in which the required threshold is a constant fraction of the total number of resources. Suppose we have $t = \gamma \cdot n$ for some constant $\gamma \in (0, 1]$. We have shown that the attacker will not play if $\frac{a \cdot \rho}{D_n} \geq \frac{1}{t-1} = \frac{1}{\gamma \cdot n - 1}$. As expected we see that if the attacker needs to compromise fewer resources, then the attacker's cost per resource needs to be greater for them not to play. It is intuitively obvious that the smaller the threshold the more likely the attacker will play (and succeed).

3.3 FlipThem$_\epsilon^{\mathcal{S}}(n, t, d, \rho)$: (n,t)-Threshold, Single Reset

So far we have set up the model such that the defender can reset the whole system regaining full control whereas the attacker compromises each resource individually. We now consider the game $\text{FlipThem}_\epsilon^S(n, t, d, \rho)$. The defender can reset a single machine at any specific time. Consider the situation at any time point where the number of resources compromised is k out of n. Assume the defender is going to reset a resource. There are multiple strategies they could employ, they could pick a resource which they have not reset recently, or pick a random resource, or pick a resource in a given secret sequence. Here we will assume the players pick resources uniformly at random. Thus the probability of resetting a compromised resource is $\frac{k}{n}$, and that of wastefully resetting a non-compromised resource $1 - \frac{k}{n}$. Letting the defender's and attacker's rate of play be μ and λ respectively, it is not hard to see that our generating matrix now becomes

$$
G = \begin{pmatrix}
-\lambda & \lambda & 0 & 0 & \cdots & 0 & 0 & 0 \\
\frac{\mu}{n} & -\frac{(\mu+(n-1)\cdot\lambda)}{n} & \frac{(n-1)\cdot\lambda}{n} & 0 & \cdots & 0 & 0 & 0 \\
0 & \frac{2\cdot\mu}{n} & -\frac{(2\cdot\mu+(n-2)\cdot\lambda)}{n} & \frac{(n-2)\cdot\lambda}{n} & \cdots & 0 & 0 & 0 \\
\vdots & \vdots & \vdots & \vdots & \ddots & \vdots & \vdots & \vdots \\
0 & 0 & 0 & 0 & \cdots & \frac{(n-1)\cdot\mu}{n} & -\frac{((n-1)\cdot\mu+\lambda)}{n} & \frac{\lambda}{n} \\
0 & 0 & 0 & 0 & \cdots & 0 & \mu & -\mu
\end{pmatrix}
$$

We then solve for the stationary distribution $\pi = (\pi_0, \pi_1, \ldots, \pi_{n-1}, \pi_n)$, by solving $\pi G = 0$, and it can be shown by induction that

$$
\pi_k = \frac{n! \cdot \lambda^k \cdot \pi_0}{(n-k)! \cdot k! \cdot \mu^k} = \frac{\binom{n}{k} \cdot \lambda^k \cdot \pi_0}{\mu^k}.
$$

Recall, that we also need to utilize the constraint $\sum_{i=0}^{n} \pi_i = 1$, which implies that we have $\pi_0 = \frac{\mu^n}{(\mu+\lambda)^n}$ so that we obtain the stationary distribution

$$
\pi = \frac{1}{(\mu+\lambda)^n}\left(\mu^n, n\cdot\lambda\cdot\mu^{n-1}, \ldots, \binom{n}{k}\cdot\mu^{n-k}\cdot\lambda^k, \ldots, n\cdot\mu\cdot\lambda^{n-1}, \lambda^n\right).
$$

Once again, this gives us the proportion of time spent in each state. We assume here that the costs and benefits have already been normalised and do not depend on n the number of resouces. Constructing these benefit functions gives

$$
\beta_D(\mu, \lambda) = 1 - \sum_{i=t}^{n} \pi_i - d\cdot\mu = 1 - \frac{1}{(\mu+\lambda)^n}\cdot\sum_{i=t}^{n}\binom{n}{i}\cdot\mu^{n-i}\cdot\lambda^i - d\cdot\mu,
$$

$$
\beta_A(\mu, \lambda) = \sum_{i=t}^{n} \pi_i - a\cdot\lambda = \frac{1}{(\mu+\lambda)^n}\cdot\sum_{i=t}^{n}\binom{n}{i}\cdot\mu^{n-i}\cdot\lambda^i - a\cdot\lambda
$$

We want to find the Nash Equilibria for these benefit functions. A point at which neither player can increase their benefit by changing their rate. We want to find pairs (μ^*, λ^*) such that

$$
\beta_D(\mu^*, \lambda^*) = \max_{\mu\in(\epsilon,\infty)} \beta_D(\mu, \lambda^*) \quad \text{and} \quad \beta_A(\mu^*, \lambda^*) = \max_{\lambda\in R_+} \beta_A(\mu^*, \lambda),
$$

where ϵ is the lowest rate we can expect the defender to play in order to ensure the stationary distributions and hence benefit functions are well defined for all valid (μ, λ). Differentiating the defender's and attacker's functions with respect to μ and λ respectively gives,

$$\frac{\partial \beta_D}{\partial \mu} = \frac{n! \cdot \mu^{n-t} \cdot \lambda^t}{(t-1)! \cdot (n-t)! \cdot (\mu + \lambda)^{n+1}} - d, \tag{10}$$

$$\frac{\partial \beta_A}{\partial \lambda} = \frac{n! \cdot \mu^{n-t+1} \cdot \lambda^{t-1}}{(t-1)! \cdot (n-t)! \cdot (\mu + \lambda)^{n+1}} - a. \tag{11}$$

Closed form solutions for μ and λ which equate to 0 are not easy to calculate directly. The second derivative of the attackers benefit with respect to λ is

$$\frac{n! \cdot \mu^{n-t+1} \cdot \lambda^{t-2}}{(t-1)! \cdot (n-t)! \cdot (\mu + \lambda)^{n+2}} \cdot [\mu \cdot (t-1) - \lambda \cdot (n+2-t)].$$

Thus, $\frac{\partial \beta_A}{\partial \lambda}$ is increasing when

$$\lambda < \frac{\mu \cdot (t-1)}{n+2-t},$$

then decreasing. Since $\frac{\partial \beta_A}{\partial \lambda}$ is $-a$ at $\lambda = 0$ and asymptotes to $-a$ as $\lambda \to \infty$ we have the derivative increasing from $-a$ to a maximum when $\lambda = \frac{\mu \cdot (t-1)}{n+2-t}$ and then decreasing back to $-a$. The maximal value of $\frac{\partial \beta_A}{\partial \lambda}$ is given by

$$\frac{n! \cdot (t-1)^{t-1}}{t^{n+1} \cdot (n+2-t)^{t-2} \cdot \mu} - a \tag{12}$$

which is positive only if μ is sufficiently small. As a function of λ, β_A therefore initially decreases (from 0), has a period of increase only if μ is sufficiently small, then decreases again. It follows that β_A has at most one non-zero maximum which occurs in the region

$$\left(\frac{\mu(t-1)}{n+2-t}, \infty \right)$$

once the derivative is decreasing, and this fixed point maximises $\beta_A(\mu, \lambda)$ on $\lambda \in [0, \infty)$ if and only if $\beta_A(\mu, \lambda) > 0$; otherwise the best response must be $\lambda = 0$. First, like the full reset case, we consider the existence of a Nash Equilibrium (μ, λ) with $\mu > \epsilon$. Since both derivatives (10) and (11) are hard to solve analytically for general n, we used a numerical method utilizing the Maple algebra system to solve for a specific n. The method for solving starts with defining the benefit functions in terms of μ and λ, we then differentiate the derivatives as above and solve for μ and λ for the defender and attacker, respectively. This provides 2 generic solutions of the form

$$\hat{\mu}(\lambda) = \text{RootOf}(f(\lambda)) \quad \text{and} \quad \hat{\lambda}(\mu) = \text{RootOf}(g(\mu))$$

where f and g are polynomials. We then put these solutions back into the derivatives to give

$$\frac{\partial \beta_D(\mu, \hat{\lambda}(\mu))}{\partial \mu} \quad \text{and} \quad \frac{\partial \beta_A(\hat{\mu}(\lambda), \lambda)}{\partial \lambda}$$

Solving these with respect to μ and λ respectively gives solutions for μ^* and λ^* with respect to the costs d and a. From this we can consider the ratio $\rho = \frac{a}{d}$ between the attacker's and defender's costs of play. A table can be constructed to show the ratios at which both the defender and attacker will and won't play for various ρ. Recall that even if the attacker is not playing, the defender must still play at some rate ϵ in order to ensure control of the system. In order to calculate the defender's benefit given a specific ρ we must calculate the lowest rate of play for the defender when the attacker is not playing. From Eq. (12), $\frac{\partial \beta_A}{\partial \lambda}$ is never positive if

$$\mu > \frac{n! \cdot (t-1)^{t-1}}{t^{n+1} \cdot (n+2-t)^{t-2} \cdot a}$$

Meaning no stationary point exists for the attackers benefit. From this we can see that by taking

$$\epsilon \geq \frac{n! \cdot (t-1)^{t-1}}{t^{n+1} \cdot (n+2-t)^{t-2} \cdot a}$$

we can ensure there is no equilibrium with $\mu^* = \epsilon$ and $\lambda \neq 0$. Recall that $\rho = \frac{a}{d}$, so that

$$\epsilon \geq \frac{n! \cdot (t-1)^{t-1}}{t^{n+1} \cdot (n+2-t)^{t-2} \cdot \rho \cdot d}$$

This shows that if ρ is large enough, ϵ will be small meaning the likely strategy for the attacker will be no play, $\lambda = 0$. So the benefit for the defender will be

$$\beta_D(\epsilon, 0) = 1 - \epsilon \cdot d = 1 - \frac{n! \cdot (t-1)^{t-1}}{t^{n+1} \cdot (n+2-t)^{t-2} \cdot \rho}.$$

However, having ρ large enough to ensure ϵ is small enough is an unrealistic assumption and choosing ϵ like this becomes a strategic choice. As it was for the full reset case, it is also very conservative and could be expensive for the defender. We therefore fix our $\epsilon > 0$ to be very small, close to zero before the game. We now want to ask the following question: Given the costs of play for both defender and attacker and a limit N for the number of resources the defender can own, what is the best set up for the defender in order to maximise their benefit function? i.e. given ρ and N we are looking for the pairs such that

$$\beta_D^*(n^*, t^*) = \max_{n \leq N, t \leq n} \beta_D^*(n, t)$$

where $\beta_D^*(n,t) = \beta_D(\mu^*, \lambda^*)$ is the Nash equilibrium for the specific number of resources n and threshold t. Recall we defined this game to be $\mathrm{Opt}_{N,\epsilon}^{\mathcal{S}}(d, \mathcal{T}, \rho)$. We turn to the method of numerical programming for this problem. Obviously, since the lowest rate of play ϵ for the defender is chosen arbitrarily before the game is played, if the equilibrium played is the trivial equilibrium then the defenders benefit is $\beta_D(\epsilon, 0) = 1 - \epsilon \cdot d$.

When running $\mathrm{Opt}_{N,\epsilon}^{\mathcal{S}}(d, \mathcal{T}, \rho)$, each round of $\mathrm{FlipThem}_{\epsilon}^{\mathcal{S}}(n, t, d, \rho)$ played has three possible outcomes.

- If ρ is so small the defender will not even play at the minimal rate ϵ.
- If ρ is "mid-size" the defender and attacker both play the non-trivial equilibrium (μ^*, λ^*).
- If ρ is large the attacker does not play and the trivial equilibrium $(\epsilon, 0)$ is played.

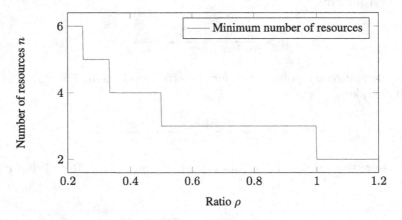

Fig. 2. Number of resources used by the defender to maximise his benefit given a specific ρ, for $\mathcal{T} = \{t \le n\}$ and $N = 7$.

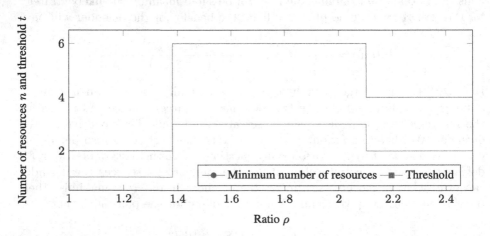

Fig. 3. Number of resources used by the defender to maximise his benefit given a specific ρ, for $\mathcal{T} = \{t < n/2\}$ and $N = 7$.

We experimentally examined two scenarios, both in which we fix $\epsilon = 0.01/d$. In the first scenario we take $\mathcal{T} = \{t \leq n\}$ and $N = 7$, in this case the function $\text{Opt}_{N,\epsilon}^{\mathcal{S}}(d, \mathcal{T}, \rho)$ outputs valid configurations for relatively small values of ρ, see Fig. 2. Interestingly the output best game for a maximum defenders benefit is always a full threshold game. In the second scenario we take $\mathcal{T} = \{t < n/2\}$, and again $N = 7$. The results are given in Fig. 3. In this case small values of ρ result in games for which the defender will not play, for larger values of ρ we end up requiring more servers.

Acknowledgements. The second author was supported by a studentship from GCHQ.This work has been supported in part by ERC Advanced Grant ERC-2010-AdG-267188-CRIPTO and by EPSRC via grant EP/I03126X.

References

1. Bedi, H.S., Shiva, S.G., Roy, S.: A game inspired defense mechanism against distributed denial of service attacks. Secur. Commun. Netw. **7**(12), 2389–2404 (2014)
2. Bowers, K.D., van Dijk, M., Griffin, R., Juels, A., Oprea, A., Rivest, R.L., Triandopoulos, N.: Defending against the unknown enemy: applying FLIPIT to system security. In: Gros80klags, J., Walrand, J. (eds.) GameSec 2012. LNCS, vol. 7638, pp. 248–263. Springer, Heidelberg (2012)
3. Collins, M.P.: A cost-based mechanism for evaluating the effectiveness of moving target defenses. In: Grossklags, J., Walrand, J. (eds.) GameSec 2012. LNCS, vol. 7638, pp. 221–233. Springer, Heidelberg (2012)
4. Das, S.K., Nita-Rotaru, C., Kantarcioglu, M. (eds.): Decision and Game Theory for Security. Lecture Notes in Computer Science, vol. 8252. Springer, Switzerland (2013)
5. Grimmett, G., Stirzaker, D.: Probability and Random Processes, 3rd edn. Oxford University Press, Oxford (2001)
6. Grossklags, J., Walrand, J.C. (eds.): Decision and Game Theory for Security. Lecture Notes in Computer Science, vol. 7638. Springer, Heidelberg (2012)
7. Laszka, A., Horvath, G., Felegyhazi, M., Buttyán, L.: FlipThem: modeling targeted attacks with FLIPIT for multiple resources. In: Poovendran, R., Saad, W. (eds.) GameSec 2014. LNCS, vol. 8840, pp. 175–194. Springer, Heidelberg (2014)
8. Laszka, A., Johnson, B., Grossklags, J.: Mitigation of targeted and non-targeted covert attacks as a timing game. In: Das, S.K., Nita-Rotaru, C., Kantarcioglu, M. (eds.) GameSec 2013. LNCS, vol. 8252, pp. 175–191. Springer, Heidelberg (2013)
9. Moayedi, B.Z., Azgomi, M.A.: A game theoretic framework for evaluation of the impacts of hackers diversity on security measures. Reliab. Eng. Syst. Saf. **99**, 45–54 (2012)
10. Nash, J.: Non-cooperative games. Ann. Math. **54**, 286–295 (1951)
11. Ostrovsky, R., Yung, M.: How to withstand mobile virus attacks (extended abstract). In: Logrippo, L. (ed.) Proceedings of the Tenth Annual ACM Symposium on Principles of Distributed Computing, Montreal, Quebec, Canada, 19–21 August 1991, pp. 51–59. ACM (1991)
12. Panaousis, E., Fielder, A., Malacaria, P., Hankin, C., Smeraldi, F.: Cybersecurity games and investments: a decision support approach. In: Saad, W., Poovendran, R. (eds.) GameSec 2014. LNCS, vol. 8840, pp. 266–286. Springer, Heidelberg (2014)

13. Pham, V., Cid, C.: Are we compromised? Modelling security assessment games. In: Grossklags, J., Walrand, J. (eds.) GameSec 2012. LNCS, vol. 7638, pp. 234–247. Springer, Heidelberg (2012)
14. Poovendran, R., Saad, W. (eds.): Decision and Game Theory for Security. Lecture Notes in Computer Science, vol. 8840. Springer, Switzerland (2014)
15. Roy, S., Ellis, C., Shiva, S.G., Dasgupta, D., Shandilya, V., Wu, Q.: A survey of game theory as applied to network security. In: Proceedings of the 43rd Hawaii International Conference on Systems Science (HICSS-43 2010), Koloa, Kauai, HI, USA, 5–8 January 2010, pp. 1–10. IEEE Computer Society (2010)
16. van Dijk, M., Juels, A., Oprea, A., Rivest, R.L.: Flipit: the game of "Stealthy Takeover". J. Cryptology 26(4), 655–713 (2013)
17. Wellman, M.P., Prakash, A.: Empirical game-theoretic analysis of an adaptive cyber-defense scenario (preliminary report). In: Saad, W., Poovendran, R. (eds.) GameSec 2014. LNCS, vol. 8840, pp. 43–58. Springer, Heidelberg (2014)
18. Zhu, Q., Başar, T.: Game-theoretic approach to feedback-driven multi-stage moving target defense. In: Das, S.K., Nita-Rotaru, C., Kantarcioglu, M. (eds.) GameSec 2013. LNCS, vol. 8252, pp. 246–263. Springer, Heidelberg (2013)

A Game Theoretic Model for Defending Against Stealthy Attacks with Limited Resources

Ming Zhang[1]($\boxtimes$), Zizhan Zheng[2], and Ness B. Shroff[1]

[1] Department of ECE and CSE, The Ohio State University, Columbus, OH, USA
{zhang.2562,shroff.11}@osu.edu
[2] Department of Computer Science, University of California Davis, Davis, CA, USA
cszheng@ucdavis.edu

Abstract. Stealthy attacks are a major threat to cyber security. In practice, both attackers and defenders have resource constraints that could limit their capabilities. Hence, to develop robust defense strategies, a promising approach is to utilize game theory to understand the fundamental trade-offs involved. Previous works in this direction, however, mainly focus on the single-node case without considering strict resource constraints. In this paper, a game-theoretic model for protecting a system of multiple nodes against stealthy attacks is proposed. We consider the practical setting where the frequencies of both attack and defense are constrained by limited resources, and an asymmetric feedback structure where the attacker can fully observe the states of nodes while largely hiding its actions from the defender. We characterize the best response strategies for both attacker and defender, and study the Nash Equilibria of the game. We further study a sequential game where the defender first announces its strategy and the attacker then responds accordingly, and design an algorithm that finds a nearly optimal strategy for the defender to commit to.

Keywords: Stealthy attacks · Resource constraints · Game theory

1 Introduction

The landscape of cyber security is constantly evolving in response to increasingly sophisticated cyber attacks. In recent years, *Advanced Persistent Threats* (APT) [1] is becoming a major concern to cyber security. APT attacks have several distinguishing properties that render traditional defense mechanism less effective. First, they are often launched by *incentive driven* entities with specific targets. Second, they are *persistent* in achieving the goals, and may involve multiple stages or continuous operations over a long period of time. Third, they are highly adaptive and *stealthy*, and often operate in a "low-and-slow" fashion [7] to avoid of being detected. In fact, some notorious attacks remained undetected for months or longer [2,6]. Hence, traditional intrusion detection and prevention

This work has been funded by QNRF fund NPRP 5-559-2-227.

© Springer International Publishing Switzerland 2015
MHR Khouzani et al. (Eds.): GameSec 2015, LNCS 9406, pp. 93–112, 2015.
DOI: 10.1007/978-3-319-25594-1_6

techniques that target one-shot and known attack types are insufficient in the face of long-lasting and stealthy attacks.

Moreover, since the last decade, it has been increasingly realized that security failures in information systems are often caused by the misunderstanding of incentives of the entities involved in the system instead of the lack of proper technical mechanisms [5,17]. To this end, game theoretical models have been extensively applied to cyber security [4,9–11,13,16,19]. Game theory provides a proper framework to systematically reason about the strategic behavior of each side, and gives insights to the design of cost-effective defense strategies. Traditional game models, however, fail to capture the persistent and stealthy behavior of advanced attacks. Further, they often model the cost of defense (or attack) as part of the utility functions of the players, while ignoring the strict resource constraints during the play of the game. For a large system with many components, ignoring such constraints can lead to either over-provision or under-provision of resources and revenue loss.

In this paper, we study a two-player non-zero-sum game that explicitly models stealth attacks with resource constraints. We consider a system with N *independent* nodes (or components), an attacker, and a defender. Over a continuous time horizon, the attacker (defender) determines when to attack (recapture) a node, subject to a unit cost per action that varies over nodes. At any time t, a node is either compromised or protected, depending on whether the player that makes the last move (i.e., action) towards it before t is the attacker or the defender. A player obtains a value for each node under its control per unit time, which again may vary over nodes. The total payoff to a player is then the total value of the nodes under its control over the entire time horizon minus the total cost incurred, and we are interested in the long-term time average payoffs.

To model stealthy attacks, we assume that the defender gets no feedback about the attacker during the game. On the other hand, the defender's moves are fully observable to the attacker. This is a reasonable assumption in many cyber security settings, as the attacker can often observe and learn the defender's behavior before taking actions. Moreover, we explicitly model their resource constraints by placing an upper bound on the frequency of moves (over all the nodes) for each player. We consider both Nash Equilibrum and Sequential Equilibrum for this game model. In the latter case, we assume that the defender is the leader that first announces its strategy, and the attacker then responds with its best strategy. The sequential setting is often relevant in cyber security, and can provide a higher payoff to the defender compared with Nash Equilibrium. To simplify the analysis, we assume that the set of nodes are independent in the sense that the proper functioning of one node does not depend on other nodes, which serves as a first-order approximation of the more general setting of interdependent nodes to be considered in our future work.

Our model is an extension of the asymmetric version of the FlipIt game considered in [15]. The FlipIt game [20] is a two-player non-zero-sum game recently proposed in response to an APT attack towards RSA Data Security [3]. In the FlipIt game, a single critical resource (a node in our model) is considered. Each player obtains control over the resource by "flipping" it subject to a cost. During the play of the game, each player obtains delayed and possibly incomplete feedback on the other player's previous moves. A player's strategy is then when to move over a time horizon, and the solution of the game heavily depends on the class of strategies adopted and the feedback structure of the game. In particular, a full analysis of Nash Equilibria has only been obtained for two special cases, when both players employ a periodic strategy [20], and when the attacker is stealthy and the defender is observable as in our model [15]. However, both works consider a single node and there is no resource constraint. The multi-node setting together with the resource constraints impose significant challenges in characterizing both Nash and Sequential Equilibria. A different multi-node extension of the FlipIt game is considered in [14] where the attacker needs to compromise either all the nodes (AND model) or a single node (OR model) to take over a system. However, only preliminary analytic results are provided.

Our game model can be applied in various settings. One example is key rotation. Consider a system with multiple nodes, e.g., multiple communication links or multiple servers, that are protected by different keys. From time to time, the attacker may compromise some of the keys, e.g., by leveraging zero-day vulnerabilities and system specific knowledge, while remaining undetected from the defender. A common practice is to periodically generate fresh keys by a trusted key-management service, without knowing when they are compromised. On the other hand, the attacker can easily detect the expiration of a key (at an ignorable cost compared with re-compromising it). Both key rotation and compromise incurs a cost, and there is a constraint on the frequency of moves at each side. There are other examples where our extension of the FlipIt game can be useful, such as password reset and virtual-machine refresh [8,15,20].

We have made following contributions in this paper.

- We propose a two-player game model with multiple independent nodes, an overt defender, and a stealthy attacker where both players have strict resource constraints in terms of the frequency of protection/attack actions across all the nodes.
- We prove that the periodic strategy is a best-response strategy for the defender against a non-adaptive *i.i.d.* strategy of the attacker, and vice versa, for general distributions of attack times.
- For the above pair of strategies, we fully characterize the set of Nash Equilibria of our game, and show that there is always one (and maybe more) equilibrium, for the case when the attack times are deterministic.
- We further consider the sequential game with the defender as the leader and the attacker as the follower. We design a dynamic programming based algorithm that identifies a nearly optimal strategy (in the sense of subgame perfect equilibrium) for the defender to commit to.

The remainder of this paper is organized as follows. We present our game-theoretic model in Sect. 2, and study best-response strategies of both players in

Sect. 3. Analysis of Nash Equilibria of the game is provided in Sect. 4, and the sequential game is studied in Sect. 5. In Sect. 6, we present numerical result, and we conclude the paper in Sect. 7.

2   Game Model

In this section, we discuss our two-player game model including its information structure, the action spaces of both attacker and defender, and their payoffs. Our game model extends the single node model in [15] to multiple nodes and includes a resource constraint to each player.

2.1   Basic Model

In our game-theoretical model, there are two players and N *independent* nodes[1]. The player who is the lawful user/owner of the N nodes is called the defender, while the other player is called the attacker. The game starts at time $t = 0$ and goes to any time $t = T$. We assume that time is continuous. A player can make a move at any time instance subject to a cost per move. At any time t, a node is under the control of the player that makes the last move towards the node before t (see Fig. 1). Each attack towards node i incurs a cost of C_i^A to the attacker, and it takes a random period of time w_i to succeed. On the other hand, when the defender makes a move to protect node i, which incurs a cost of C_i^D, node i is recovered immediately even if the attack is still in process. Each node i has a value r_i that represents the benefit that the attacker receives from node i per unit of time when node i is compromised.

In addition to the move cost, we introduce a strict resource constraint for each player, which is a practical assumption but has been ignored in most prior works on security games. In particular, we place an upper bound on the average amount of resource that is available to each player at any time (to be formally defined below). As typical security games, we assume that r_i, C_i^A, C_i^D, the distribution of w_i, and the budget constraints are all common knowledge of the game, that is, they are known to both players. For instance, they can be learned from history data and domain knowledge. Without loss of generality, all nodes are assumed to be protected at time $t = 0$. Table 1 summarizes the notations used in the paper.

As in [15], we consider an asymmetric feedback model where the attacker's moves are *stealthy*, while the defenders' moves are *observable*. More specifically, at any time, the attacker knows the full history of moves by the defender, as well as the state of each node, while the defender has no idea about whether a node is compromised or not. Let $\alpha_{i,k}$ denote the time period the attacker waits from the latest time when node i is recovered, to the time when the attacker starts its k-th attack against node i, which can be a random variable in general. The attacker's action space is then all the possible selections of $\{\alpha_{i,k}\}$. Since the set of nodes are independent, we can assume $\alpha_{i,k}$ to be independent across i without

---

[1] The terms "components" and "nodes" are interchangeable in this paper.

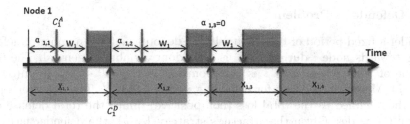

Fig. 1. Game model

Table 1. List of notations

Symbol	Meaning
T	Time horizon
N	Number of nodes
r_i	Value per unit of time of compromising node i
w_i	Attack time for node i
C_i^A	Attacker's move cost for node i
C_i^D	Defender's move cost for node i
$\alpha_{i,k}$	Attacker's waiting time in its k-th move for node i
$X_{i,k}$	Time between the (k−1)-th and the k-th defense for node i
B	Budget to the defender, greater than 0
M	Budget to the attacker, greater than 0
m_i	Frequency of defenses for node i
p_i	Probability of immediate attack on node i once it recovers
L_i	Number of defense moves for node i

loss of generality. However, they may be correlated across k in general, as the attacker can employ a time-correlated strategy. On the contrary, the defender's strategy is to determine the time intervals between its $(k-1)$-th move and k-th move for each node i and k, denoted as $X_{i,k}$.

In this paper, we focus on *non-adaptive (but possibly randomized) strategies*, that is, neither the attacker nor the defender changes its strategy based on feedback received during the game. Therefore, the values of $\alpha_{i,k}$ and $X_{i,k}$ can be determined by the corresponding player before the game starts. Note that assuming non-adaptive strategies is not a limitation for the defender since it does not get any feedback during the game anyway. Interestingly, it turns out not to be a big limitation on the attacker either. As we will show in Sect. 3, periodic defense is a best-response strategy against any non-adaptive *i.i.d.* attacks (formally defined in Definition 2) and vice versa. Note that when the defender's strategy is periodic, the attacker can predict defender's moves before the game starts so there is no need to be adaptive.

2.2 Defender's Problem

Consider a fixed period of time T and let L_i denote the total number of defense moves towards node i during T. L_i is a random variable in general. The total amount of time when node i is compromised is then $T - \sum_{k=1}^{L_i} \min(\alpha_{i,k} + w_i, X_{i,k})$. Moreover, the cost for defending node i is $L_i C_i^D$. The defender's payoff is then defined as the total loss (non-positive) minus the total defense cost over all the nodes. Given the attacker's strategy $\{\alpha_{i,k}\}$, the defender faces the following optimization problem:

$$\max_{\{X_{i,k}\}, L_i} E\left[\sum_{i=1}^{N} \frac{-\left(T - \sum_{k=1}^{L_i} \min(\alpha_{i,k} + w_i, X_{i,k})\right) \cdot r_i - L_i C_i^D}{T}\right]$$

$$s.t. \sum_{i=1}^{N} \frac{L_i}{T} \leq B \text{ w.p.1} \tag{1}$$

$$\sum_{k=1}^{L_i} X_{i,k} \leq T \text{ w.p.1 } \forall i$$

The first constraint requires that the average number of nodes that can be protected at any time is upper bounded by a constant B. The second constraint defines the feasible set of $X_{i,k}$. Since T is given, the expectation in the objective function can be moved into the summation in the numerator.

2.3 Attacker's Problem

We again let L_i denote the total number of defense moves towards node i in T. The total cost of attacking i is then $(\sum_{k=1}^{L_i} \mathbf{1}_{\alpha_{i,k} < X_{i,k}}) \cdot C_i^A$, where $\mathbf{1}_{\alpha_{i,k} < X_{i,k}} = 1$ if $\alpha_{i,k} < X_{i,k}$ and $\mathbf{1}_{\alpha_{i,k} < X_{i,k}} = 0$ otherwise. It is important to note that when $\alpha_{i,k} \geq X_{i,k}$, the attacker actually gives up its k-th attack against node i (this is possible as the attacker can observe when the defender moves). Given the defender's strategy, the attacker's problem can be formulated as follows, where M is an upper bound on the average number of nodes that the attacker can attack at any time instance.

$$\max_{\alpha_{i,k}} E\left[\sum_{i=1}^{N} \frac{(T - \sum_{k=1}^{L_i} \min(\alpha_{i,k} + w_i, X_{i,k})) \cdot r_i - (\sum_{k=1}^{L_i} \mathbf{1}_{\alpha_{i,k} < X_{i,k}}) \cdot C_i^A}{T}\right]$$

$$s.t. E\left[\sum_{i=1}^{N} \frac{1}{T} \int_0^T v_i(t) dt\right] \leq M \tag{2}$$

where $v_i(t) = 1$ if the attacker is attacking node i at time t and $v_i(t) = 0$ otherwise. Note that we make the assumption that the attacker has to keep consuming resources when the attack is in progress instead of making an instantaneous move like the defender; hence it has a different form of budget constraint. On the other hand, we assume that C_i^A captures the total cost for each attack on node i, which is independent of the attack time. We further have the following equation:

$$\int_0^T v_i(t)dt = \sum_{k=1}^{L_i} (\min(\alpha_{i,k} + w_i, X_{i,k}) - \min(\alpha_{i,k}, X_{i,k})) \qquad (3)$$

Putting (3) into (2) and moving the expectation inside, the attacker's problem becomes

$$\max_{\alpha_{i,k}} \sum_{i=1}^N \frac{T \cdot r_i - E[\sum_{k=1}^{L_i} \min(\alpha_{i,k} + w_i, X_{i,k})] \cdot r_i - E[\sum_{k=1}^{L_i} P(\alpha_{i,k} < X_{i,k})] \cdot C_i^A}{T}$$

$$s.t. \sum_{i=1}^N \frac{E[\sum_{k=1}^{L_i} \min(\alpha_{i,k} + w_i, X_{i,k}) - \min(\alpha_{i,k}, X_{i,k})]}{T} \le M. \qquad (4)$$

3 Best Responses

In this section, we analyze the best-response strategies for both players. Our main result is that when the attacker employs a non-adaptive *i.i.d.* strategy, a periodic strategy is a best response for the defender, and vice versa. To prove this result, however, we have provided characterization of best responses in more general settings. In this and following sections, we have omitted most proofs to save space. All the missing proofs can be found in our online technical report [21].

3.1 Defender's Best Response

We first show that for the defender's problem (1), an optimal deterministic strategy is also optimal in general. We then provide a sufficient condition for a deterministic strategy to be optimal against any non-adaptive attacks. Finally, we show that periodic defense is optimal against non-adaptive *i.i.d.* attacks.

Lemma 1. *Suppose $X_{i,k}^\star$ and $L_i^\star$ are the optimal solutions of (1) among all deterministic strategies, then they are also optimal among all the strategies including both deterministic and randomized strategies.*

According to the lemma, it suffices to consider defender's strategies where both $X_{i,k}$ and $L_{i,k}$ are deterministic.

Definition 1. *For a given L_i, we define a set $\mathcal{X}_i$ including all deterministic defense strategies with the following properties:*

1. $\sum_{k=1}^{L_i} X_{i,k} = T$;
2. $F_{\alpha_{i,k}+w_i}(X_{i,k}) = F_{\alpha_{i,j}+w_i}(X_{i,j}) \quad \forall k, j,$

where $F_{\alpha_{i,k}+w_i}(\cdot)$ is the CDF of r.v. $\alpha_{i,k} + w_i$.

Note that $\mathcal{X}_i$ can be an empty set in general due to the randomness of $\alpha_{i,k} + w_i$. The following lemma shows that when $\mathcal{X}_i$ is non-empty for all i, any strategy that belongs to $\mathcal{X}_i$ is the defender's best deterministic strategy against a non-adaptive attacker.

Lemma 2. *For any given set of $\{L_i\}$ with $\sum_{i=1}^{N} \frac{L_i}{T} \leq B$, if $\mathcal{X}_i \neq \emptyset\ \forall i$, then any set of $\{X_{i,k}\}$ that belongs to $\mathcal{X}_i$ is the defender's best deterministic strategy.*

Lemma 2 gives a sufficient condition for a deterministic defense strategy to be optimal. The main idea of the proof is to show that the defender's payoff for each node i is concave with respect to $X_{i,k}$. The optimality then follows from the KKT conditions. Intuitively, the defender tries to equalize its expected loss in each period in a deterministic way, which gives the defender the most stable system to avoid a big loss in any particular period. We then show that a periodic defense is sufficient when the attacker employs a non-adaptive *i.i.d.* strategy formally defined below.

Definition 2. *An attack strategy is called non-adaptive i.i.d. if it is non-adaptive, and $\alpha_{i,k}$ is independent across i and is i.i.d. across k.*

Theorem 1. *A periodical strategy is the best response for the defender if the attacker employs a non-adaptive i.i.d. strategy.*

According to the theorem, the periodic strategy gives the defender the most stable system when the attacker adopts the non-adaptive *i.i.d.* strategy. Since the attacker's waiting time $\alpha_{i,k}$ does not change with time, a fixed defense interval provides the same expected payoff between every two consecutive moves. Moreover, since the defender's problem is a convex optimization problem, the optimal defending frequency for a given attack strategy can be easily determined by solving the convex program.

3.2   Attacker's Best Response

We first analyze the attacker's best response against any deterministic defense strategies, then show that the non-adaptive *i.i.d.* strategy is the best response against periodic defense.

Lemma 3. *When defense strategies are deterministic, the attacker's best response (among non-adaptive strategies) must satisfy the following condition*

$$\alpha_{i,k}^{*} = \begin{cases} 0 & w.p.\ p_{i,k} \\ \geq X_{i,k} & w.p.\ 1 - p_{i,k} \end{cases} \tag{5}$$

Proof Sketch: The main idea of the proof is to divide the problem (4) into $\sum_{i=1}^{N} L_i$ independent sub-problems, one for each node and a single period, where each subproblem has a similar target function and a budget $M_{i,k}$ where $\sum_{i=1}^{N} \sum_{k=1}^{L_i} M_{i,k} = M$. Due to the independence of nodes, it suffices to prove the lemma for any of these sub-problems.

Lemma 3 implies that for each node i, the attacker's best strategy is to either attack node i immediately after it realizes the node's recovery, or gives up the attack until the defender's next move. There is no incentive for the attacker to wait a small amount of time to attack a node before the defender's next move. The constraint M actually determines the probability that the attacker will attack immediately. If M is large enough, the attacker will never wait after defender's each move. We then find the attacker's best responses when the defender employs the periodic strategy.

Theorem 2. *When the defender employs periodical strategy, the non-adaptive i.i.d. strategy is the attacker's best response among all non-adaptive strategies.*

3.3 Simplified Optimization Problems

According to Theorems 1 and 2, periodic defense and non-adaptive *i.i.d.* attack can form a pair of best-response strategies with respect to each other. Consider such pair of strategies. Let $m_i \triangleq \frac{L_i}{T} = \frac{1}{X_{i,k}}$, and let p_i denote the probability that $\alpha_{i,k} = 0, \forall k$. The optimization problems to the defender and the attacker can then be simplified as follows.

Defender's problem:

$$\max_{m_i} \sum_{i=1}^{N} \left[\left(E[\min(w_i, \frac{1}{m_i})]p_i r_i - C_i^D \right) \cdot m_i - p_i r_i \right]$$

$$s.t. \sum_{i=1}^{N} m_i \leq B \tag{6}$$

Attacker's problem:

$$\max_{p_i} \sum_{i=0}^{N} p_i \cdot \left(r_i(1 - E[\min(w_i, \frac{1}{m_i})] \cdot m_i) - C_i^A m_i \right)$$

$$s.t. \sum_{i=0}^{N} E[\min(w_i, \frac{1}{m_i})] \cdot m_i \cdot p_i \leq M \tag{7}$$

We observe that the defender's problem is a continuous convex optimization problem (see the discussion in Sect. 3.1), and the attacker's problem is a fractional knapsack problem. Therefore, the best response strategy of each side can be easily determined. Also, the time period T disappears in both problems.

4 Nash Equilibria

In this section, we study the set of Nash Equilibria of the simplified game as discussed in Sect. 3.3 where the defender employs a periodic strategy, and the attacker employs a non-adaptive *i.i.d.* strategy. We further assume that the attack time w_i is deterministic for all i. We show that this game always has a Nash equilibrium and may have multiple equilibria of different values.

We first observe that for deterministic w_i, when $m_i \geq \frac{1}{w_i}$, the defender's payoff becomes $-m_i C_i^D$, which is maximized when $m_i = \frac{1}{w_i}$. Therefore, it suffices to consider $m_i \leq \frac{1}{w_i}$. Thus, the optimization problems to the defender and the attacker can be further simplified as follows.

For a given p, the defender aims at maximizing its payoff:

$$\max_{m_i} \sum_{i=1}^{N} [m_i(r_i w_i p_i - C_i^D) - p_i r_i]$$

$$s.t. \sum_{i=1}^{N} m_i \leq B \tag{8}$$

$$0 \leq m_i \leq \frac{1}{w_i}, \forall i$$

On the other hand, for a given m, the attacker aims at maximizing its payoff:

$$\max_{p_i} \sum_{i=1}^{N} p_i[r_i - m_i(r_i w_i + C_i^A)]$$

$$s.t. \sum_{i=1}^{N} m_i w_i p_i \leq M \tag{9}$$

$$0 \leq p_i \leq 1, \forall i$$

For a pair of strategies (m, p), the payoff to the defender is $U_d(m, p) = \sum_{i=1}^{N} [m_i(p_i r_i w_i - C_i^D) - p_i r_i]$, while the payoff to the attacker is $U_a(m, p) = \sum_{i=1}^{N} p_i[r_i - m_i(r_i w_i + C_i^A)]$. A pair of strategies (m^*, p^*) is called a (pure strategy) *Nash Equilibrium (NE)* if for any pair of strategies (m, p), we have $U_d(m^*, p^*) \geq U_d(m, p^*)$ and $U_a(m^*, p^*) \geq U_a(m^*, p)$. In the following, we assume that $C_i^A > 0$ and $C_i^D > 0$. The cases where $C_i^A = 0$ or $C_i^D = 0$ or both exhibit slightly different structures, but can be analyzed using the same approach. Without loss of generality, we assume $r_i > 0$ and $\frac{C_i^D}{r_i w_i} \leq 1$ for all i. Note that if $r_i = 0$, then node i can be safely excluded from the game, while if $\frac{C_i^D}{r_i w_i} > 1$, the coefficient of m_i in U_d (defined below) is always negative and there is no need to protect node i.

Let $\mu_i(p) \triangleq p_i r_i w_i - C_i^D$ denote the coefficient of m_i in U_d, and $\rho_i(m) \triangleq \frac{r_i - m_i(r_i w_i + C_i^A)}{m_i w_i}$. Note that for a given p, the defender tends to protect more a component with higher $\mu_i(p)$, while for a given m, the attacker will attack a component more frequently with higher $\rho_i(m)$. When m and p are clear from the context, we simply let μ_i and ρ_i denote $\mu_i(p)$ and $\rho_i(m)$, respectively.

To find the set of NEs of our game, a key observation is that if there is a full allocation of defense budget B to m such that $\rho_i(m)$ is a constant for all i, any full allocation of the attack budget M gives the attacker the same payoff. Among these allocations, if there is further an assignment of p such that $\mu_i(p)$ is a constant for all i, then the defender also has no incentive to deviate from m; hence (m, p) forms an NE. The main challenge, however, is that such an assignment of p does not always exist for the whole set of nodes. Moreover, there are NEs that do not fully utilize the defense or attack budget as we show below. To characterize the set of NEs, we first prove the following properties satisfied by any NE of the game. For a given strategy (m, p), we define $\mu^*(p) \triangleq \max_i \mu_i(p)$, $\rho^*(m) \triangleq \min_i \rho_i(m)$, $F(p) \triangleq \{i : \mu_i(p) = \mu^*(p)\}$, and $D(m, p) \triangleq \{i \in F : \rho_i(m) = \rho^*(m)\}$. We omit m and p when they are clear from the context.

Lemma 4. *If (m, p) is an NE, we have:*

1. $\forall i \notin F, m_i = 0, p_i = 1, \rho_i = \infty$;
2. $\forall i \in F \backslash D, m_i \in [0, \frac{r_i}{w_i r_i + C_i^A}], p_i = 1$;
3. $\forall i \in D, m_i \in [0, \frac{r_i}{w_i r_i + C_i^A}], p_i \in [\frac{C_i^D}{r_i w_i}, 1]$.

Lemma 5. *If (m, p) forms an NE, then for $i \in D, j \in F \backslash D$ and $k \notin F$, we have* $r_i w_i - C_i^D \geq r_j w_j - C_j^D > r_k w_k - C_k^D$.

According to the above lemma, to find all the equilibria of the game, it suffices to sort all the nodes by a non-increasing order of $r_i w_i - C_i^D$, and consider each F_h consisting of the first h nodes such that $r_h w_h - C_h^D > r_{h+1} w_{h+1} - C_{h+1}^D$, and each subset $D_k \subseteq F_h$ consisting of the first $k \leq h$ nodes in the list. In the following, we assume such an ordering of nodes. Consider a given pair of F and $D \subseteq F$. By Lemma 4 and the definitions of F and D, the following conditions are satisfied by any NE with $F(p) = F$ and $D(m, p) = D$.

$$m_i = 0, p_i = 1, \forall i \notin F; \tag{10}$$

$$m_i \in [0, \frac{r_i}{w_i r_i + C_i^A}], p_i = 1, \forall i \in F \backslash D; \tag{11}$$

$$m_i \in [0, \frac{r_i}{w_i r_i + C_i^A}], p_i \in [\frac{C_i^D}{r_i w_i}, 1], \forall i \in D; \tag{12}$$

$$\sum_{i \in F} m_i \leq B, \sum_{i \in F} m_i w_i p_i \leq M; \tag{13}$$

$$\mu_i = \mu^*, \forall i \in F; \qquad \mu_i < \mu^*, \forall i \notin F; \tag{14}$$

$$\rho_i = \rho^*, \forall i \in D; \qquad \rho_i > \rho^*, \forall i \notin D. \tag{15}$$

The following theorem provides a full characterization of the set of NEs of the game.

Theorem 3. *Any pair of strategies (m, p) with $F(p) = F$ and $D(m, p) = D$ is an NE iff it is a solution to one of the following sets of constraints in addition to (10) to (15).*

1. $\sum_{i \in F} m_i = B; \rho^* = 0;$
2. $\sum_{i \in F} m_i = B; \rho^* > 0; \sum_{i \in F} m_i w_i p_i = M;$
3. $\sum_{i \in F} m_i = B; \rho^* > 0; p_i = 1, \forall i \in F;$
4. $\sum_{i \in F} m_i < B; \mu^* = 0; F = F_N; \rho^* = 0;$
5. $\sum_{i \in F} m_i < B; \mu^* = 0; F = F_N; \rho^* > 0; \sum_{i \in F} m_i w_i p_i = M;$
6. $\sum_{i \in F} m_i < B; \mu^* = 0; F = F_N; \rho^* > 0; p_i = 1, \forall i \in F.$

In the following, NEs that fall into each of the six cases considered above are named as Type 1–Type 6 NEs, respectively. The next theorem shows that our game has at least one equilibrium and may have more than one NE.

Theorem 4. *The attacker-defender game always has a pure strategy Nash Equilibrium, and may have more than one NE of different payoffs to the defender.*

Proof. The proof of the first part is given in [21]. To show the second part, consider the following example with two nodes where $r_1 = r_2 = 1, w_1 = 2, w_2 = 1, C_1^D = 1/5, C_2^D = 4/5, C_1^A = 1, C_2^A = 7/2, B = 1/3$, and $M = 1/5$. It is easy to check that $m = (1/6, 1/6)$ and $p = (3/20, 9/10)$ is a Type 2 NE, and $m = (1/3, 0)$ and $p = (p_1, 1)$ with $p_1 \in [1/5, 3/10]$ are all Type 1 NEs, and all these NEs have different payoffs to the defender.   □

5   Sequential Game

In this section, we study a sequential version of the simplified game considered in the last section. In the simultaneous game we considered in the previous section, neither the defender nor the attacker can learn the opponent's strategy in advance. While this is a reasonable assumption for the defender, an advanced attacker can often observe and learn defender's strategy before launching attacks. It therefore makes sense to consider the setting where the defender first commits to a strategy and makes it public, the attacker then responds accordingly. Such a sequential game can actually provide defender higher payoff comparing to a Nash Equilibrium since it gives the defender the opportunity of deterring the attacker from moving. We again focus on non-adaptive strategies, and further assume that at $t = 0$, the leader (defender) has determined its strategy, and the follower (attacker) has learned the defender's strategy and determined its own strategy in response. In addition, the players do not change their strategies thereafter. Our objective is to identify the best sequential strategy for the defender to commit to, in the sense of subgame perfect equilibrium [18] defined as follows. We again focus on the case where w_i is deterministic for all i.

Definition 3. *A pair of strategies (m^*, p^*) is a subgame perfect equilibrium of the simplified game (8) and (9) if m^* is the optimal solution of*

$$\max_{m_i} \sum_{i=1}^{N} [m_i(r_i w_i p_i^* - C_i^D) - p_i^* r_i]$$

$$s.t. \quad \sum_{i=1}^{N} m_i \leq B \tag{16}$$

$$0 \leq m_i \leq \frac{1}{w_i}, \forall i$$

where $p_i^\star$ is the optimal solution of

$$\max_{p_i} \sum_{i=1}^{N} p_i[r_i - m_i(r_i w_i + C_i^A)]$$

$$s.t. \sum_{i=1}^{N} m_i w_i p_i \leq M \qquad (17)$$

$$0 \leq p_i \leq 1, \forall i$$

Note that in a subgame perfect equilibrium, $p_i^\star$ is still the optimal solution of (9) as in a Nash Equilibrium. However, defender's best strategy $m_i^\star$ is not necessarily optimal with respect to (8). Due to the multi-node setting and the resource constraints, it is very challenging to identify an exact subgame perfect equilibrium strategy for the defender. To this end, we propose a dynamic programming based algorithm that finds a nearly optimal defense strategy.

Remark 1. Since for any given defense strategy $\{m_i\}$, the attacker's problem (17) is a fractional knapsack problem, the optimal $p_i, \forall i$ has the following form: Sort the set of nodes by $\rho_i(m_i) = \frac{r_i - m_i(r_i w_i + C_i^A)}{m_i w_i}$ non-increasingly, then there is an index k such that $p_i = 1$ for the first k nodes, and $p_i \leq 1$ for the $k+1$-th node, and $p_i = 0$ for the rest nodes. However, if $\rho_i = \rho_j$ for some $i \neq j$, the optimal attack strategy is not unique. When this happens, we assume that the attacker always breaks ties in favor of the defender, a common practice in Stackelberg security games [12].

Before we present our algorithm to the problem, we first establish the following structural properties on the subgame perfect equilibria of the game.

Lemma 6. *In any subgame perfect equilibrium (m, p), the set of nodes can be partitioned into the following four disjoint sets according to the attack and defense strategies applied:*

1. $F = \{i | m_i > 0, \ p_i = 1\}$
2. $D = \{i | m_i > 0, \ 0 < p_i < 1\}$;
3. $E = \{i | m_i > 0, \ p_i = 0\}$;
4. $G = \{i | m_i = 0, \ p_i = 1\}$.

Moreover, they satisfy the following properties:

1. $F \cup D \cup E \cup G = \{i | i = 1, ..., n\}$ and $|D| \leq 1$
2. $\rho_i \geq \rho_k \geq \rho_j$ for $\forall i \in F, \ k \in D, \ j \in E$

Since the set D has at most one element, we use m_d to represent $m_i, i \in D$ for simplicity, and let $\rho_d = \rho(m_d)$. If D is empty, we pick any node i in F with minimum ρ_i and treat it as a node in D.

Lemma 7. *For any given nonnegative ρ_d, the optimal solution for (16)–(17) satisfy the following properties:*

1. $r_i w_i - C_i^D > 0 \ \forall i \in F \cup E \cup D$
2. $m_i \leq \overline{m}_i \ \forall i \in F$
3. $m_j = \overline{m}_j \ \forall j \in E$
4. $\overline{m}_i \leq \frac{1}{w_i} \ \forall i$
5. $B - \sum_{i \in E} \overline{m}_i - m_d > 0.$

where $\overline{m}_i = m_i(\rho_d)$ and $m_i(\cdot)$ is the reverse function of $\rho_i(\cdot)$.

Remark 2. If $\rho_d < 0$, the defender can give less budget to the corresponding node to bring ρ_d down to 0. In any case, the payoffs from nodes in set D and E are 0 since the attacker will give up attacking the nodes in set D and E. Thus, the defender has more budget to defend the nodes in set F and G which brings him more payoffs. Therefore we only need to consider nonnegative ρ_d.

Lemma 8. *For any nonnegative ρ_d, there exists an optimal solution for (16)–(17) such that $\forall i \in F$, there are at most two $m_i < \overline{m}_i$ and all the other $m_i = \overline{m}_i$*

From the above lemmas, we can establish the following results about the structure of the optimal solution for (16)–(17).

Proposition 1. *For any nonnegative ρ_d, there exists an optimal solution $\{m_i\}_{i=1}^n$ such that*

1. *$\forall i \in F$, there are at most two $m_i < \overline{m}_i$ and all the other $m_i = \overline{m}_i$;*
2. *$m_d = \overline{m}_d$;*
3. *$\forall i \in E, \ m_i = \overline{m}_i$;*
4. *$\forall i \in G, \ m_i = 0$.*

According to Proposition 1, for any nonnegative ρ_d, once the set allocation is determined, the value of m_i can be immediately determined for all the nodes except the two fractional nodes in set F. Further, for the two fractional nodes, their m_i can be found using linear programming as discussed below. From these observations, we can convert (16), (17) to (18) for any given nonnegative ρ_d, d, f_1 and f_2.

$$
\max_{p, m_{f_1}, m_{f_2}, E, F, G} \sum_{i \in F \setminus \{f_1, f_2\}} [\overline{m}_i(r_i w_i - C_i^D) - r_i] + \sum_{j=1}^{2} [m_{f_j}(r_{f_j} w_{f_j} - C_{f_j}^D) - r_{f_j}]
$$

$$
- \sum_{i \in G} r_i - \sum_{i \in E} \overline{m}_i C_i^D + m_d(p r_d w_d - C_d^D) - p r_d
$$

$$
s.t. \quad \sum_{i \in F \setminus \{f_1, f_2\}} \overline{m}_i + m_{f_1} + m_{f_2} + \sum_{i \in E} \overline{m}_i + m_d \leq B
$$

$$
\sum_{i \in F \setminus \{f_1, f_2\}} w_i \overline{m}_i + w_{f_1} m_{f_1} + w_{f_2} m_{f_2} + p w_d m_d \leq M
$$

$$
0 \leq m_{f_1} \leq \overline{m}_1, \ 0 \leq m_{f_2} \leq \overline{m}_2, \ 0 \leq p \leq 1 \tag{18}
$$

Note that, the set allocation is part of the decision variables in (18).

We then propose the following algorithm to the defender's problem (see Algorithm 1). The algorithm iterates over nonnegative ρ_d (with a step size ρ_{step}) (lines 3–10). For each ρ_d, it iterates over all possible node d in set D, and all possible nodes f_1, f_2 with fractional assignment in set F (lines 5–8). Given ρ_d, d, f_1, f_2, the best set allocation (together with m_i for all i and p) are determined using dynamic programming as explained below (lines 6–7), where we first assume that $B, M, \overline{m}_i$ and w_i have been rounded to integers for all i. The loss of performance due to rounding will be discussed later.

Consider any ρ_d, node d is in set D, and nodes f_1, f_2 with frictional assignment in set F. Let $SEQ(i, b, m, d, f_1, f_2, ind)$ denote the maximum payoff of the defender considering only node 1 to node i (excluding nodes d, f_1 and f_2), for given budgets b and m for the two constraints in (18), respectively. The ind is a boolean variable that indicates whether the second constraint of (18) is tight for node 1 to i. If ind is $True$, it means all the budget m is used up for node 1 to i. ind is $False$ meaning that there is still budget m available for the attacker. Here, $0 \leq b \leq B$ and $0 \leq m \leq M$. The value of $SEQ(i, b, m, d, f_1, f_2, ind)$ is determined recursively as follows. If $b < 0$ or $m < 0$, the value is set to $-\infty$. If node i is one of d, f_1 and f_2, we simply set $SEQ(i, b, m, d, f_1, f_2, ind) = SEQ(i - 1, b, m, d, f_1, f_2, ind)$. Otherwise, we have the following recurrence equation, where the three cases refer to the maximum payoff when putting nodes i in set F, E, and G, respectively.

$$SEQ(i, b, m, d, f_1, f_2, ind)$$
$$= \max \left\{ SEQ(i - 1, b - \overline{m}_i, m - w_i\overline{m}_i, d, f_1, f_2, ind) + \overline{m}_i(r_iw_i - C_i^D) - r_i, \right.$$
$$\left. SEQ(i - 1, b - \overline{m}_i, m, d, f_1, f_2, ind) - \overline{m}_iC_i^D, SEQ(i - 1, b, m, d, f_1, f_2, ind) - r_i \right\} \quad (19)$$

Meanwhile, if ind is $False$, node i can be allocated to set E only if $r_i - \overline{m}_i(r_iw_i + C_i^A) \leq 0$. Otherwise, there is still available budget for the attacker to attack other nodes with reward greater than 0 which violates the structure of the greedy solution for (17). Also, if ind is $False$, it means m is not used up. Thus we should return $-\infty$ if ind is $False$, $i > 0$ and $m = 0$.

Moreover, we let $SEQ(0, b, m, d, f_1, f_2, ind)$ denote the maximum defense payoff when only nodes in d, f_1, and f_2 are considered. If ind is $True$, the following linear program in (20) determines the optimal values of p, m_{f_1} and m_{f_2} for given budgets b and m:

$$\max_{m_{f_i}, m_{f_2}} \sum_{j=1}^{2} [m_{f_j}(r_{f_j}w_{f_j} - C_{f_j}^D) - r_{f_j}] + m_d(pr_dw_d - C_d^D) - pr_d$$
$$s.t. m_{f_1} + m_{f_2} + m_d \leq b$$
$$m_{f_1}w_{f_1} + m_{f_2}w_{f_2} \leq m \quad (20)$$
$$m_{f_1} \leq \overline{m}_{f_1}, \quad m_{f_2} \leq \overline{m}_{f_2}$$
$$p = \frac{m - m_{f_1}w_{f_1} - m_{f_2}w_{f_2}}{w_dm_d} \leq 1$$

If ind is $False$, we must have $p = 1$. The optimal values of m_{f_1} and m_{f_2} are determined by (21):

$$\max_{m_{f_i}, m_{f_2}} \sum_{j=1}^{2} [m_{f_j}(r_{f_j} w_{f_j} - C_{f_j}^D) - r_{f_j}] + m_d(r_d w_d - C_d^D) - r_d$$

$$s.t. m_{f_1} + m_{f_2} + m_d \le b \tag{21}$$

$$m_{f_1} w_{f_1} + m_{f_2} w_{f_2} \le m - w_d m_d$$

$$m_{f_1} \le \overline{m}_{f_1}, \quad m_{f_2} \le \overline{m}_{f_2}$$

Algorithm 1. Sequential Strategy for Defender

1: Initialize ρ_{step}
2: $\rho_{max} \leftarrow \min\{\rho : \sum_{i=1}^{n} w_i m_i(\rho) \le M\}$
3: **for** $\rho_d \leftarrow 0$ to ρ_{max} with step size ρ_{step} **do**
4: $\quad \overline{m}_i \leftarrow m_i(\rho_d)$ for all i
5: $\quad$ **for** $d, f_1, f_2 \leftarrow 1$ to n **do**
6: $\quad\quad val_{d,f_1,f_2} \leftarrow SEQ(n, B, M, d, f_1, f_2, True)$
7: $\quad\quad val'_{d,f_1,f_2} \leftarrow SEQ(n, B, M, d, f_1, f_2, False)$
8: $\quad$ **end for**
9: $\quad C_{dp}(\rho_d) \leftarrow \max_{d,f_1,f_2}\{val_{d,f_1,f_2}, val'_{d,f_1,f_2}\}$
10: **end for**
11: $C_{alg}^{\star} \leftarrow \max_{\rho_d}\{C_{dp}(\rho_d)\}$

Since the dynamic program searches for all the possible solutions that satisfy Proposition 1, $C_{dp}(\rho_d)$ gives us the optimal solution of (16)–(17) for any given nonnegative ρ_d. Algorithm 1 then computes the optimal solution by searching all the nonnegative ρ_d. Note that d, f_1 and f_2 can be equal to include the case that there is only one or zero node in set F. The minimum possible value of ρ is 0 (explained in Remark 2). The maximum possible value of ρ is $\min\{\rho : \sum_{i=1}^{n} w_i m_i(\rho) \le M\}$. For larger ρ, the sum of all $w_i \overline{m}_i$ will be less than M. In this case, all the nodes will be in set F and $p_i = 1$ $\forall i$, which makes (16)–(17) a simple knapsack problem that can be easily solved.

Additionally, since the dynamic program searches over all feasible integer values, we use a simple rounding technique to guarantee it is implementable. Before the execution of $SEQ(n, B, M, d, f_1, f_2, ind)$, we set $\overline{m}_i \leftarrow \lfloor \frac{\overline{m}_i}{\delta} \rfloor$, $w_i \leftarrow \lfloor \frac{w_i}{\delta} \rfloor$ for all i and $B \leftarrow \lfloor \frac{B}{\delta} \rfloor$, $M \leftarrow \lfloor \frac{M}{\delta} \rfloor$ where δ is an adjustable parameter. Intuitively, by making δ and ρ_{step} small enough, Algorithm 1 can find a strategy that is arbitrarily close to the subgame perfect equilibrium strategy of the defender. Formally, we can establish the following result.

Theorem 5. *Let C_{alg} denote the payoffs of the strategy found by Algorithm 1, and $C^{\star}$ the optimal payoffs. Then for any $\epsilon > 0$, Algorithm 1 can ensure that $\frac{|C_{alg}|}{|C^{\star}|} \le 1 + \epsilon$ with a total time complexity of $O(\frac{n^8 BM}{\epsilon^3})$, where B and M are values before rounding.*

Note that both C_{alg} and C^* are non-positive. The details can be found in our online technical report [21].

6 Numerical Result

In this section, we present numerical results for our game models. For the illustrations, we assume that all the attack times w_i are deterministic as in Sects. 4 and 5. We study the payoffs of both attacker and defender and their strategies in both Nash Equilibrium and subgame perfect equilibrium in a two-node setting, and study the impact of various parameters including resource constraints B, M, and the unit value r_i. We further study the payoffs and strategies for both players in subgame perfect equilibrium in a five-node setting, and study the impact of various parameters.

We first study the impact of the resource constraints M, B, and the unit value r_1 on the payoffs for the two node setting in Fig. 2. In the figure, we have plotted both Type 1 and Type 5 NE[2] and subgame perfect equilibrium. Type 5 NE only occurs when M is small as shown in Fig. 2(a), while Type 1 NE appears when B is small as shown in Fig. 2(b), which is expected since B is fully utilized in a Type 1 NE while M is fully utilized in a Type 5 NE. When the defense budget B becomes large, the summation of m_i does not necessarily equal to B and thus Type 1 NE disappears. Similarly, the Type 5 NE disappears for large attack budget M. In Fig. 2(c) and (d), we vary the unit value of node 1, r_1. At the beginning, the defender protects node 2 only since $w_2 > w_1$. As r_1 becomes larger and larger, the defender starts to change its strategy by protecting node 1 instead of node 2 in NE Type 1. On the other hand, since node 1 is fully protected by the defender and the defender gives up defending node 2, the attacker begins to attack node 2 with probability 1, and uses the rest budget to attack node 1 with probability less than 1, due to the high defending frequency and limited resources M. We further observe that in both the simultaneous game and the sequential game, the value of m_1 increases along with the increase of r_1, while the value of m_2 decreases at the same time. This implies that the defender tends to protect the nodes with higher values more frequently. In addition, the subgame perfect equilibrium always bring the defender higher payoffs compared with Nash Equilibrium, which is expected.

Moreover, it interesting to observe that under the Type 5 NE, the attacker's payoff decreases for a larger M as shown in Fig. 2(a). This is because the defender's budget B is not fully utilized in Type 5 NE, and the defender can use more budget to protect both nodes when M increases. The increase of the attacker's payoff by having a larger M is canceled by the increase of the defender's move frequency m_1 and m_2. We also note that the Type 5 NE is less preferable for the defender in Fig. 2(c) when r_1 is small and favors defender as r_1 increases, which tells us that the defender may prefer different types of NEs under different scenarios and so does the attacker.

---

[2] There are also Type 2 NE, which are omitted for the sake of clarify.

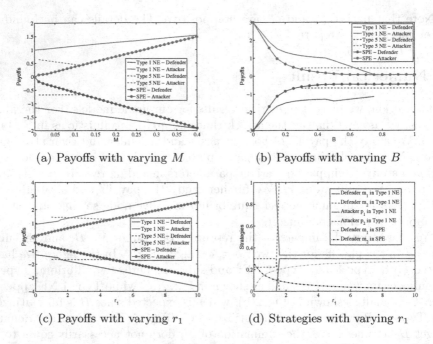

Fig. 2. The effects of varying resource constraints, where in all the figures, $r_2 = 1, w_1 = 1.7, w_2 = 1.6, C_1^D = 0.5, C_2^D = 0.6, C_1^A = 1, C_2^A = 1.5$, and $r_1 = 2$ in (a) and (b), $B = 0.3$ in (a), (c), and (d), and $M = 0.1$ in (b), (c), and (d).

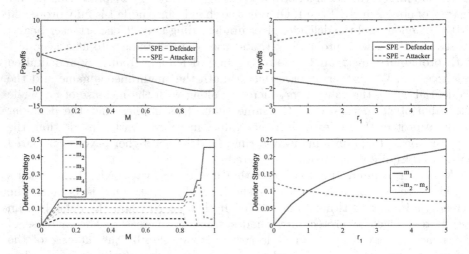

Fig. 3. The effects of varying resource constraints and r_1, where $w = [2\ 2\ 2\ 2\ 2]$, $C^D = C^A = [1\ 1\ 1\ 1\ 1]$, $B = 0.5$, $r = [5\ 4\ 3\ 2\ 1]$ in (a), $r = [r_1\ 1\ 1\ 1\ 1]$ and $M = 0.3$ in (b).

We then study the effects of varying M and r_1 on both players' payoffs and strategies in the sequential game for the five-node setting. In Fig 3(a), the parameters of all the nodes are the same except r_i. We vary the attacker's budget M from 0 to 1. When $M = 0$, the defender can set m_i for all i to arbitrary small (but positive) values, so that the attacker is unable to attack any node, leading to a zero payoff for both players. As M becomes larger, the attacker's payoff increases, while the defender's payoff decreases, and the defender tends to defend the nodes with higher values more frequently, as shown in Fig. 3(a)(lower). After a certain point, the defender gives up some nodes and protects higher value nodes more often. This is because with a very large M, the attacker is able to attack all the nodes with high probability, so that defending all the nodes with small m_i is less effective than defending high value nodes with large m_i. This result implies that the attacker's resource constraint has a significant impact on the defender's behavior and when M is large, protecting high value nodes more frequently and giving up several low value nodes is more beneficial for the defender compared to defending all the nodes with low frequency.

In Fig. 3(b), we vary r_1 while setting other parameters to be the same for all the nodes. Since all the nodes other than node 1 are identical, they have the same m_i as shown in Fig. 3(b)(lower). We observe that the defender protects node 1 less frequently when r_1 is smaller than the unit value of other nodes. When r_1 becomes larger, the defender defends node 1 more frequently, which tells us the defender should protect the nodes with higher values more frequently in the subgame perfect equilibrium when all the other parameters are the same.

7   Conclusion

In this paper, we propose a two-player non-zero-sum game for protecting a system of multiple components against a stealthy attacker where the defender's behavior is fully observable, and both players have strict resource constraints. We prove that periodic defense and non-adaptive $i.i.d.$ attack are a pair of best-response strategies with respect to each other. For this pair of strategies, we characterize the set of Nash Equilibria of the game, and show that there is always one (and maybe more) equilibrium, for the case when the attack times are deterministic. We further study the sequential game where the defender first publicly announces its strategy, and design an algorithm that can identify a strategy that is arbitrarily close to the subgame perfect equilibrium strategy for the defender.

References

1. Advanced persistent threat. http://en.wikipedia.org/wiki/Advanced_persistent_threat
2. ESET and Sucuri Uncover Linux/Cdorked.A: The Most Sophisticated Apache Backdoor (2013). http://www.eset.com/int/about/press/articles/article/eset-and-sucuri-uncover-linuxcdorkeda-apache-webserver-backdoor-the-most-sophisticated-ever-affecting-thousands-of-web-sites/

3. Coviello, A.: Open letter to RSA customers, 17 March 2011. http://www.rsa.com/node.aspx?id=3872
4. Alpcan, T., Başar, T.: Network Security: A Decision and Game-Theoretic Approach. Cambridge University Press, Cambridge (2010)
5. Anderson, R.: Why information security is hard - an economic perspective. In: Proceedings of ACSAC (2001)
6. Bencsáth, B., Pék, G., Buttyán, L., Félegyházi, M.: The cousins of stuxnet: duqu, flame, and gauss. Future Internet 4, 971–1003 (2012)
7. Bowers, K.D., Dijk, M.E.V., Juels, A., Oprea, A.M., Rivest, R.L., Triandopoulos, N.: Graph-based approach to deterring persistent security threats. US Patent 8813234 (2014)
8. Bowers, K.D., van Dijk, M., Griffin, R., Juels, A., Oprea, A., Rivest, R.L., Triandopoulos, N.: Defending against the unknown enemy: applying flipIt to system security. In: Walrand, J., Grossklags, J. (eds.) GameSec 2012. LNCS, vol. 7638, pp. 248–263. Springer, Heidelberg (2012)
9. Buttyan, L., Hubaux, J.-P.: Security and Cooperation in Wireless Networks: Thwarting Malicious and Selfish Behavior in the Age of Ubiquitous Computing. Cambridge University Press, New York (2007)
10. Gueye, A., Marbukh, V., Walrand, J.C.: Towards a metric for communication network vulnerability to attacks: a game theoretic approach. In: Krishnamurthy, V., Zhao, Q., Huang, M., Wen, Y. (eds.) GameNets 2012. LNICST, vol. 105, pp. 259–274. Springer, Heidelberg (2012)
11. Kearns, M., Ortiz, L.E.: Algorithms for interdependent security games. In: Proceedings of NIPS (2003)
12. Korzhyk, D., Yin, Z., Kiekintveld, C., Conitzer, V., Tambe, M.: Stackelberg vs. Nash in security games: an extended investigation of interchangeability, equivalence, and uniqueness. J. Artif. Intell. Res. 41, 297–327 (2011)
13. Kunreuther, H., Heal, G.: Interdependent security. J. Risk Uncertainty 26(2–3), 231–249 (2003)
14. Laszka, A., Horvath, G., Felegyhazi, M., Buttyán, L.: Flipthem: modeling targeted attacks with flipit for multiple. In: Saad, W., Poovendran, R. (eds.) GameSec 2014. LNCS, vol. 8840, pp. 175–194. Springer, Heidelberg (2014)
15. Laszka, A., Johnson, B., Grossklags, J.: Mitigating covert compromises: a game-theoretic model of targeted and non-targeted covert attacks. In: Chen, Y., Immorlica, N. (eds.) WINE 2013. LNCS, vol. 8289, pp. 319–332. Springer, Heidelberg (2013)
16. Manshaei, M.H., Zhu, Q., Alpcan, T., Başar, T.: Game theory meets network security and privacy. ACM Comput. Surv. (2012)
17. Moore, T., Anderson, R.: Economics and internet security: a survey of recent analytical, empirical and behavioral research (2011). ftp://ftp.deas.harvard.edu/techreports/tr-03-11.pdf
18. Osborne, M.J., Rubinstein, A.: A Course in Game Theory. The MIT Press, Cambridge (1994)
19. Tambe, M.: Security and Game Theory: Algorithms, Deployed Systems. Cambridge University Press, New York (2011)
20. van Dijk, M., Juels, A., Oprea, A., Rivest, R.L.: FlipIt: the game of "Stealthy Takeover". J. Cryptology 26(4), 655–713 (2013)
21. Zhang, M., Zheng, Z., Shroff, N.B.: A game theoretic model for defending against stealthy attacks with limited resources. Tehnical Report. http://arxiv.org/abs/1508.01950

Passivity-Based Distributed Strategies for Stochastic Stackelberg Security Games

Phillip Lee[1], Andrew Clark[2], Basel Alomair[3], Linda Bushnell[1(✉)], and Radha Poovendran[1]

[1] Network Security Lab, Department of Electrical Engineering, University of Washington, Seattle, WA 98195, USA
{leep3,lb2,rp3}@uw.edu
[2] Department of Electrical and Computer Engineering, Worcester Polytechnic Institute, Worcester, MA 01609, USA
aclark@wpi.edu
[3] National Center for Cybersecurity Technology, King Abdulaziz City for Science and Technology (KACST), Riyadh, Saudi Arabia
alomair@kacst.edu.sa

Abstract. Stackelberg Security Games (SSGs) model scenarios where a defender implements a randomized security policy, while an attacker observes the policy and selects an optimal attack strategy. Applications of SSG include critical infrastructure protection and dynamic defense of computer networks. Current work focuses on centralized algorithms for computing stochastic, mixed-strategy equilibria and translating those equilibria into security policies, which correspond to deciding which subset of targets (e.g., infrastructure components or network nodes) are defended at each time step. In this paper, we develop *distributed* strategies for multiple, resource-constrained agents to achieve the same equilibrium utility as these centralized policies. Under our approach, each agent moves from defending its current target to defending a new target with a precomputed rate, provided that the current target is not defended by any other agent. We analyze this strategy via a passivity-based approach and formulate sufficient conditions for the probability distribution of the set of defended targets to converge to a Stackelberg equilibrium. We then derive bounds on the deviation between the utility of the system prior to convergence and the optimal Stackelberg equilibrium utility, and show that this deviation is determined by the convergence rate of the distributed dynamics. We formulate the problem of selecting a minimum-mobility security policy to achieve a desired convergence rate, as well as the problem of maximizing the convergence rate subject to mobility constraints, and prove that both formulations are convex. Our approach is illustrated and compared to an existing integer programming-based centralized technique through a numerical study.

This work was supported by ONR grant N00014-14-1-0029, NSF grant CNS-1446866 and a grant from the King Abdulaziz City for Science and Technology (KACST).

© Springer International Publishing Switzerland 2015
MHR Khouzani et al. (Eds.): GameSec 2015, LNCS 9406, pp. 113–129, 2015.
DOI: 10.1007/978-3-319-25594-1_7

1   Introduction

Intelligent and persistent adversaries typically observe a targeted system and its security policies over a period of time, and then mount efficient attacks tailored to the weaknesses of the observed policies. These attacks have been analyzed within the framework of Stackelberg Security Games (SSG), where the defender (leader) selects a policy in order to maximize its utility under the best response strategy of the adversary (follower) [1,2]. Applications of SSGs include defense of critical infrastructures [3,4] and intrusion detection in computer networks [5]. In both of these applications, the security policy corresponds to defending a set of targets, including ports, checkpoints, or computer network nodes.

The security of the system targeted in an SSG can be further improved through randomized policies, in which the set of nodes or locations that are guarded varies over time with a probability distribution that is chosen by the defender [2–4,6]. An attacker with knowledge of the probability distribution, but not the outcome of the randomized policy at each time step, will have greater uncertainty of the system state and reduced effectiveness of the attack.

Current work in SSGs focuses on centralized computation of the Stackelberg equilibria against different types of attackers, including rational, min-max, and bounded rational [6] attackers, under complete, incomplete, or uncertain information. In scenarios including patrolling and intrusion defense, however, security policies are implemented by distributed agents (e.g., multi-robot patrols, or malware filters in intrusion detection). These agents have limitations on computation, communication, and ability to move between targets. Currently, however, computationally efficient distributed strategies for resource-constrained defenders to achieve the same Stackelberg equilibria as centralized mechanisms are lacking.

In this paper, we developed distributed strategies for multiple defenders that guarantee convergence to a stochastic Stackelberg equilibrium distribution while minimizing the cost of movement. We propose a distributed strategy in which each defender first checks if a neighboring target is undefended, and then transitions to defending that with a certain probability if it is undefended. Since each defender only needs to know whether the neighboring targets are defended, the proposed policy can be implemented with only local communication. We analyze our approach by introducing nonlinear continuous dynamics, where each state variable is equal to the probability that a corresponding target is guarded by at least one defender, that approximate our proposed strategy. We show that, under this mapping, the Stackelberg equilibrium is achieved if and only if the continuous dynamics converge to a fixed point corresponding to the Stackelberg equilibrium. We develop sufficient conditions for convergence of these nonlinear dynamics via a passivity-based approach.

We derive bounds on the utility of an adversary with partial information as a function of the convergence rate of the dynamics, which we characterize as a passivity index. We then formulate the problem of maximizing the convergence rate, subject to mobility constraints, and prove that the formulation is convex, leading to efficient algorithms for computing the optimal policy. Our approach is validated and compared with an existing integer programming-based approach via numerical study.

The paper is organized as follows. In Sect. 2, we review related works on Stackelberg security games. In Sect. 3, the defenders and attacker models are introduced, and a zero-sum game is formulated between multiple defenders and an attacker. In Sect. 4, we propose a distributed defender strategy and prove convergence to the desired Stackelberg equilibrium. Section 5 bounds the utility of the attacker using the convergence rate of the dynamics and presents a convex optimization approach for maximizing the convergence rate. Section 6 presents our simulation results. Section 7 concludes the paper.

2   Related Work

Stackelberg Security Games (SSGs) have been gaining increasing attention in the security community in application including the defense of critical infrastructures such as airports [3,7], large interconnected computer networks [5,8] and protection of location privacy [9,10]. In particular, stochastic Stackelberg games have been used to design randomized security policies instead of deterministic policies that can be learned by the attacker with certainty.

Computing the Stackelberg equilibria has been studied in the existing literatures [11,12]. Computation of mixed-strategy Stackelberg equilibria against a worst-case (minimax or zero-sum) attacker was considered in [7]. Randomized security policies against bounded rational adversaries were proposed in [11]. When the defender has partial or uncertain information on the adversary's goals and capabilities, a repeated Stackelberg framework was proposed to model the learning and adaptation of the defender strategy over time [12]. In [13], a human adversary with bounded rationality was modeled as the quantal response (QR) in which the rationality of the adversary is characterized by a positive parameter λ, with perfect rationality and worst-case (minimax) behavior as the two extremes. Games when the defender is uncertain about the behavioral models of the attacker has been studied. In [6], a monotonic maximin solution was proposed that guarantees utility bound for the defender against a class of QR adversaries. These existing works focus on computing the Stackelberg equilibria, where optimization framework including mixed-integer programming has been used for the computation.

Centralized algorithms for choosing which targets to defend over time to achieve a Stackelberg equilibrium have received significant recent attention [14, 15], leading to deployment in harbor patrols [4] and mass transit security [3,16]. In [14], randomized patrolling of a one-dimensional perimeter by multiple robots was considered, where all robots are governed by a parameter p determining to move forward or back. In [15], a game when the attacker not only has the knowledge of the randomized policy but also the current location of the defender was analyzed, leading to attacker's strategy being function of the defense policy and the previous moves of the defender. In these works, mixed integer linear programming techniques were proposed to compute the defender strategy, which provide guaranteed optimality but require a centralized entity with worst-case exponential complexity in the number of defenders, time steps, and targets.

In the present paper, we instead consider a set of defenders who choose their strategies in a distributed manner in order to approximate the equilibrium of a one-shot SSG.

3   Model and Game Formulation

In this section, we present the defender and adversary models. We then formulate a Stackelberg game modeling the behavior of the adversary and defenders.

3.1   Defender Model

We assume that there are n targets and m defenders where $m \leq n$. The targets are represented as nodes on a complete graph, and each defender is located at one node in the graph at each time t. We model the constrained mobility of defenders and physical distances between nodes by assigning a cost d_{ij} of traversing from target i to target j. The cost of traversing may not be symmetric $(d_{ij} \neq d_{ji})$. Each defender is able to communicate with other defender to obtain information regarding whether any target is currently occupied by another defender. We define S_t to be the set of targets that is defended at time t.

3.2   Adversary Model

We consider an adversary whose goal is to successfully penetrate the system by attacking one or more targets over time. If the adversary attacks target i at time t, the adversary will collect the reward $r_i \geq 0$ if no defender is present at the target at time t. If at least one defender is present at target i at time t, the adversary will pay the cost $c_i \geq 0$. Both reward and cost values are known to the defenders and the adversary.

We consider two types of adversaries with different levels of available information. The first type of adversary is able to observe the fraction of time that a target is occupied by at least one defender for all targets but is unable to observe the current locations of defenders. The second type of adversary is able to observe exact location of one or more defenders at a sequence of times $t_1 < t_2 < \cdots < t_k$ and plan the attack strategy at time $t > t_k$ based on these observations.

3.3   Game Formulation

We consider a Stackelberg game where the defenders first choose the fraction of time that each target will be occupied by at least one defender. The adversary then observes the chosen fraction of time and decides to either attack a specific target, or not attack any target. The goal of the adversary is to maximize its expected utility, defined as the expected reward minus the expected cost of detection. The goal of the defender is to minimize the best-case expected utility of the adversary, leading to a zero-sum formulation.

To formally define the game, we denote x_i as the fraction of time that target i is occupied by at least one defender. If the adversary decides to attack target i, then the expected utility of attacking i, denoted $U_{adv}(i)$, is given as

$$U_{adv}(x_i) = (1 - x_i)r_i - x_i c_i = -(r_i + c_i)x_i + r_i \qquad (1)$$

Let z_i be the adversary's chosen probability of attacking target i. Writing $\mathbf{x}$ and $\mathbf{z}$ as the vectors of defender and adversary probabilities, respectively, the expected utility of the adversary can be written as

$$U_{adv}(\mathbf{x}, \mathbf{z}) = -\mathbf{x}^T (C + R)\mathbf{z} + \mathbf{1}^T R \mathbf{z} \qquad (2)$$

where C and R are $n \times n$ diagonal matrices with $C_{ii} = c_i$ and $R_{ii} = r_i$. Given $\mathbf{x}$, the adversary obtains the best-response strategy $\mathbf{z}$ by solving the linear program

$$
\begin{aligned}
& \text{maximize} \ -\mathbf{x}^T(C+R)\mathbf{z} + \mathbf{1}^T R\mathbf{z} \\
& \mathbf{y} \\
& \text{s.t.} \qquad \mathbf{1}^T \mathbf{z} \le 1, 0 \le z_i \le 1, \ i = 1,\dots,n
\end{aligned}
\qquad (3)
$$

We note that the adversary can maximize its utility by selecting $z_i = 1$ for some i satisfying

$$i \in \arg\max \{ (\mathbf{x}^T(C + R) + \mathbf{1}^T R)_j : j = 1,\dots,n \}$$

and $z_j = 0$ otherwise. Hence, without loss of generality we assume that the adversary selects a best-response strategy $\mathbf{z}^*$ with this structure, implying that the expected utility of the adversary is given by

$$U_{adv}^*(\mathbf{x}) = \max\{ \max_{i=1,\dots,n} \{ -(r_i + c_i)x_i + r_i \}, 0 \} \qquad (4)$$

which is a piecewise linear function in $\mathbf{x}$.

The Stackelberg equilibrium $\mathbf{x}^*$ of the defender can then be obtained as the solution to the optimization problem

$$
\begin{aligned}
& \text{minimize} \ U_{adv}^*(\mathbf{x}) \\
& \mathbf{x} \\
& \text{s.t.} \qquad \mathbf{1}^T \mathbf{x} \le m, x_i \in [0,1]
\end{aligned}
\qquad (5)
$$

where the constraint $\mathbf{1}^T \mathbf{x} \le m$ reflects the fact that there are m defenders. Equation (5) is a piecewise linear optimization problem, and hence is convex. In the following section, we will discuss how to design the mobility patterns of defenders to achieve the computed $\mathbf{x}^*$ in a distributed manner.

4 Passivity-Based Distributed Defense Strategy

In this section, we present the proposed distributed patrolling strategy of the defenders. We define continuous dynamics that approximate the probability that each target is defended at time t, and show that convergence of the continuous dynamics to the distribution $\mathbf{x}^*$ is equivalent to convergence of the time-averaged defender positions to the Stackelberg equilibrium. We formulate sufficient conditions for convergence of the continuous dynamics via a passivity-based approach.

4.1   Distributed Defender Strategy

Our proposed distributed patrolling strategy is as follows. Each defender decides whether to move to a different target according to an i.i.d. Poisson process with rate γ. At time t, the defender at target i selects a target $j \neq i$ uniformly at random and sends a query message to determine if there is already a defender at target j. If so, then the defender remains at target i. If not, the defender moves to target j with probability p_{ij}.

This defender strategy can be modeled via nonlinear continuous dynamics. Let $x_i(t)$ denote the probability that at least one defender guards target i at time t. For $\delta > 0$ sufficiently small, we then have

$$x_i(t + \delta) = x_i(t) + (1 - x_i(t)) \sum_{j \neq i} \gamma \delta p_{ji} x_j(t) - \sum_{j \neq i} \gamma \delta p_{ij} x_i(t)(1 - x_j(t)).$$

This approximation makes the simplifying assumption that the events $i \in S_t$ and $j \notin S_t$ are independent for $i \neq j$. Dividing by δ and taking the limit as $\delta \to 0$ yields

$$\dot{x}_i(t) = (1 - x_i(t)) \sum_{j \neq i} Q_{ji} x_j(t) - x_i(t) \sum_{j \neq i} Q_{ij}(1 - x_j(t)), \qquad (6)$$

where $Q_{ij} = p_{ij}\gamma$. The following lemma establishes that under the dynamics (6), the number total expected number of defended targets is equal to m at each time step, and the probability that each target is defended is within the interval $[0,1]$.

Lemma 1. *If $x_i(0) \in [0,1]$ for all i and $\mathbf{1}^T \mathbf{x}(0) = m$, then $x_i(t) \in [0,1]$ and $\mathbf{1}^T \mathbf{x}(t) = m$ for all $t \geq 0$.*

Proof. To show that $x_i(t) \in [0,1]$ for all $t \geq 0$ when $x_i(0) \in [0,1]$, let

$$t^* = \inf \{t : x_i(t) \notin [0,1] \text{ for some } i\}.$$

By continuity, $x_i(t^*) \in \{0,1\}$ for some i and $x_j(t) \in [0,1]$ for all $j \neq i$. Suppose without loss of generality that $x_i(t^*) = 0$. Then

$$\dot{x}_i(t^*) = \sum_{j \neq i} Q_{ji} x_j(t) \geq 0,$$

implying that $x_i(t) \in [0,1]$ within a neighborhood of t^* and contradicting the definition of t^*. Hence $x_i(t) \in [0,1]$ for all i and $t \geq 0$.

Now, we have that

$$\mathbf{1}^T \dot{\mathbf{x}}(t) = \sum_{i=1}^{n} \left[(1 - x_i(t)) \sum_{j \neq i} Q_{ji} x_j(t) - x_i(t) \sum_{j \neq i} Q_{ij}(1 - x_j(t)) \right]$$

$$= \sum_{i=1}^{n} \left[\sum_{j \neq i} (Q_{ji} x_j(t) - Q_{ij} x_i(t)) + \sum_{j \neq i} (Q_{ij} x_i(t) x_j(t) - Q_{ji} x_i(t) x_j(t)) \right] = 0,$$

implying that $\mathbf{1}^T \mathbf{x}(t)$ is constant.

4.2   Passivity-Based Convergence Analysis

We now derive conditions on the matrix Q to ensure that, for any initial distribution $\mathbf{x}(0)$, the dynamics (6) satisfy $\lim_{t \to \infty} \mathbf{x}(t) = \mathbf{x}^*$. If this condition holds, then the time-averaged distribution satisfies $\frac{1}{T} \int_0^T \mathbf{x}(t) \, dt \to \mathbf{x}^*$, and hence the Stackelberg equilibrium is achieved.

By inspection of (6), convergence to $\mathbf{x}^*$ occurs only if

$$(1 - x_i^*) \sum_{j \neq i} Q_{ji} x_j^* = x_i^* \sum_{j \neq i} Q_{ij}(1 - x_j^*)$$

for all i. Defining D^* to be a diagonal matrix with $D_{ii}^* = x_i^*$, this necessary condition can be written in matrix form as

$$(D^*(Q - Q^T) + Q^T)\mathbf{x}^* = D^* Q \mathbf{1}. \tag{7}$$

In order to develop sufficient conditions for convergence to $\mathbf{x}^*$, we introduce a decomposition of the dynamics (6) into a negative feedback interconnection between two passive dynamical systems. Recall that a dynamical system Σ is *output feedback passive* if there exists a positive semidefinite function V such that

$$\dot{V}(t) \leq \rho y(t)^T y(t) + u(t)^T y(t) \tag{8}$$

for all input u and output y for all time t. If $\rho = 0$, then the system is called passive, and the system is called strictly passive if $\rho < 0$. The parameter ρ is defined as the output feedback passivity index of the system [17].

Define $\hat{\mathbf{x}}(t) = \mathbf{x}(t) - \mathbf{x}^*$, and let two input-output dynamical systems be given by

$$(\Sigma_1) \quad \begin{cases} \dot{\hat{x}}_i(t) = -(R_{in}(i) + R_{out}(i))\hat{x}_i(t) + u_i^{(1)}(t) \\ y_i^{(1)}(t) = \hat{x}_i(t) \end{cases} \tag{9}$$

$$(\Sigma_2): \quad \mathbf{y}^{(2)}(t) = -(D^*(Q - Q^T) + Q^T)\mathbf{u}^{(2)}(t) \tag{10}$$

where $R_{in}(i) = \sum_{j \in N(i)} Q_{ji} x_j(t)$ and $R_{out}(i) = \sum_{j \in N(i)} Q_{ij}(1 - x_i(t))$. By inspection, the trajectory of $\hat{x}_j(t)$ in the negative feedback interconnection between (Σ_1) and (Σ_2), shown in Fig. 1, is equivalent to the trajectory of $\hat{x}_j(t)$ under the dynamics (6).

The decomposition of Fig. 1 can be interpreted as follows. The top block represents the change in the probability that each target i is defended, based on the current probability that target i is defended. The input signal from the bottom block can be interpreted as the rate at which defenders from other targets move to target i.

A standard result states that the negative feedback interconnection between two strictly passive systems is globally asymptotically stable [17], which in this case implies that $\mathbf{x}(t)$ converges asymptotically to $\mathbf{x}^*$. Hence, it suffices to derive conditions under which systems (Σ_1) and (Σ_2) are strictly passive. We now present sufficient conditions for strict passivity of (Σ_1) and (Σ_2), starting with (Σ_1).

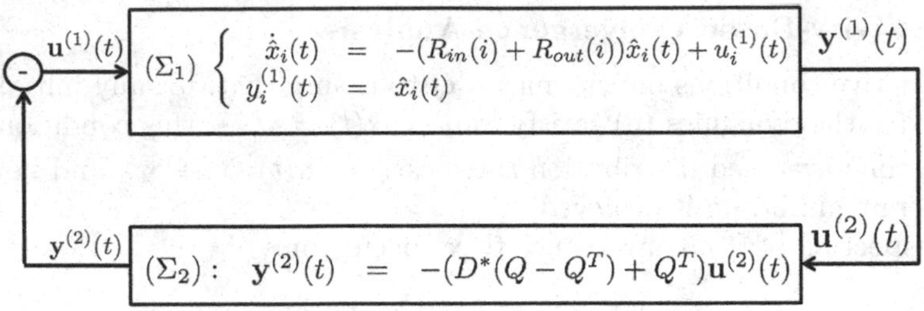

Fig. 1. Decomposition of the patrol dynamics as negative feedback interconnection between passive systems.

Proposition 1. *The system* (Σ_1) *is passive from input* $\mathbf{u}^{(1)}(t)$ *to output* $\mathbf{y}^{(1)}(t)$. *If* $\max_j \{\min \{Q_{ji}, Q_{ij}\}\} > 0$ *for all* i, *then* (Σ_1) *is strictly passive.*

Proof. Consider the storage function $V(\hat{\mathbf{x}}) = \frac{1}{2}\hat{\mathbf{x}}^T \hat{\mathbf{x}}$. We have

$$\dot{V}(\hat{\mathbf{x}}) = -\sum_i (R_{in}(i) + R_{out}(i))\hat{x}_i^2 + (\mathbf{u}^{(1)})^T \hat{\mathbf{x}}.$$

Since the output $\mathbf{y}^{(1)}$ is given by $\mathbf{y}^{(1)}(t) = \hat{\mathbf{x}}$, it suffices to show that $R_{in}(i) + R_{out}(i) > 0$ for all feasible $\mathbf{x}$. We have

$$R_{in}(i) + R_{out}(i) = \sum_i [Q_{ji}x_j + Q_{ij}(1 - x_j)]. \tag{11}$$

Since $x_j \in [0, 1]$, each term of (11) is bounded below by $\min \{Q_{ji}, Q_{ij}\} \geq 0$. Hence the system (Σ_1) satisfies $\dot{V}(\hat{\mathbf{x}}) \leq (\mathbf{u}^{(1)})^T \mathbf{y}$, implying passivity. Furthermore, if the condition $\max_j \{\min \{Q_{ji}, Q_{ij}\}\} =: k > 0$ holds for all i, then

$$\dot{V}(\hat{\mathbf{x}}) < -k\hat{\mathbf{x}}^T \hat{\mathbf{x}} + (\mathbf{u}^{(1)})^T \mathbf{y},$$

implying strict passivity.

The condition $\max_j \{\min \{Q_{ji}, Q_{ij}\}\} > 0$ implies that, for target i, there exists at least one target j such that defenders will transition to target i from target j, and vice versa, with positive probability.

For the system (Σ_2), define matrix $K = (D^*(Q - Q^T) + Q^T)$, so that $\mathbf{y}^{(2)} = -K\mathbf{u}^{(2)}$. If $-\mathbf{u}^T K\mathbf{u} \geq 0$ for all $\mathbf{u}$, then passivity of the bottom block would be guaranteed. On the other hand, since the diagonal entries of K are all 0, the matrix K is neither positive- nor negative-definite. The following proposition gives a weaker sufficient condition.

Proposition 2. *Define* $P = I - \frac{1}{n}\mathbf{1}\mathbf{1}^T$. *If* $PKP \leq 0$ *for all* $\mathbf{u}$, *then the system* (Σ_2) *satisfies* $\mathbf{u}^T \mathbf{y} \geq 0$ *for all* $\mathbf{u}$ *satisfying* $\mathbf{1}^T \mathbf{u} = 0$.

Proof. Suppose that $\mathbf{1}^T \mathbf{u} = 0$. Then $P\mathbf{u} = \mathbf{u}$, since P projects any vector onto the subspace orthogonal to $\mathbf{1}$, and hence $\mathbf{u}^T K\mathbf{u} = \mathbf{u}^T PKP\mathbf{u}$. The inequality $PKP \leq 0$ then implies that $\mathbf{u}^T \mathbf{y} = \mathbf{u}^T K\mathbf{u} \leq 0$.

Combining the conditions for passivity of (Σ_1) and (Σ_2) with the fact that $\mathbf{1}^T \hat{\mathbf{x}}(t) = 0$ (Lemma 1) yields the following sufficient condition for convergence to the desired distribution $\mathbf{x}^*$.

Theorem 1. *If the conditions*

$$K\mathbf{x}^* = D^* Q \mathbf{1} \tag{12}$$

$$\max_j \{\min \{Q_{ji}, Q_{ij}\}\} > 0 \ \forall i \tag{13}$$

$$P^T \frac{K + K^T}{2} P \leq 0 \tag{14}$$

hold, then the vector of probabilities $\mathbf{x}(t)$ *converges to* $\mathbf{x}^*$ *as* $t \to \infty$. *There exists at least one realization of* Q *with* $Q_{ij} \geq 0$ *for all* $i \neq j$ *and* $Q_{ii} = 0$ *that satisfies (12)–(14).*

Proof. Condition (12) implies that the equilibrium of the dynamics (6) corresponds to the Stackelberg equilibrium $\mathbf{x}^*$. Conditions (13) and (14) establish strict passivity of (Σ_1) (Proposition 1) and passivity of (Σ_2) (Proposition 2), respectively, when the trajectory satisfies $\mathbf{1}^T \hat{\mathbf{x}}(t) = 0$ and $x_i(t) \in [0, 1]$ for all i and t, which is guaranteed by Lemma 1. Hence the overall system is globally asymptotically stable with equilibrium $\mathbf{x}^*$. It remains to show that there is a feasible matrix Q that satisfies the conditions (12)–(14).

The proof constructs a matrix Q such that $\frac{K+K^T}{2} = \zeta(\frac{1}{n}\mathbf{1}\mathbf{1}^T - I)$ for some $\zeta \geq 0$. By construction, $\frac{1}{2}P(K + K^T)P = -\zeta P^3 \leq 0$, since $P \geq 0$.

For this choice of $\frac{K+K^T}{2}$, the identities $\frac{K+K^T}{2} = \zeta(\frac{1}{n}\mathbf{1}\mathbf{1}^T - I)$ and $K\mathbf{x}^* = D^* Q \mathbf{1}$ are equivalent to

$$x_i^* Q_{ij} + (1 - x_j^*) Q_{ij} + x_j^* Q_{ji} + (1 - x_i^*) Q_{ji} = \zeta \ \forall i \neq j \tag{15}$$

$$\sum_j x_i^*(1 - x_j^*) Q_{ij} = \sum_j x_j^*(1 - x_i^*) Q_{ji} \ \forall i \tag{16}$$

Define

$$\tau_{ij} = \frac{1}{1 - x_j^*} + \frac{1}{x_i^*} + \frac{1}{1 - x_i^*} + \frac{1}{x_j^*},$$

and let $Q_{ij} = \frac{\zeta}{\tau_{ij} x_i^*(1 - x_j^*)}$. Substitution of Q_{ij} and Q_{ji} into (15) yields

$$\frac{x_i^* \zeta}{\tau_{ij} x_i^*(1 - x_j^*)} + \frac{(1 - x_j^*)\zeta}{\tau_{ij} x_i^*(1 - x_j^*)} + \frac{x_j^* \zeta}{\tau_{ij} x_j^*(1 - x_i^*)} + \frac{(1 - x_i^*)\zeta}{\tau_{ij} x_j^*(1 - x_i^*)} = \zeta,$$

implying that (15) holds. Furthermore,

$$x_i^*(1 - x_j^*) Q_{ij} = \frac{\gamma}{\tau_{ij}} x_j^*(1 - x_i^*) Q_{ji},$$

and hence (16) holds as well.

Observe that under this choice of Q, $Q_{ij} \geq 0$ for all i, j, and condition (13) is satisfied as well.

While there may be multiple matrices Q satisfying conditions (12)–(14), and hence guaranteeing convergence to $\mathbf{x}^*$, the corresponding dynamics of each defender may lead to a high cost associated with moving between distant targets. The problem of selecting the values of Q that minimize the total movement can be formulated as

$$\begin{aligned}
\underset{Q,K}{\text{minimize}} \quad & \sum_{i=1}^{n}\sum_{j=1}^{n} d_{ij}Q_{ij}x_i^*(1-x_j^*) \\
\text{s.t.} \quad & K = D^*(Q - Q^T) + Q^T \\
& P(K + K^T)P \leq 0 \\
& K\mathbf{x}^* = D^*Q\mathbf{1} \\
& Q_{ij} \geq 0 \ \forall i \neq j, \ Q_{ii} = 0 \ \forall i \\
& \max_j\{\min\{Q_{ji}, Q_{ij}\}\} > 0 \ \forall i
\end{aligned} \tag{17}$$

The objective function $\sum_{i=1}^{n}\sum_{j=1}^{n} d_{ij}Q_{ij}x_i^*(1-x_j^*)$ can be interpreted as the total movement cost to maintain the Stackelberg equilibrium $\mathbf{x}^*$ once the equilibrium is reached. Equation (17) can be reformulated as a standard-form semidefinite program and solved in polynomial time. Furthermore, the procedure described in Theorem 1 can be used to construct a feasible solution to (17) in $O(n^2)$ time when the number of targets is large.

5   Mitigating Side Information of Adversary

In this section, we analyze the performance of our approach against an adversary with knowledge of the defender positions at a previous time period. We first bound the deviation between the utility of an adversary with partial information and the Stackelberg equilibrium utility. Our bound is a function of the convergence rate of the dynamics (6). We then formulate the problem of maximizing the convergence rate subject to mobility constraints, as well as the problem of selecting the least-costly patrolling strategy to achieve a desired convergence rate.

5.1   Deviation from Stackelberg Equilibrium

An adversary who observes the defender positions at time t' can estimate the probability $x_i(t)$ that target i is defended at time $t > t'$ via the dynamics (6). The adversary then computes the optimal strategy $\mathbf{z}(t)^*$, where $z_i(t)^*$ is the probability of attacking target i at time t, by solving the optimization problem $\max\{-\mathbf{x}(t)^T(C + R)\mathbf{z} + \mathbf{1}^T R\mathbf{z} : \mathbf{1}^T\mathbf{z} = 1, \mathbf{z} \geq 0\}$.

The deviation of the resulting utility from the Stackelberg equilibrium is given by

$$E(t) = \sum_j [z_j(t)^*(c_j x_j(t) + (1 - x_j(t))r_j) - z_j^*(x_j^* c_j + (1 - x_j^*)r_j)].$$

The following theorem provides an upper bound on $E(t)$ as a function of the convergence rate.

Theorem 2. *The expression $E(t)$ satisfies*

$$E(t) \leq 2 \max_j \{|c_j||x_j(t) - x_j^*| + |r_j||x_j(t) - x_j^*|\} + \max_j |c_j - r_j| \sum_j |x_j(t) - x_j^*|.$$

(18)

Proof. Letting $\alpha_j(x_j(t)) = c_j x_j(t) + r_j(1 - x_j(t))$,

$$E(t) = \sum_j [\alpha_j(x_j(t))(z_j(t)^* - z_j^* + z_j^*) - z_j^* \alpha_j(x_j^*)]$$

$$= \sum_j [\alpha_j(x_j(t))(z_j(t)^* - z_j^*) + z_j^*(\alpha_j(x_j(t)) - \alpha_j(x_j^*))].$$ (19)

Considering the two terms of the inner summation in (19) separately, we first have that $\sum_j \alpha_j(x_j(t))(z_j(t)^* - z_j^*)$ is equal to $\alpha_j(x_j(t)) - \alpha_i(x_i(t))$, where j is the target attacked by the adversary in the best-response to distribution $\mathbf{x}(t)$ and i is the target attacked by the adversary in the best-response to $\mathbf{x}^*$. We then have

$$\alpha_j(x_j(t)) - \alpha_i(x_i(t)) = c_j x_j(t) + r_j(1 - x_j(t)) - c_i x_i(t) - r_i(1 - x_i(t))$$

$$= c_j \hat{x}_j(t) - r_j \hat{x}_j(t) - c_i \hat{x}_i(t) + r_i \hat{x}_i(t)$$

$$+ c_j x_j^* + r_j(1 - x_j^*) - c_i x_i^* - r_i(1 - x_i^*)$$

$$\leq c_j \hat{x}_j(t) - r_j \hat{x}_j(t) - c_i \hat{x}_i(t) + r_i \hat{x}_i(t) \quad (20)$$

$$\leq |c_j||x_j - x_j^*| + |r_j||x_j - x_j^*| \quad (21)$$

$$+ |c_i||x_i - x_i^*| + |r_i||x_i - x_i^*|$$

where (20) follows from the fact that i is a best-response to $\mathbf{x}^*$ and (21) follows from the triangle inequality. Taking an upper bound over i and j yields the first term of (18).

Now, consider the second term of $E(t)$. We have

$$\alpha_j(x_j(t)) - \alpha_j(x_j^*) = c_j x_j(t) + (1 - x_j(t))r_j - c_j x_j^* - r_j(1 - x_j^*) = (c_j - r_j)(x_j(t) - x_j^*).$$

Hence

$$\sum_j z_j^*(\alpha_j(x_j(t)) - \alpha(x_j^*)) = \sum_j z_j^*(c_j - r_j)(x_j(t) - x_j^*)$$

$$\leq \max_i |c_i - r_i| \sum_j |x_j(t) - x_j^*|,$$

the second term of (18).

Theorem 1 implies that the deviation between the optimal adversary utility at time t and the Stackelberg equilibrium is determined by the convergence rate. The convergence rate can be bounded via a Lyapunov-type argument. As a preliminary, we have the following standard result.

Proposition 3. *[17] Let $V(x)$ be a continuously differentiable function such that*

$$c_1||x||^a \leq V(x) \leq c_2||x||^a \tag{22}$$
$$\dot{V}(x) \leq -c_3||x||^a \tag{23}$$

over a domain $D \subset \mathbb{R}^n$. Suppose $\dot{x} = f(x)$ satisfies $f(0) = 0$. Then

$$||x(t)|| \leq \left(\frac{c_2}{c_1}\right)^{1/a} \exp\left(-\frac{c_3}{c_2 a}\right)||x(0)||.$$

A bound on the convergence rate can then be derived via the passivity analysis of Sect. 4.

Proposition 4. *Define $K_p = P^T(\frac{K+K^T}{2})P$, where $P = (I - \frac{1}{n}\mathbf{1}\mathbf{1}^T)$, and suppose that $K_p \leq 0$. Denote the eigenvalues of K_p as $0 \geq -\lambda_1 \geq \cdots \geq -\lambda_{n-1}$ and associated eigenvector of λ_i as q_i. Then, the deviation $||\mathbf{x}(t) - \mathbf{x}^*||_2$ satisfies*

$$||\mathbf{x}(t) - \mathbf{x}^*||_2 \leq \exp(-\lambda_1). \tag{24}$$

Proof. Let $V(\hat{\mathbf{x}}) = \frac{1}{2}\hat{\mathbf{x}}^T\hat{\mathbf{x}}$. In the notation of Proposition 3, we have $a = 2$ and $c_1 = c_2 = \frac{1}{2}$. We will bound $\dot{V}(\hat{\mathbf{x}})$ as a function of $||\hat{\mathbf{x}}||^2$. Any $\hat{\mathbf{x}}$ such that $\mathbf{1}^T\hat{\mathbf{x}} = 0$ satisfies $\hat{\mathbf{x}} = P\hat{\mathbf{x}}$. Then, from the passivity analysis in Proposition 1, we have

$$\dot{V}(\hat{\mathbf{x}}) \leq \hat{\mathbf{x}}^T K\hat{\mathbf{x}} = \hat{\mathbf{x}}^T P^T \frac{K + K^T}{2} P\hat{\mathbf{x}} = \hat{\mathbf{x}}^T K_p\hat{\mathbf{x}}$$

which can be upper bounded as

$$\hat{\mathbf{x}}^T K_p\hat{\mathbf{x}} \overset{(a)}{=} \sum_{i=1}^{n-1} -\lambda_i(q_i^T\hat{\mathbf{x}})^2 \leq -\lambda_1 \sum_{i=1}^{n-1} \hat{\mathbf{x}}^T q_i q_i^T \hat{\mathbf{x}}$$

$$\overset{(b)}{=} -\lambda_1 \sum_{i=1}^{n-1} \hat{\mathbf{x}}^T(I - \frac{1}{n}\mathbf{1}\mathbf{1}^T)\hat{\mathbf{x}} = -\lambda_1\hat{\mathbf{x}}^T P\hat{\mathbf{x}}$$

$$\overset{(c)}{=} -\lambda_1\hat{\mathbf{x}}^T P^T P\hat{\mathbf{x}} = -\lambda_1||\hat{\mathbf{x}}||^2$$

where (a) is from eigen decomposition, (b) is from the orthogonality of eigenvectors for symmetric matrices, and (c) is from the idempotent property of the projection matrix. Substituting $-\lambda_1$ as c_3 from Proposition 3, we obtain the desired bound.

The proof of Proposition 4 implies that $\dot{V}(\hat{\mathbf{x}}) \leq -\lambda_1\hat{\mathbf{x}}^T\hat{\mathbf{x}}$, implying that λ_1 is a *passivity index* [17] for the system (Σ_1). Proposition 4 shows that maximizing over the convergence rate is equivalent to maximizing $|\lambda_1|$, which will be considered in the following section.

5.2   Optimizing the Convergence Rate

The problem of maximizing the convergence rate subject to the mobility constraint can be formulated as

$$\begin{aligned}
&\text{maximize } s\\
&Q, K, s\\
&\text{s.t.} \quad K = D^*(Q - Q^T) + Q^T\\
&\qquad\quad K\mathbf{x}^* = D^*Q\mathbf{1}\\
&\qquad\quad Q_{ij} \geq 0 \; \forall i \neq j, \; Q_{ii} = 0 \; \forall i\\
&\qquad\quad \sum_{i=1}^{n}\sum_{j=1}^{n} d_{ij}Q_{ij} \leq d\\
&\qquad\quad \max_j \{\min\{Q_{ji}, Q_{ij}\}\} > 0 \; \forall i\\
&\qquad\quad P\left(\frac{K+K^T}{2}\right)P + sP \leq 0, s \geq 0
\end{aligned} \tag{25}$$

The first four constraints are from (17). The last constraint ensures the negative semi-definiteness of the matrix $P(K+K^T)P$ and maximization of $|\lambda_1|$, as shown in the following proposition.

Proposition 5. *Denote the eigenvalues of $P(K + K^T)P$ as $0, \lambda_1, \ldots, \lambda_{n-1}$ ordered such that $\lambda_1 \geq \lambda_2 \geq \cdots \geq \lambda_{n-1}$, and let q_i denote the eigenvector associated with eigenvalue λ_i. If $P(K + K^T)P + sP \leq 0$, then $\lambda_1 \leq -s$.*

Proof. Let $K_P = P(K + K^T)P$. Then the matrix $K_P + sP$ can be rewritten as

$$K_P + sP = PK_PP + sPIP = P(K_P + sI)P \tag{26}$$

by the idempotent property of P. If $P(K_P+sI)P \leq 0$, then $\mathbf{x}^T P(K_P+sI)P\mathbf{x} \leq 0$ for all $\mathbf{x}$. Letting $\hat{\mathbf{x}} = P\mathbf{x}$, we have

$$\hat{\mathbf{x}}^T(K_P + sI)\hat{\mathbf{x}} \leq 0$$

for all $\hat{\mathbf{x}}$ that satisfies $\mathbf{1}^T\hat{\mathbf{x}} = 0$. In particular, choose $\hat{\mathbf{x}} = q_1$, which satisfies the condition $\mathbf{1}^T q_1$ from the orthogonality of eigenvectors of a symmetric matrix. Then $q_1^T(K_P + I)q_1 = \lambda_1 + s \leq 0$, and hence $\lambda_1 \leq -s$.

By Proposition 5, the constraints $P(K + K^T)P + sP$ and $s \geq 0$ ensure the negative semidefiniteness of $P(K + K^T)P$ and maximizing s will result in $s^* = |\lambda_1|$. The formulated optimization problem is a semidefinite program and can be solved efficiently in polynomial time as in the case of (17).

An alternative optimization is minimizing the patrol cost for a given convergence rate λ. This optimization problem can be formulated as

$$\begin{aligned}
&\text{minimize } \sum_{i=1}^{n}\sum_{j=1}^{n} d_{ij}Q_{ij}x_i^*(1 - x_j^*)\\
&Q, K\\
&\text{s.t.} \quad K = D^*(Q - Q^T) + Q^T\\
&\qquad\quad P\left(\frac{K+K^T}{2}\right)P + \lambda P \leq 0\\
&\qquad\quad K\mathbf{x}^* = D^*Q\mathbf{1}\\
&\qquad\quad Q_{ij} \geq 0 \; \forall i \neq j, \; Q_{ii} = 0 \; \forall i
\end{aligned} \tag{27}$$

which is also convex. This optimization problem is always feasible by the same argument given in Theorem 1, since given a $\lambda > 0$, one can set $\zeta = \lambda$ in the proof of Theorem 1 and construct a matrix Q that satisfies the constraint of (27). This optimization problem returns the least costly patrolling policy given a security constraint of achieving a desired convergence rate to the Stackelberg equilibrium.

6   Numerical Study

In this section, we conduct a numerical study via Matlab on a patrolling application. The formulated optimization problems were solved using cvx. We consider a network with 30 targets deployed uniformly at random in a square of size 10. The mobility cost d_{ij} was set as the Euclidean distance between target i and j. The number of defenders was set to 5. The diagonal reward and cost matrices R and C were randomly generated where the reward and cost values r_i and c_i were chosen uniformly in the interval $(0, 10)$.

We first obtained a Stackelberg equilibrium $\mathbf{x}^*$ by solving the convex optimization problem (5), and solved for Q for a set of convergence rates λ by solving the optimization problem (27) where the movement cost is minimized for a given convergence rate. The adversary's utility at the Stackelberg equilibrium was 3.56.

Convergence of $\mathbf{x}(t)$ to the Stackelberg equilibrium $\mathbf{x}^*$ under the continuous dynamics (6) is shown in Fig. 2(a). The initial positions were chosen at random among 30 targets. We observe that $\mathbf{x}(t)$ converges to $\mathbf{x}^*$ exponentially with differing convergence rates as shown in Proposition 4. Figure 2(b) shows the maximum

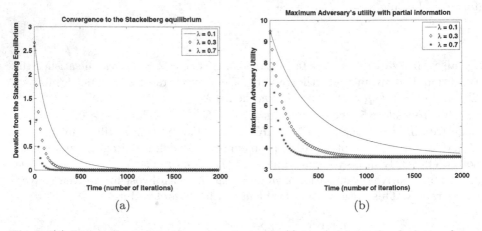

(a) (b)

Fig. 2. (a) Figure illustrating the convergence of $\mathbf{x}(t)$ to $\mathbf{x}^*$. Metric for deviation from the Stackelberg equilibrium was $||\mathbf{x}(t) - \mathbf{x}^*||$ with Q matrices obtained with varying λ by solving optimization problem (27). (b) Maximum adversary's utility with information of the initial locations of defenders. The maximum utility of the adversary decays exponentially, with the maximum utility being the reward value of the target that is not covered by a defender initially.

utility of the adversary over time when the adversary observes the positions of defenders at time $t = 0$. The maximum utility of the adversary at time $t - 0$ is shown to be 9.5 which is the maximum reward value of targets that are not guarded by defender at time $t = 0$. Maximum adversary's utility converges to the defender's utility at Stackelberg equilibrium. The maximum utility of the adversary also decays exponentially with higher convergence rate of (6) offering faster decay of the adversary's utility as observed in Theorem 2.

Our proposed approach is compared with the integer programming-based technique, denoted Raptor, for centralized computation of patrol routes developed in [16] as shown in Fig. 3. Each data point represents an average over 15 independent and random trials with different cost and reward matrices, as well as target locations. The number of defenders was set to 3. For our approach, the minimum patrolling cost was obtained from the optimization problem (27), while the movement cost of Raptor is the minimum cost to transition between two sets of patroller locations sampled randomly with distribution $\mathbf{x}^*$. Our approach is able to achieve comparable mobility cost to Raptor with a convergence rate of $\lambda = 10^{-3}$. We observe that under our approach, as the number of targets increases, the minimum movement cost increases, with the rate of increase proportional to the convergence rate while Raptor's minimum patrolling cost stays relatively constant as the number of targets increase.

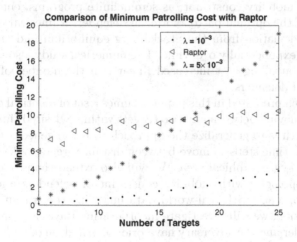

Fig. 3. Minimum patrolling cost with different convergence rate λ and Raptor [16]. The number of defenders was set to 3. It is shown that our approach is able to achieve comparable mobility cost to Raptor with a convergence rate of $\lambda = 10^{-3}$. Under our approach, the minimum movement cost grows in a linear manner as the number of targets grows, and the slope of the line is proportional to the convergence rate λ. Raptor's minimum patrolling cost remains relatively constant as the number of targets grows.

7   Conclusions and Future Work

Stackelberg security games are a modeling framework for scenarios in which a defender chooses a randomized security policy, and an adversary observes the distribution of the randomized policy and selects an attack accordingly. In this paper, we developed a strategy for a team of defenders to implement a stochastic Stackelberg equilibrium security policy. Under our proposed strategy, each defender selects a target according to a precomputed probability distribution at each time step and moves to that target if the target is currently unoccupied. We formulated sufficient conditions, via a passivity-based approach, for a chosen probability distribution to guarantee convergence to the desired Stackelberg equilibrium.

We analyzed the behavior of an intelligent adversary who observes the previous positions of the set of defenders and selects an attack strategy based on these positions and the knowledge of the defender strategies. We proved that the additional impact of the attack provided by knowledge of the defender positions can be bounded as a function of the convergence rate of the defenders to the Stackelberg equilibrum. Under the passivity framework, this convergence rate is interpreted as a passivity index. We formulated the problem of selecting the minimum-cost (in terms of defender movement) strategy to achieve a desired convergence rate, as well as the problem of selecting the fastest-converging defender strategy under mobility constraint, as semidefinite programs, enabling efficient computation of the optimal patrols for each defender. Numerical results verified that both the deviation from the Stackelberg equilibrium and the adversary's utility decayed exponentially over time. The numerical study also suggested that the minimum patrolling cost increased linearly in the number of targets for a fixed number of defenders.

The approach presented in this paper assumes a set of identical defenders that are capable of moving between any two targets within a desired time. A direction of future research is to generalize the approach to heterogeneous defenders who require multiple time steps to move between distant targets, reflecting a deployment over a wide geographical area. We will also extend the proposed approach to arbitrary topologies with mobility constraint of defenders and numerically evaluate the approach with real-world data including the transit network used in [16]. In addition, we will investigate incorporating Bayesian framework where both the defender and the adversary have prior distribution of each other's utility and initial locations and develop approximation algorithms to solve the Bayesian Stackelberg game.

References

1. Paruchuri, P., Pearce, J.P., Marecki, J., Tambe, M., Ordonez, F., Kraus, S.: Playing games for security: an efficient exact algorithm for solving Bayesian Stackelberg games. In: Proceedings of the 7th International Joint Conference on Autonomous Agents and Multiagent Systems, vol. 2, pp. 895–902 (2008)

2. Manshaei, M.H., Zhu, Q., Alpcan, T., Bacşar, T., Hubaux, J.-P.: Game theory meets network security and privacy. ACM Comput. Surv. (CSUR) **45**(3), 25 (2013)
3. Pita, J., Jain, M., Marecki, J., Ordóñez, F., Portway, C., Tambe, M., Western, C., Paruchuri, P., Kraus, S.: Deployed ARMOR protection: the application of a game theoretic model for security at the Los Angeles international airport. In: Proceedings of the 7th International Joint Conference on Autonomous Agents and Multiagent Systems: Industrial Track, pp. 125–132 (2008)
4. Shieh, E., An, B., Yang, R., Tambe, M., Baldwin, C., DiRenzo, J., Maule, B., Meyer, G.: Protect: a deployed game theoretic system to protect the ports of the United States. In: Proceedings of the 11th International Conference on Autonomous Agents and Multiagent Systems, vol. 1, pp. 13–20 (2012)
5. Chen, L., Leneutre, J.: A game theoretical framework on intrusion detection in heterogeneous networks. IEEE Trans. Inf. Forensics Secur. **4**(2), 165–178 (2009)
6. Jiang, A.X., Nguyen, T.H., Tambe, M., Procaccia, A.D.: Monotonic maximin: a robust stackelberg solution against boundedly rational followers. In: Das, S.K., Nita-Rotaru, C., Kantarcioglu, M. (eds.) GameSec 2013. LNCS, vol. 8252, pp. 119–139. Springer, Heidelberg (2013)
7. Brown, G., Carlyle, M., Salmerón, J., Wood, K.: Defending critical infrastructure. Interfaces **36**(6), 530–544 (2006)
8. Zonouz, S., Khurana, H., Sanders, W.H., Yardley, T.M.: Rre: a game-theoretic intrusion response and recovery engine. IEEE Trans. Parallel Distrib. Syst. **25**(2), 395–406 (2014)
9. Shokri, R., Theodorakopoulos, G., Troncoso, C., Hubaux, J.-P., Le Boudec, J.-Y.: Protecting location privacy: optimal strategy against localization attacks. Conf. Comput. Commun. Secure. 617–627 (2012)
10. Shokri, R.: Privacy games: optimal user-centric data obfuscation. Proc. Priv. Enhancing Technol. **2015**(2), 1–17 (2015)
11. Yang, R., Ordonez, F., Tambe, M.: Computing optimal strategy against quantal response in security games. In: Proceedings of the 11th International Conference on Autonomous Agents and Multiagent Systems, vol. 2, pp. 847–854 (2012)
12. Kiekintveld, C., Marecki, J., Tambe, M.: Approximation methods for infinite bayesian stackelberg games: modeling distributional payoff uncertainty. In: The 10th International Conference on Autonomous Agents and Multiagent Systems, vol. 3, pp. 1005–1012 (2011)
13. McKelvey, R.D., Palfrey, T.R.: Quantal response equilibria for normal form games. Games Econ. Behav. **10**(1), 6–38 (1995)
14. Agmon, N., Kraus, S., Kaminka, G. et al.: Multi-robot perimeter patrol in adversarial settings. In: International Conference on Robotics and Automation, pp. 2339–2345 (2008)
15. Vorobeychik, Y., An, B., Tambe, M., Singh, S.: Computing solutions in infinite-horizon discounted adversarial patrolling games. In: International Conference on Automated Planning and Scheduling (2014)
16. Varakantham, P., Lau, H.C., Yuan, Z.: Scalable randomized patrolling for securing rapid transit networks. In: Proceedings of the Twenty-Fifth Innovative Applications of Artificial Intelligence Conference
17. Khalil, H.K.: Nonlinear Systems. Prentice Hall Upper Saddle River (2002)

Combining Online Learning and Equilibrium Computation in Security Games

Richard Klíma[1,4(✉)], Viliam Lisý[1,2], and Christopher Kiekintveld[3]

[1] Department of Computer Science, FEE, Czech Technical University in Prague,
Prague, Czech Republic
rklima@liverpool.ac.uk
[2] Department of Computing Science, University of Alberta, Edmonton, Canada
viliam.lisy@agents.fel.cvut.cz
[3] Computer Science Department, University of Texas at El Paso, El Paso, USA
cdkiekintveld@utep.edu
[4] Department of Computer Science, University of Liverpool, Liverpool, UK

Abstract. Game-theoretic analysis has emerged as an important method for making resource allocation decisions in both infrastructure protection and cyber security domains. However, static equilibrium models defined based on inputs from domain experts have weaknesses; they can be inaccurate, and they do not adapt over time as the situation (and adversary) evolves. In cases where there are frequent interactions with an attacker, using learning to adapt to an adversary revealed behavior may lead to better solutions in the long run. However, learning approaches need a lot of data, may perform poorly at the start, and may not be able to take advantage of expert analysis. We explore ways to combine equilibrium analysis with online learning methods with the goal of gaining the advantages of both approaches. We present several hybrid methods that combine these techniques in different ways, and empirically evaluated the performance of these methods in a game that models a border patrolling scenario.

Keywords: Game theory · Security games · Online learning · Stackelberg game · Stackelberg equilibrium · Nash equilibrium · Border patrol · Multi-armed bandit problem

1 Introduction

Game theory has become an important paradigm for modeling resource allocation problems in security [23]. Deciding how to deploy limited resources is a core problem in security, and game theoretic models are particularly useful for finding randomized policies that make it difficult for attackers to exploit predictable patterns in the security. There are several examples of successful decision support systems that have been developed using this methodology, including the ARMOR system for airport security [19], the IRIS tool for scheduling Federal Air Marshals [24], and the PROTECT system for scheduling Coast Guard

© Springer International Publishing Switzerland 2015
MHR Khouzani et al. (Eds.): GameSec 2015, LNCS 9406, pp. 130–149, 2015.
DOI: 10.1007/978-3-319-25594-1_8

patrols [22]. All of these examples focus primarily on terrorism, where attacks are very infrequent, the adversaries are highly sophisticated, and the stakes of individual events are extremely high. These factors all lead to constructing game models based mostly on inputs from domain experts.

There are many other security domains that are characterized by much more frequent interactions with lower stakes for individual events. These types of domains include border security, cyber security, and urban policing. When there is enough observable data about the actual behavior of attackers, it makes sense to use this data to construct and continually improve the models used for decision making.

However, pure learning/data-driven approaches also have drawbacks: they are entirely reactive, and cannot anticipate adversaries' reactions, they cannot easily incorporate additional information from experts or intelligence, and they can suffer from very poor initial performance during the initial data collection/exploration phase.

We introduce *hybrid* methods that seek to combine the best features of model-based equilibrium analysis and data-driven machine learning for security games with frequent interactions. By using analysis of (imperfect) game models we can warm-start the learning process, avoiding problems with initial poor performance. Using learning allows us to achieve better long-term performance because we are not limited by inaccuracies in a specified model, and we can also adapt to changes in adversary behaviors over time.

The primary motivating domain for our approach is border security, though we believe that our methods are relevant to many other domains with similar features. Border security is a major national security issue in the United States and many other countries around the world. The Customs and Border Protection agency (CBP) is charged with enforcing border security in the United States. The U.S. has thousands of miles of land and sea borders, so CBP faces a very large-scale resource allocation problem when they decide how to allocate infrastructure and patrolling resources to detect and apprehend illegal border crossings. They also have a large amount of data available to inform these resource allocation decisions; in particular, detailed information is available about all apprehensions, including times and precise locations. In principle, this data allows CBP to identify patterns of activity and adopt risk-based resource allocation policies that deploy mobile resources to the areas with the greatest threat/activity levels. The shift to a more data-driven, risk-based strategy for deploying border patrol resources is a major element of the most CBP strategy plan [1].

We study a game with repeated interactions between an attacker and a defender that is designed to capture several of the main features of the border patrol problem. For this model we introduce several hybrid solution techniques that combine Stackelberg equilibrium analysis with online learning methods drawn from the literature on multi-armed bandits. We also introduce variations of these methods for the realistic case where the defender is allowed to allocate multiple patrolling resources in each round (similar to the case of combinatorial multi-armed bandits). We perform an empirical evaluation of our hybrid methods to show the tradeoffs between equilibrium methods and learning methods, and how our hybrid methods can mitigate these tradeoffs.

2 Related Work

There are several lines of work in the area of security games that acknowledge that the game models and assumption about adversary behaviors are only approximations. These models typically focus on finding equilibrium solutions that are robust to some type approximation error. For example, several works have focused on robustness to errors in estimating payoffs, including both Bayesian and interval models of uncertainty [15,16]. Other works have focused on uncertainty about the surveillance capabilities of the attacker [2,3,9,28], or about the behavior of humans who may act with bounded rationality [20,21,27]. Finally, some recent works have combined multiple types of uncertainty in the same model [18].

Our approach is not focused on simply making equilibrium solutions more robust to modeling error, but on integrating equilibrium models with learning methods based on repeated interactions with an attacker. The learning methods we use are drawn from the literature on online learning in multi-armed bandits (MAB), where the focus is on balancing exploration and exploitation. One well-known method for learning a policy for a MAB with fixed distributions is UCB [4], which has also been modified into Sliding-window UCB [13] for situations with varying underlying distributions. The algorithms that most closely fit our setting are for the adversarial MAB problem, where there are no assumptions about the arms having a fixed distribution of rewards, but instead an adversary can arbitrarily modify the rewards. The EXP3 method is one of the most common learning methods for this case [5]. There have been several other recent works that have considered using learning in the context of security games [6,17,26,30], but these have not considered combining learning with equilibrium models. The most closely related work that considers combining learning and equilibrium models is in Poker, where implicit agent models have been proposed that adopt online learning to select among a portfolio of strategies [7,8].

3 Game Model

We introduce a game model that captures several important features of resource allocation for border patrol [1]. The core of the model is similar to the standard Stackelberg security game setting [14,23]. The border is represented by a set of K distinct zones, which represent the possible locations where an attacker can attempt to enter the country illegally. There is a defender (i.e., border patrol), denoted by Θ, who can allocate d resources to patrol a subset of the K zones; there are not enough resources to patrol every area all of the time. The attackers, denoted by Ψ, attempt to cross the border without being detected by avoiding the patrolling agents.

An important difference between our model and the standard security game model is that we consider this a repeated game between the attacker and defender that plays out in a series of rounds. This models the frequent interactions over time between CBP and organized criminal groups that smuggle people, drugs, and other contraband across the border. Each round $t \in 1 \ldots N$ in the game corresponds to a time period (e.g., a shift) for which the defender assigns resources to protect d out of the K zones. Attackers attempt to cross in each round, and each individual attacker selects a single zone to cross at.

The utilities in the game are not zero-sum, but follow the standard security game assumption that the defender prefers to patrol zones that are attacked and the attacker prefers to cross in zones that are not patrolled. More precisely, we assume that for any zone the defender receives payoff $x_c^\Theta = 1$ if an attacker chooses the zone and it is patrolled by a resource, and $x_u^\Theta = 0$ if it is not selected or not patrolled. We assume that the attacker has a zone preference vector $x_u^\Psi = \langle v_1^\Psi, \ldots, v_K^\Psi \rangle$, which describes his payoff for crossing a zone if it is not patrolled. This vector can represent how easy/difficult it is to cross a zone because of specific conditions in the terrain (i.e., without the risk of being caught, an attacker would prefer an easy crossing near a city, rather than a dangerous crossing over miles of open desert). If the attacker is apprehended in zone j, he suffers penalty of $\pi^\Psi = 0.5$; hence, his payoff is $x_{c,j}^\Psi = v_j^\Psi - 0.5$. The goal of each player is to maximize the sum of payoffs obtained over all rounds of the game.

An important characteristic of the border patrol domain is limited observability. In particular, the border patrol only gathers reliable information about the illegal entrants they actually apprehend; they do not observe the complete strategy of all attackers.[1] In our model, we capture this by revealing to the defender only the attackers that are apprehended (i.e., the attacker chooses a zone where the defender currently has a resource patrolling). The defender does not observe the attackers that choose zones that are not patrolled. This leads to a classic exploration vs. exploitation problem, since the defender may need to explore zones that appear to be suboptimal based on the current information to learn more about the attacker's strategy. In a long run of the game we overcome the possibility of high, unnoticed immigrant flows in an unpatrolled zone by an extra exploration, which we use in the defender strategies.

As a simplifying assumption, we assume that the attacker observes the whole patrol history of the defender in all zones but does not know the defender strategy vector. At time t, the attacker knows the number of previous rounds in which the defender protected zone j further denoted $h_j^t = c_j^t * (t-1)$. This can be justified in part by the domain, since border patrol agents are more easily observable (i.e., they are uniformed, drive in marked vehicles, etc.), and smuggling organizations are known to use surveillance to learn border patrolling strategies. We also assume the attackers to cooperate and form a gang or a cartel and thus share fully the gained information about the patrols. However, it also allows us to more easily define a simple but realistic adaptive strategy for the attackers to follow in our model based on fictitious play. We describe this behavior in more detail later on.

We do not generally assume that the defender knows the attacker's payoffs (i.e., zone preferences). However, when we consider equilibrium solutions we will

---

[1] This is sometimes described as the problem of estimating the total flow of traffic, rather than just the known or observed flow based on detections and apprehensions.

assume that the defender is able to estimate with some uncertainty the payoffs for the attacker. Formally, the defender will have an approximation of the attacker's preference vector v'^{Ψ}, such that $|v_j^{\Psi} - v_j'^{\Psi}| < \epsilon$ for each j and some ϵ known to both players.

In Fig. 1 we present an example of border patrol game, where the defender chooses first a zone to patrol and then the attacker chooses a zone to cross without knowing which specific zone the defender is currently patrolling. There is the zone preference vector v^{Ψ} and patrol history vector h^{100} at round 100. In this example the attacker is apprehended because she chose the same zone as the defender. If the attacker had chosen zone 1 he would have successfully crossed the border because the defender does not patrol it.

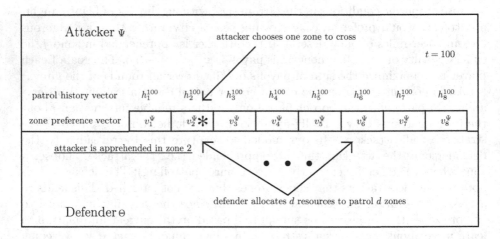

Fig. 1. Border patrol game example

3.1   Attacker Behavior Model

Our main focus in this work will be on designing effective policies for the defender against an adaptive adversary. While there are many ways that the attackers can be modeled as learning and adapting to the defender policy, here we will focus on one simple but natural strategy. We assume that the attackers adapt based on a version of the well-known fictitious play learning policy (e.g., [12]). In fictitious play the player forms beliefs about the opponent strategy and behaves rationally with respect to these beliefs. The standard model of fictitious play assumes the opponent plays a stationary mixed strategy, so the player forms his beliefs about opponent's strategy based on the empirical frequencies of the opponent's play. We define an *adversarial attacker* as an attacker who attacks the zone that maximizes his expected payoff under the assumption that the defender plays a mixed strategy corresponding his normalized patrol history: $j^t = \arg\max_j (v_j^{\Psi} - \pi^{\Psi} * c_j^t)$. This type of the attacker strategy can be seen as the worst-case strategy compared to some naive attacker strategies, which we also successfully tested our proposed algorithm against. Algorithms for minimizing regret in the adversarial bandit setting are designed to be efficient against any adversary and therefore we expect the proposed combined algorithms to be effective against any attacker's strategy.

We also want to evaluate robustness of the designed strategies to rapid changes of the attacker's behavior. In the real world, these can be introduced by new criminal organization starting operations in the area, or by changes in the demand or tactics used by an organization, such as the adoption of a new smuggling route. Therefore, we introduce also an *adversarial attacker with changes*, which differs form the basic adversarial attacker in having variable preference vector x_u^Ψ, that rapidly changes at several points in the game. The defender is not informed about these changes or the time when it happens.

4 Background

4.1 Stackelberg Security Game

Our model of the border patrolling problem is similar to the standard Stackelberg security game model, as described in [14]. The game has two players, the defender Θ and the attacker Ψ. In our model the defender represents the Office of Border Patrol (OBP) and the attacker represents a group of illegal immigrants or a criminal smuggling organization. In security games we usually do not have individuals playing against each other but rather groups of people who have similar or same goal. These groups can represent terrorists, hackers, etc. on the attacker side and officers, authorities, security units etc. on the defender side. These groups use a joint strategy so we can think of the group as an individual player with several resources. The defender has a set of pure strategies, denoted $\sigma_\Theta \in \Sigma_\Theta$ and the attacker has a set of pure strategies, denoted $\sigma_\Psi \in \Sigma_\Psi$. We consider a mixed strategy, which allows playing a probability distribution over all pure strategies, denoted $\delta_\Theta \in \Delta_\Theta$ for the defender and $\delta_\Psi \in \Delta_\Psi$ for the attacker. We define payoffs for the players over all possible joint pure strategy outcomes by $\Omega_\Theta : \Sigma_\Psi \times \Sigma_\Theta \to \mathbb{R}$ for the defender and $\Omega_\Psi : \Sigma_\Theta \times \Sigma_\Psi \to \mathbb{R}$ for the attacker. The payoffs for the mixed strategies are computed based on the expectations over pure strategy outcomes.

An important concept in Stackelberg security games is the idea of a leader and a follower. This concept is the main difference from the normal-form game. The defender is considered to be the leader and the attacker is the follower. The leader plays first, and then the attacker is able to fully observe the defender strategy before acting. This is quite a strong assumption and it represents very adversarial and intelligent attacker who can fully observe the defender's strategy before deciding how to act. In our model we assume less intelligent attacker who does not know the exact defender strategy as described in Sect. 3. Formally we can describe the attacker's strategy as a function which chooses a mixed distribution over pure strategies for any defender's strategy: $F_\Psi : \Delta_\Theta \to \Delta_\Psi$.

4.2   Stackelberg Equilibrium

Stackelberg equilibrium is a strategy profile where no player can gain by unilaterally deviating to another strategy for the case where the leader moves first, and the follower plays a best response. We follow the standard definition of Stackelberg equilibrium (SE) for security games [14]. This version of Stackelberg equilibrium is known as *strong Stackelberg equilibrium*. The strong SE assumes that in cases of indifference between targets the follower chooses the optimal strategy for the leader. A strong SE exists in every Stackelberg game. The leader can motivate the desired strong equilibrium by choosing a strategy, which is arbitrary close to the equilibrium. This makes the follower strictly better off for playing the preferred strategy.

4.3   Nash Equilibrium

Nash equilibrium is a basic concept in game theory for players who move simultaneously. A profile of strategies form a Nash equilibrium if the defender plays a best response s^* that holds $x^\Theta(s_i^*, s_{-i}) \geq x^\Theta(s_i, s_{-i})$ for all strategies $s_i \in S_i^\Theta$ and the attacker plays a best response s^* that holds $x^\Psi(s_i^*, s_{-i}) \geq x^\Psi(s_i, s_{-i})$ for all strategies $s_i \in S_i^\Psi$.

The relationship between strong Stackelberg equilibrium and Nash equilibrium is described in detail in [29]. The authors show that Nash equilibria are interchangeable in security games, avoiding equilibrium selection problems. They also prove that under the SSAS (Subsets of Schedules Are Schedules) restriction on security games, any Stackelberg strategy is also a Nash equilibrium strategy; and furthermore, this strategy is unique in a class of real-world security games.

5   Defender Strategies

The problem the defender faces closely resembles the multi-armed bandit problem, in which each arm represents one of the zones. Therefore, we first explain the online learning algorithms designed for this problem and then we explain how we combine them with game-theoretic solutions.

5.1   Online Learning with One Resource

First we focus on the problem with a single defender resource ($d = 1$). The defender's problem then directly corresponds to the adversarial multi-armed bandit problem. A standard algorithm for optimizing cumulative reward in this setting is *Exponential-weight algorithm for Exploration and Exploitation* (EXP3), which was introduced in [5]. The algorithm estimates the cumulative sum $s(i)$ of all past rewards the player could have received in each zone using the important sampling correction. If zone i is selected with probability p_i and reward r is received, the estimate of the sum is updated by $s(i) = s(i) + \frac{r}{p_i}$. This ensures that $s(i)$ is an unbiased estimate of the real cumulative sum for that zone.

The defender then chooses actions proportionally to the exponential of this cumulative reward estimate. We use the numerically more stable formulation introduced by [11]. Formally, a given zone i is protected with probability:

$$p_i^\Theta = \frac{1 - \gamma}{\sum\limits_{j \in K} e^{(s(j) - s(i))\frac{\gamma}{K}}} + \frac{\gamma}{K}, \tag{1}$$

where γ represents the amount of random exploration in the algorithm.

5.2 Online Learning with Multiple Resources

When computing the strategy for multiple defenders ($d > 1$), we could consider each combination of allocations of the resources to be a separate action in a multi-armed bandit problem. It would require the algorithm to learn the quality of each of the exponentially many allocations independently. However, thanks to the clear structure of payoffs from individual zones, the problem can be solved more efficiently as a combinatorial multi-armed bandit problem. We solve it using the COMB-EXP-1 algorithm introduced in [10] and presented here as Algorithm 1.

COMB-EXP-1 algorithm

Initialization: Start with the distribution $q_0(i) = \frac{1}{K}$ and set $\eta = \sqrt{\frac{2d \log K}{KN}}$

for $t = 1, ..., N$ do

    1. Sample d actions from vector $p_{t-1} = dq_{t-1}$.

    2. Obtain the reward vector $X_i(t)$ for all chosen actions i.

    3. Set $\bar{X}_i(t) = \frac{1 - X_i(t)}{p_{t-1}(i)}$ for all chosen actions i and $\bar{X}_i(t) = 0$ for all other not chosen actions.

    4. Update $\bar{q}_t(i) = q_{t-1}(i) \exp(-\eta \bar{X}_i(t))$.

    5. Compute q_t as a projection of $\bar{q}_t$ to $\mathcal{P} = \{x \in \mathbb{R}^K : \sum_i x_i = 1, x_i \le \frac{1}{d}\}$ using KL divergence.

end

Algorithm 1. Combinatorial EXP3 learning algorithm

The algorithm starts with a uniform distribution over all zones q_0. In each round, it samples d distinct zones from this distribution using the Algorithm 1, so that the probability of protecting each zone is p_i (line 1). It protects the selected zones and receives reward for each of the selected zones (line 2). It computes the loss vector rescaled by importance sampling (line 3) and updates the probability of protecting individual zones using the exponential weighting (line 4). After the update, vector q_t may not represent a correct probability and not sum to one. Therefore, it must be projected back to the simplex of valid probability distributions (line 5).

Similar to the non-combinatorial EXP3 algorithm, the COMB-EXP-1 algorithm can be numerically unstable if some zone is played with very small probability ($p_t(i) \to 0$). We prevent this instability in our implementation by adding an uniform vector with very small values (10^{-7}) to the strategy vector q, which bounds the scaled losses $\bar{X}$.

Combinatorial Sampling. On line 1, Algorithm 1 samples d zones so that each zone i is protected with probability $p(i)$. We use combinatorial sampling as introduced in [25]. From vector p we create a new cumulative sum vector. For each integer $j \in (1, K)$, let $S_j = \sum_{i<j} p_i$. Based on that we define a disjoint partition of interval $[0, d)$ as $I_j = [S_j, S_j + p_j)$. Interval I_j represents zone j. To sample d zones, we generate single random number y from interval $[0, 1)$ uniformly at random. The selected zones correspond to the intervals that contain points $y, y+1, ..., y+(d-1)$. Since each zone is covered with probability at most 1, no two of these points will be part of the same interval and the probability of hitting interval i is p_i. In Fig. 2 there is an example of combinatorial sampling, where we have 6 zones $z_1, ..., z_6$ and 3 resources (defenders). We generate a random number y and sample the intervals created by cumulation from the probability vector p.

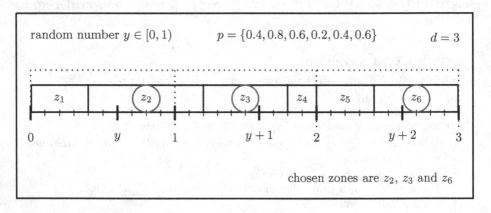

Fig. 2. Combinatorial sampling example

Projection Heuristic. The COMB-EXP-1 algorithm requires projection using KL-divergence on line 5. This projection defines a distribution $q \in \mathcal{P}$ which has the minimal KL-divergence from vector $\bar{q}$.

$$q = \arg\min_{p \in \mathcal{P}} KL(p, \bar{q}) \qquad KL(p, q) = \sum_{i \in 1...K} p(i) \log \frac{p(i)}{q(i)} \qquad (2)$$

We are not aware of a computationally efficient algorithm for computing such projection. Therefore, we propose a heuristic algorithm H_1, where we decrease all values greater than $1/d$ to $1/d$ and normalize all other values in the vector to sum to $(1 - a/d)$, where a is the number of values in the original vector greater than $1/d$ and d is the number of resources.

We compare our heuristic to another heuristic H_2 where we redistribute the difference value from value in a vector greater than $1/d$ uniformly among other values. In 10000 experiments with randomly generated vectors and with different numbers of resources d, heuristic H_1 was always better than H_2. Further, we tried to randomly perturb the vectors returned by H_1 by small amounts in individual zones, while still ensuring the perturbed vector belongs to $\mathcal{P}$ and we were able to find a better value in less than 1 % of cases. We conclude that H_1 is a good approximation of the projection and we use it in the experimental evaluation.

6   Combined Algorithms

In this section we propose four algorithms that combine the online learning algorithms described above with a (possibly inaccurate) game-theoretic solution. The main idea is to start the learning algorithm with some prior information, but allow the algorithms to learn close to optimal solutions even if the initial information is inaccurate.

6.1   Combined Algorithm 1

The EXP3 algorithm described in Sect. 5.1 computes the values $s(j)$ that estimate the cumulative rewards for each zone j. We can initialize the EXP3 algorithm by initializing these values. If both players knew the exact preference vector of the attacker, their optimal static strategy would be the Nash equilibrium (NE) of the game. Due to the security games utility restrictions, this equilibrium is unique [29]. If the attacker was playing an equilibrium attacking zone j with probability NE_j for τ rounds, the cumulative rewards obtainable in individual zones would be $s^{init}(j) = NE_j * \tau$. By using this initialization for the values $s(j)$ in EXP3, it starts from a state similar to the state where it has played τ rounds of the game against the optimal attacker. Since the defender does not have access to the exact preference vector, we compute the approximate Nash equilibrium strategy based on his inaccurate estimate. Algorithm COMB1 than uses it for initialization of $s(i)$ as described above, but otherwise runs the standard EXP3 algorithm.

Combinatorial COMB1. Combinatorial version of the COMB1 algorithms is also based on the intuition of initialization by the estimated equilibrium play. Since COMB-EXP-1 uses the current strategy vector instead of cumulative rewards, we use the defender's strategy for initialization. The algorithms for computing the equilibrium for security games with multiple defender resources, such as [14], directly output the strategy in the form of a coverage vector representing the probability that each zone will be covered. Let Stackelberg equilibrium (SE) be this coverage vector, than the initial distribution for COMB-EXP-1 is:

$$q_0(i) = \frac{\tau}{d} SE + \frac{1 - \tau}{K} \tag{3}$$

where τ is the parameter which sets how confident we are about the Stackelberg equilibrium strategy. The basic setting is $\tau = 0.9$. After initializing the online learning algorithm we continue playing standard combinatorial EXP3.

6.2 Combined Algorithm 2

In this combined algorithm, instead of initializing the learning algorithm as if it played based on the equilibrium strategy before the games starts, it actually plays the estimated equilibrium strategy for the first T rounds of the game. Even though the actions are selected based on the equilibrium in these rounds, EXP3 learns form the observed apprehensions. In order to also learn about the zones that are never played in the equilibrium strategy, we add 10 % uniform exploration to the strategy.

EXP3 learns by computing the vector of estimates s. This vector is computed from the beginning of the game no matter which strategy the defender uses. For finding the point where to switch from first stage to the second we compute the EXP3 payoff virtually while playing the estimated Nash equilibria. Virtual EXP3 payoff is computed using the importance sampling correction. It gives higher payoff for a strategy with higher probability of visiting a particular zone. If the probability of EXP3 protecting a particular zone with positive payoff is higher than the probability in Nash equilibrium vector, we get a relatively higher payoff for EXP3 than for the NE strategy. In this manner we prioritize the strategy that has the higher estimated payoff. The defender gets covered payoff 1 and uncovered payoff 0 and virtual EXP3 defender covered and uncovered payoff is

$$\bar{x}_c^{\Theta}(t) = \frac{e_i^t}{n_i^t} * 1 \qquad \bar{x}_u^{\Theta}(t) = \frac{e_i^t}{n_i^t} * 0 \qquad (4)$$

where e_i^t is the probability of playing zone i in round t by playing EXP3 and n_i^t is the probability of playing zone i in round t by the estimated Nash equilibrium strategy.

We compute the total payoff for both strategies as the sum over all rounds played so far. The algorithm switches to the EXP3 learning algorithm if the cumulative payoff of virtually playing EXP3 exceeds the actual cumulative reward obtained by playing the Nash equilibrium with the additional exploration.

Combinatorial COMB2. Combinatorial COMB2 algorithm is analogous to the standard COMB2 algorithm. We use the estimated Stackelberg equilibrium strategy for multiple resources with 10 % extra exploration and combinatorial EXP3 algorithm. We start with SE strategy and compute virtually expected payoff for playing EXP3. Once the virtual EXP3 payoff becomes greater than actual payoff by playing SE with extra exploration we switch to EXP3 algorithm and use the standard updates.

6.3   Combined Algorithm 3

The third combined algorithm is based on a similar concept to the previous one, but in this case we continually switch between two strategies based on which one has the higher current estimated payoff. One of these strategies is based on the estimated equilibrium, and the other is a learning policy. For the strategy we are currently playing we store the total actual payoff and for the other strategy we compute the payoff in the same way we did for virtual play of EXP3 in the previous algorithm. Similar to above, for virtually playing NE strategy the defender gets covered and uncovered payoff

$$\bar{x}_c^\Theta(t) = \frac{n_i^t}{e_i^t} * 1 \qquad \bar{x}_u^\Theta(t) = \frac{n_i^t}{e_i^t} * 0 \tag{5}$$

Let $\bar{X}_{Alg}^\Theta$ be the estimated cumulative payoff of an algorithms, COMB3 plays the estimated Nash equilibria with exploration if $\bar{X}_{EXP3}^\Theta < \bar{X}_{NE}^\Theta$ or we play EXP3 if $\bar{X}_{EXP3}^\Theta > \bar{X}_{NE}^\Theta$.

The EXP3 algorithm learns using the expected payoff vector s from all previously played rounds including those rounds when the defender played the NE strategy with exploration.

Combinatorial COMB3. Analogously to the non-combinatorial COMB3 algorithm, combinatorial COMB3 algorithm is a generalization of previous combinatorial COMB2 algorithm. In this COMB3 algorithm we enable the switching between the two strategies arbitrary according to the highest payoff. We compute the virtual SE strategy payoff while playing combinatorial EXP3 algorithm and vice versa.

6.4   Combined Algorithm 4

With this algorithm, the defender uses several estimated Nash equilibria corresponding to random modifications of the attacker preference vector by at most ϵ. This models the scenario of building a model based on the input of multiple domain experts, rather than a single expert. There is extra exploration of 10 % added to each estimated Nash equilibrium. The main idea is that some of these random variations may be a more accurate estimate of the true preference vector and the algorithm can learn which one from the interaction. COMB4 starts playing with one of the strategies and in parallel computes the expected payoffs for the other estimated Nash strategies and for the EXP3 learning algorithm. In each round, we select an action based on the strategy with the highest current estimate of the cumulative payoff. In our model we did experiments with 3 estimated Nash equilibria (NE).

Combinatorial COMB4. The combinatorial version of this algorithm is practically the same as the non-combinatorial version. The only difference is that the equilibria are computed for multiple defender resources and the learning algorithm is also combinatorial.

7   Experiments

If not otherwise specified, we consider a game model where $K = 8$ (8 zones) and $N = 1000$ (1000 rounds). In the border patrol domain we can consider 1 round as 1 day, so a 1000 round game represents approximately 3 years. All the experiments are run 1000 times to get reliable results. In each of these runs, we generated a new preference vector for the attacker. Each value is i.i.d. and comes from range $(0, 1)$. We compute the estimated preference vector known to the defender by adding a random number from interval $(-\epsilon, \epsilon)$ to each entry. The exploration parameter γ for the learning algorithms has been hand-tuned to $\gamma = 0.2$, i.e., 20 % exploration.

7.1   Imprecise Stackelberg Equilibrium Strategy

We test the influence of different levels of error (ϵ) in the zone preference
vector on the performance of the estimated SSE. In Fig. 3 we show apprehension rates for different levels of error.
We observe the performance of the SSE
strategy for $\epsilon \in [0, 0.2]$. The adversarial
attacker can learn the strategy and over
time the apprehension rate decreases. In
particular, for higher values of ϵ there is
a large decrease in performance. For $\epsilon \geq$
0.15 we get even worse performance than
for playing a random defender strategy,
which has the expected payoff 12.5 %.
For SSE with no error the performance
is still very good even after the attacker

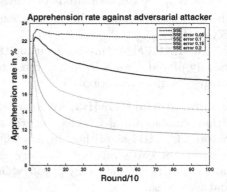

Fig. 3. SSE strategies with different levels of error against an adversarial attacker

learns the strategy. In our further experiments we focus on error 0.1, for which the game theoretic strategy is better than random, but there is still room for improvement. The widest mean 95 % confidence interval in these experiments is ±0.56 % for error 0.1.

7.2   Performance of Combined Algorithms with One Resource

We compare the performance of the EXP3 learning algorithm, Stackelberg equilibrium strategy (SSE), and Stackelberg equilibrium strategy with error (which is used in the COMB algorithms). For each graph we compute a 95 % confidence interval and provide a mean interval width across all rounds.

In Fig. 4 we use two styles of result visualization to better understand the behavior of the algorithms. One is a moving average of apprehensions in 20 rounds (a,b) and the other is the mean apprehension rate from the beginning of the game (c,d). The moving average better represents the immediate perfor-mance of the algorithm and the cumulative performance captures the overall

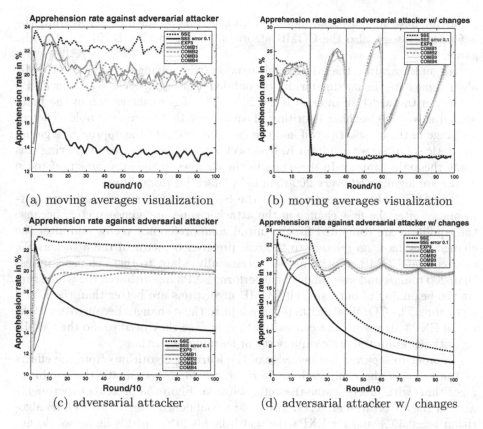

Fig. 4. COMB algorithms with 0.1 error against adversarial attacker

performance of the algorithm. Figure 4a shows the same experiment as Fig. 4b and c shows the same experiment as Fig. 4d. The COMB algorithms use the imprecise game-theoretic solution with error $\epsilon = 0.1$.

In Fig. 4c the COMB algorithms have the widest confidence interval $\pm 0.39\%$ and for EXP3 algorithm the width of interval is $\pm 0.30\%$. The mean reward of SSE with error decreases with the attacker learning the strategy. SSE without error gives a very good, stable performance. COMB1 has better but similar performance to EXP3. This comes from the nature of COMB1 algorithm, which is an initialized EXP3. COMB2 algorithm starts with playing SSE with error plus some extra exploration and then switches permanently to EXP3. We can see that this switch occurs close to the intersection of SSE with error and EXP3 algorithm which is a desired feature of COMB2 algorithm. COMB3 outperforms COMB2, which is caused by better adaptability to the intelligent attacker. COMB4 has the best performance out of all COMB algorithms and also outperforms EXP3 algorithm. COMB2, COMB3 and especially COMB4 algorithms have very good performance for the first half of the game (up to round 500) and outperform EXP3 and SSE with error. At the end of our game COMB algorithms and EXP3 algorithm have similar performance, which is caused by the attacker learning the defender strategy, also the COMB algorithms tend to play EXP3 later in the game.

In games against the adversarial attacker with changes in Fig. 4d COMB algorithms have the maximal width of confidence interval $\pm0.32\%$ and for EXP3 algorithm the width of interval is $\pm0.26\%$. This figure shows one of the main advantages of the learning algorithm. If we assume that we are not able to detect a change in the attacker payoff and therefore to compute the appropriate game-theoretic solution, we can intuitively expect a poor performance by playing this game-theoretic strategy. In these figures the changes in the attacker's preference vector are highlighted every 200 rounds by black horizontal lines.

The SSE with error strategy and the SSE strategy have almost same performance after the first change in the attacker zone preference vector, because the equilibria are computed for the initial zone preference vector and after the change they have no relation to the real preference vector of the attacker. We can see that COMB algorithms can successfully adapt to these changes in less than 200 rounds and even slightly outperform EXP3 algorithm in the whole run. At the beginning of our game all COMB algorithms are better than the EXP3 algorithm. The COMB algorithms can adapt to these changes because they make use of EXP3 algorithm and can switch to it in case they need to. So the COMB algorithms retain the desired property of learning algorithms.

In order to separate the behavior of the learning algorithms from the effects of the error in the computed equilibrium, we further evaluate the combined algorithms with precise game-theoretic solution. Figure 5a presents experiments against the adversarial attacker. The widest confidence interval for COMB algorithm is $\pm0.39\%$ and for EXP3 the width is $\pm0.29\%$. In this figure we do not visualize COMB4 since it is identical to COMB3 in this case. The COMB2 and COMB3 algorithms get even better than the SSE strategy, because for the attacker it is more difficult to learn the defender strategy if it is not static.

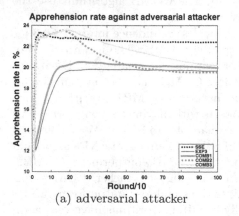

(a) adversarial attacker

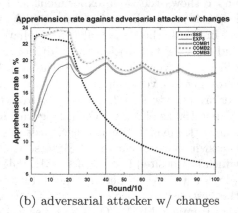

(b) adversarial attacker w/ changes

Fig. 5. COMB algorithms with no error against adversarial attacker

This is partly caused by the extra exploration in the COMB algorithm playing SSE, which can confuse the attacker. The attacker learns quite fast against a static defender strategy vector SSE. One can observe that even though COMB2 and COMB3 outperform the SSE strategy for a short period of time, it then drops substantially in performance due to the attacker eventually learning the strategy. The apprehension rate of the COMB algorithms decreases under the SSE strategy even though they use this SSE strategy, because there is the extra 10 % exploration added to SSE strategy. Nevertheless we can see that COMB algorithms significantly outperform EXP3 algorithm for the first half of the game and then they all converge to a similar performance.

In Fig. 5b we test the COMB algorithms using the precise game-theoretic solution against the adversarial attacker with changes. For COMB algorithms the widest interval is $\pm 0.32\,\%$ and for EXP3 algorithm the width of interval is $\pm 0.26\,\%$. The COMB algorithms can react well to changes in the attacker strategy because of the learning algorithm part. If the defender has a precise SSE strategy he might prefer playing it instead of any other strategy in the case of the adversarial attacker however if there are some changes in the attacker payoff matrix the defender would be better off by playing some more sophisticated algorithm like EXP3 or preferably one of the proposed COMB algorithms, because these can adapt to the changes in the attacker behavior over time.

Now we focus on the convergence of the algorithms in a substantially longer time window. Figure 6 presents the COMB algorithms using game-theoretic solution with error against adversarial attacker for 10000 rounds. This experiment is done 100 times for each setting. The maximal mean width of confidence intervals for COMB algorithms is 0.99 % and the width of confidence interval for EXP3 is 0.92 %. We can see that COMB algorithms and EXP3 algorithm converge to the same performance quite quickly. Playing precise Stackelberg equilibrium strategy has the best performance however the SSE strategy with 0.1

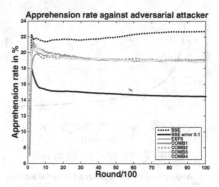

Fig. 6. Convergence of COMB algorithms against adversarial attacker

error gives quite poor results. The precise SSE strategy performance increases during the time, which is caused by the attacker learning more precisely the defender strategy and therefore there are more ties in the attacker strategy which the attacker breaks in favor of the defender.

7.3   Combinatorial Combined Algorithms

In this section we focus on the combinatorial case where the defender uses multiple resources so he can patrol d zones in each round where $d > 1$. We test combinatorial variants of COMB algorithms which use combinatorial variant of

EXP3 as described in Sect. 5.2. For brevity, we continue to refer to the combinatorial variant as EXP3. The experiments are done for a larger game model
with 20 zones ($K = 20$). We compare the strategies in models with 2, 4, 6 and
8 defender resources ($d = 2, 4, 6, 8$). These experiments are run 1000 times for
each setting and each game has 1000 rounds.

In Fig. 7 there are 4 COMB algorithms, SSE with error and SSE without error
strategies. The widest mean confidence interval in all the figures is $\pm 0.36\%$. We
observe in Fig. 7a that EXP3 outperforms the COMB algorithms, which is caused
by poor performance of the SSE with error strategy. The EXP3 algorithm gives
almost 2 times better performance than SSE with error strategy, because there
are too few defenders for too many zones and even a small error in the SSE
strategy causes a low apprehension rate. Due to this fact, the COMB algorithms
have worse performance than EXP3.

When we increase the number of defenders to 4 in Fig. 7b, SSE with error does
better and so do the COMB algorithms. COMB3 outperforms EXP3 algorithm
after the half of the game and COMB4 does even better than COMB3, which

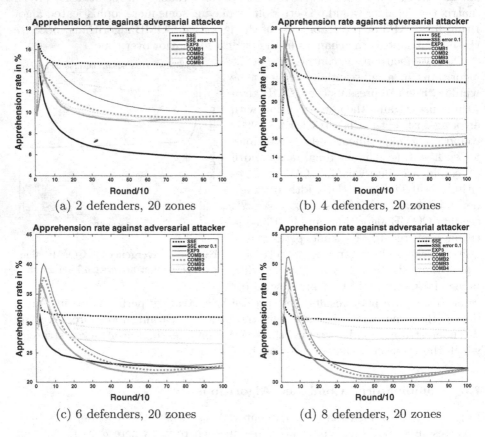

(a) 2 defenders, 20 zones

(b) 4 defenders, 20 zones

(c) 6 defenders, 20 zones

(d) 8 defenders, 20 zones

Fig. 7. COMB algorithms against adversarial attacker, 0.1 error in SSE, varying
number of defenders

comes from the nature of the algorithms. One can observe interesting peaks of the performance curves at the beginning of the game, which are caused by increasing the number of defenders. The attacker needs time to learn effectively against multiple defenders and at the beginning he plays poorly. However by the steepness of the algorithms curves we can see that the attacker learns very quickly after playing very badly at the beginning. These described features are even stronger with increasing number of defenders in Fig. 7c and d.

The SSE strategy with error approaches even more closely the performance of EXP3 algorithm because the more defenders there are, the less the error in the SSE strategy vector matters. The defender still chooses the zones with high probabilities even though there are some errors, because these 0.1 errors cannot decrease the real values too much to not be chosen. For the last figure with 8 defenders the SSE with error strategy even outperforms EXP3 algorithm. Nevertheless COMB3 and especially COMB4 algorithms have very strong performance and approach to SSE strategy performance. COMB1 and COMB2 have obvious drawbacks in the limited use of SSE with error strategy. COMB1 use the game-theoretic strategy only to initialize EXP3 and then cannot make use of it anymore and similarly for COMB2 algorithm, which uses the game-theoretic strategy at the beginning and then permanently switches to EXP3 algorithm.

8   Conclusion

We argue that security games with frequent interactions between defenders and attackers require a different approach than the more classical application domains for security games, such as preventing terrorist attacks. Game theoretic models generally require a lot of assumptions about the opponent's motivations and rationality, which are inherently imprecise, and may even change during the course of a long-term interaction. Therefore, it may be more efficient to learn the optimal strategy from the interaction. However, the standard methods for online learning in adversarial environment do not provide ways to incorporate the possibly imprecise knowledge available about the domain.

We propose learning algorithms that are able to take into consideration imprecise domain knowledge that can even become completely invalid at any point during the game. We further show how to efficiently extend these algorithms to allow for the combinatorial case with multiple defender resources. We show that these algorithms achieve significant improvement on the performance of learning algorithms in the initial stages of the game as well as significant improvement to using only an imprecise game theoretic model in the long run. On the other hand, especially in the combinatorial case, it may be better to use the EXP3 learning algorithm without any knowledge if we expect the performance of imprecise game theoretic solution to be very low. With increasing quality of this solution it is quickly beneficial to use the proposed COMB3 or COMB4 algorithm. Even in the cases where EXP3 outperforms the COMB algorithms, the COMB algorithms still have a very good performance due to using EXP3 as their main component. In a sufficiently long time period all of the

COMB algorithms converge to long-run performance of the EXP3 algorithm, and they retain the theoretical guarantees that make EXP3 attractive in adversarial settings.

Future work could focus on a formal analysis of the proposed combined algorithms. For example, it may be possible to derive and prove improved regret bounds which would provide further guarantees on the algorithm performance. Another direction for future work is bringing defender action preferences into the game model, which would better reflect real-world applications.

Acknowledgements. This research was supported by the Office of Naval Research Global (grant no. N62909-13-1-N256).

References

1. 2012–2016 border patrol strategic plan. U.S. Customs and Border Protection (2012)
2. An, B., Brown, M., Vorobeychik, Y., Tambe, M.: Security games with surveillance cost and optimal timing of attack execution. In: AAMAS, pp. 223–230 (2013)
3. An, B., Kiekintveld, C., Shieh, E., Singh, S., Tambe, M., Vorobeychik, Y.: Security games with limited surveillance. In: AAAI, pp. 1241–1248 (2012)
4. Auer, P., Cesa-Bianchi, N., Fischer, P.: Finite-time analysis of the multi-armed bandit problem. Mach. Learn. **47**, 235–256 (2002)
5. Auer, P., Cesa-Bianchi, N., Freund, Y., Schapire, R.E.: The non-stochastic multi-armed bandit problem. SIAM J. Comput. **32**(1), 48–77 (2001)
6. Balcan, M.-F., Blum, A., Haghtalab, N., Procaccia, A.D.: Commitment without regrets: online learning in stackelberg security games. In: ACM Conference on Economics and Computation (EC-2015), pp. 61–78 (2015)
7. Bard, N., Johanson, M., Burch, N., Bowling, M.: Online implicit agent modelling. In: AAMAS, pp. 255–262 (2013)
8. Bard, N., Nicholas, D., Szepesvari, C., Bowling, M.: Decision-theoretic clustering of strategies. In: AAMAS, pp. 17–25 (2015)
9. Blum, A., Nika, H., Procaccia, F.D.: Lazy defenders are almost optimal against diligent attackers. In: AAAI, pp. 573–579 (2014)
10. Combes, R., Lelarge, M., Proutiere, A., Talebi, M.S.: Stochastic and adversarial combinatorial bandits (2015). arXiv:1502.03475
11. Cowling, P.I., Powley, E.J., Whitehouse, D.: Information set monte carlo tree search. IEEE Trans. Comput. Intell. AI Games **4**, 120–143 (2012)
12. Fudenberg, D., Levine, D.K.: The Theory of Learning in Games. The MIT Press, Cambridge (1998)
13. Garivier, A., Moulines, E.: On upper-confidence bound policies for non-stationary bandit problems. In: ALT, pp. 174–188 (2011)
14. Kiekintveld, C., Jain, M., Tsai, J., Pita, J., Ordonez, F., Tambe, M.: Computing optimal randomized resource allocations for massive security games. In: AAMAS, pp. 689–696 (2009)
15. Kiekintveld, C., Kreinovich, V.: Efficient approximation for security games with interval uncertainty. In: AAAI, pp. 42–45 (2012)
16. Kiekintveld, C., Marecki, J., Tambe, M.: Approximation methods for infinite Bayesian Stackelberg games: modeling distributional payoff uncertainty. In: AAMAS, pp. 1005–1012 (2011)

17. Klima, R., Kiekintveld, C., Lisy, V.: Online learning methods for border patrol resource allocation. In: GAMESEC, pp. 340–349 (2014)
18. Nguyen, T.H., Jiang, A., Tambe, M.: Stop the compartmentalization: unified robust algorithms for handling uncertainties in security games. In: AAMAS, pp. 317–324 (2014)
19. Pita, J., Jain, M., Ordonez, F., Portway, C., Tambe, M., Western, C., Paruchuri, P., Kraus, S.: ARMOR security for los angeles international airport. In: AAAI, pp. 1884–1885 (2008)
20. Pita, J., Jain, M., Ordonez, F., Tambe, M., Kraus, S.: Robust solutions to stackelberg games: addressing bounded rationality and limited observations in human cognition. Artif. Intell. J. **174**(15), 1142–1171 (2010)
21. Pita, J., John, R., Maheswaran, R., Tambe, M., Kraus, S.: A robust approach to addressing human adversaries in security games. In: European Conference on Artificial Intelligence (ECAI), pp. 660–665 (2012)
22. Shieh, E., An, B., Yang, R., Tambe, M., Baldwin, C., Direnzo, J., Meyer, G., Baldwin, C.W., Maule, B.J., Meyer, G.R.: PROTECT : a deployed game theoretic system to protect the ports of the United States. In: AAMAS, pp. 13–20 (2012)
23. Tambe, M.: Security and Game Theory: Algorithms, Deployed Systems, Lessons Learned. Cambridge University Press, Cambridge (2011)
24. Tsai, J., Rathi, S., Kiekintveld, C., Ordóñez, F., Tambe, M.: IRIS - a tools for strategic security allocation in transportation networks. In: AAMAS, pp. 37–44 (2009)
25. Tsai, J., Yin, Z., Kwak, J.-Y., Kempe, D., Kiekintveld, C., Tambe, M.: Urban security: game-theoretic resource allocation in networked physical domains. In: AAAI, pp. 881–886 (2010)
26. Yang, R., Ford, B., Tambe, M., Lemieux, A.: Adaptive resource allocation for wildlife protection against illegal poachers. In: AAMAS, pp. 453–460 (2014)
27. Yang, R., Kiekintvled, C., Ordonez, F., Tambe, M., John, R.: Improving resource allocation strategies against human adversaries in security games: an extended study. Artif. Intell. J. (AIJ) **195**, 440–469 (2013)
28. Yin, Z., Jain, M., Tambe, M., Ordonez, F.: Risk-averse strategies for security games with execution and observational uncertainty. In: AAAI, pp. 758–763 (2011)
29. Yin, Z., Korzhyk, D., Kiekintveld, C., Conitzer, V., Tambe, M.: Stackelberg vs. nash in security games: interchangeability, equivalence, and uniqueness. In: AAMAS, pp. 1139–1146 (2010)
30. Zhang, C., Sinha, A., Tambe, M.: Keeping pace with criminals: designing patrol allocation against adaptive opportunistic criminals. In: AAMAS, pp. 1351–1359 (2015)

Interdependent Security Games Under Behavioral Probability Weighting

Ashish R. Hota and Shreyas Sundaram[⊠]

School of Electrical and Computer Engineering,
Purdue University, West Lafayette, USA
{ahota,sundara2}@purdue.edu

Abstract. We consider a class of *interdependent security games* where the security risk experienced by a player depends on her own investment in security as well as the investments of other players. In contrast to much of the existing work that considers risk neutral players in such games, we investigate the impacts of behavioral probability weighting by players while making security investment decisions. This weighting captures the transformation of objective probabilities into perceived probabilities, which influence the decisions of individuals in uncertain environments. We show that the Nash equilibria that arise after incorporating probability weightings have much richer structural properties and equilibrium risk profiles than in risk neutral environments. We provide comprehensive discussions of these effects on the properties of equilibria and the social optimum when the players have homogeneous weighting parameters, including comparative statics results. We further characterize the existence and uniqueness of pure Nash equilibria in *Total Effort* games with heterogeneous players.

1 Introduction

Interdependent security games are a class of strategic games where multiple selfish players choose personal investments in security, and the security risk faced by a player depends on the investments of other players in the society [18,20]. These games serve as abstract frameworks that capture various forms of risk externalities that users face in networked environments. There is a large body of literature on this class of problems starting from the early works by Varian [30] and Kunreuther and Heal [18]; a comprehensive recent survey can be found in [20].

The risk faced by individuals in these settings is often manifested as the probability of a successful attack, and this probability is a function of the investment by the individual and the externalities due to the investments by other interacting individuals. The system-wide landscape of security investments in this setting will be a function of the decisions that individuals make under this notion of risk. Much of the work in interdependent security games considers players who are risk neutral, or are risk averse in the sense of classical expected utility theory [20]. On the other hand, there is a rich literature in decision theory and behavioral economics which concerns itself with decision making under risk with findings

© Springer International Publishing Switzerland 2015
MHR Khouzani et al. (Eds.): GameSec 2015, LNCS 9406, pp. 150–169, 2015.
DOI: 10.1007/978-3-319-25594-1_9

that suggest consistent and significant deviations in human behavior from the predictions of classical expected utility theory [2,5]. One of the most important deviations is the way individuals perceive the probability of an uncertain outcome (e.g., cyber attack). In particular, empirical studies show that individuals tend to overweight small probabilities and underweight large probabilities. Thus, the objective (i.e., true) probabilities are typically transformed in a highly nonlinear fashion into *perceived* probabilities, which are then used for decision making [8,16].

While there have been some studies highlighting the significance of biases and irrationalities in human decision making in information security domains [1,6], theoretical analyses of relevant deviations from classical notions of rational behavior are scarce in the literature on interdependent security games.[1] Empirical investigations [7,27] are also limited in this context.

In this paper, our goal is to study the effects of behavioral probability weighting of players on their equilibrium strategies in three fundamental interdependent security game models: *Total Effort*, *Weakest Link* and *Best Shot* games. Each of these games captures a certain manifestation of risk (i.e., probability of successful attack) as a function of investment by the players, and is motivated by practical scenarios as discussed in [9]. In the *Total Effort* game, the probability of a successful attack depends on the average of the security investments by the players. As an example, an attacker might want to slow down the transfer of a file in a peer-to-peer file sharing system, while the speed is determined by the sum of efforts of several participating machines. In the *Weakest Link* game, the society is only as secure as the least secure player, while in the *Best Shot* game, the player with the maximum investment must be successfully attacked for the attack on other players to be successful. Weakest link externalities are prevalent in computer security domains; successful breach of one subsystem often increases the vulnerability of other subsystems, such as by giving the attacker increasing access to otherwise restricted parts. Best shot externalities often arise in cyber systems with built in redundancies. In order to disrupt a certain functionality of the target system, the attack must successfully compromise all the entities that are responsible for maintaining that functionality.

These game-theoretic models were first introduced in [30], and were subsequently extended in [9,10] to cases where players can also purchase (static) cyber insurance in addition to investing in security. All three models are instances where the nature of externalities is positive[2] as the investment by a player

---

[1] In a related class of security games known as Stackelberg security games with two players, one attacker and one defender, there have been recent studies [4,13,17,31,32] that incorporate behavioral decision theoretic models, including prospect theory and quantal response equilibrium. However, this class of games is very different from interdependent security games [20], which is the focus of the current work.

[2] Both positive and negative externalities have been studied in the literature. Negative externalities capture settings where more investment by others makes a player more vulnerable, and this is usually the case where the attack is targeted towards individuals who have invested less in security. Most of the literature in security games has focused on positive externalities [20].

(weakly) improves the security of the everyone in the society (by reducing the attack probability). The fact that the players' utility functions in these games are coupled through the shared probability of successful attack motivates our focus on studying the effects of behavioral probability weighting.[3]

We model the nonlinear probability weightings of players using the weighting function due to Prelec [26], whose properties we discuss in the next section. We first characterize the pure Nash equilibria (PNE) in *Total Effort* games; we compare the (structural) properties of the equilibrium under behavioral probability weighting to the equilibria that arise under risk neutral [9] and classical expected utility maximization behavior [15]. We then examine how the intensity of probability weighting affects the probability of successful attack at equilibrium. We carry out a similar analysis for the social welfare maximizing solution in the *Total Effort* game, and Nash equilibria in *Weakest Link* and *Best Shot* games. Subsequently, we prove general existence and uniqueness results in *Total Effort* games when the probability weighting parameters and cost parameters are heterogeneous (player-specific) under certain conditions on the number of players.

2   Probability Weighting

As discussed in the previous section, our focus in this paper will be on understanding the effects of nonlinear weighting of objective probabilities by individuals while making decisions under risk. Such weightings have been comprehensively studied in the behavioral economics and psychology literature [5], and have certain fundamental characteristics, including (i) *possibility effect*: overweighting of probabilities very close to 0, (ii) *certainty effect*: underweighting of probabilities very close to 1, and (iii) *diminishing sensitivity* from the end points 0 and 1. These characteristics are usually captured by an inverse S-shaped weighting function. After the initial advancements in the development of prospect theory and rank dependent utility theory, various parametric forms of weighting functions were proposed, most prominently by Kahneman and Tversky [29], Gonzalez and Wu [8], and Prelec [26]. All of these parametric weighting functions exhibit the qualitative and analytical characteristics (i) to (iii) described above.

In this paper, we consider the one parameter probability weighting function due to Prelec.[4] If the objective (i.e., true) probability of an outcome is x, the weighting function is given by

$$w(x) = \exp(-(-\ln(x))^{\alpha}), \qquad x \in [0, 1], \qquad (1)$$

---

[3] There are also various behavioral characteristics that affect the perceived *values* of gains and losses [12,16]. However, as the values of the gains and losses are not strategy-dependent in the games that we consider here, behavioral value functions would not affect the equilibria that arise.

[4] While we focus on the Prelec weighting function here, many of our results will also hold under a broader class of weighting functions with similar qualitative properties as the Prelec weighting function.

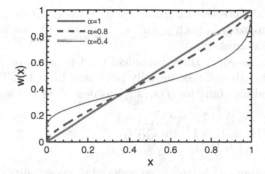

Fig. 1. Shape of the probability weighting function. The quantity x is the objective probability of failure, and $w(x)$ is the corresponding perceived probability of failure.

where exp() is the exponential function. The parameter $\alpha \in (0, 1]$ controls the curvature of the weighting function as illustrated in Fig. 1. For $\alpha = 1$, we have $w(x) = x$, i.e., the weighting function is linear. As α decreases away from 1, $w(x)$ becomes increasingly nonlinear, with an inverse S-shape. For smaller α, the function $w(x)$ has a sharper overweighting of low probabilities and underweighting of high probabilities.

Remark 1. The probability weighting function $w(x)$ in (1) has the following properties.

1. $w(0) = 0$, $w(1) = 1$, and $w(\frac{1}{e}) = \frac{1}{e}$.
2. $w(x)$ is concave for $x \in [0, \frac{1}{e}]$, and convex for $x \in [\frac{1}{e}, 1]$.
3. $w'(x)$ attains its minimum at $x = \frac{1}{e}$. In other words, $w''(x) = 0$ at $x = \frac{1}{e}$. The minimum value of $w'(x)$ is $w'(\frac{1}{e}) = \alpha$.
4. $w'(\epsilon) \to \infty$ as $\epsilon \to 0$, and $w'(1 - \epsilon) \to \infty$ as $\epsilon \to 0$.

3   Interdependent Security Games

As discussed in the introduction, in interdependent security games, a player makes her security investment decision independently, while the probability of successful attack on the player depends on the strategies of other players. We denote the number of players by n, and denote the investment in security by player i as s_i, where $s_i \in [0, 1]$. Following the conventional game theoretic notation, we use s_{-i} to denote the investment profile of all players other than i. The formulation that we consider here has the following characteristics. The objective probability of a successful attack on the system is given by $f(s_i, s_{-i}) \in [0, 1]$, for some function f. Player i incurs a cost-per-unit of security investment of $b_i \in \mathbb{R}_{\geq 0}$, and if the system experiences a successful attack, player i incurs a loss of $L_i \in \mathbb{R}_{>0}$. The expected utility of a player (under the true probabilities of successful attack) is then

$$Eu_i = -L_i f(s_i, s_{-i}) - b_i s_i. \tag{2}$$

In settings that model positive externalities, as is the focus in the current paper, $f(s_i, s_{-i})$ is nonincreasing in both s_i and s_{-i}, and usually assumed to be convex in s_i for analytical tractability.

In this work, we consider three canonical models of interdependent security games with positive externalities, initially presented in [9,10]. The models differ in the attack probability function $f(s_i, s_{-i})$ as described below.

- Total Effort: $f(s_i, s_{-i}) = 1 - \frac{1}{N}(\sum_{i=1}^{n} s_i)$.
- Weakest Link: $f(s_i, s_{-i}) = 1 - \min_{i=1}^{n} s_i$.
- Best Shot: $f(s_i, s_{-i}) = 1 - \max_{i=1}^{n} s_i$.

In [9,10], the authors additionally considered the possibility of players investing in insurance to reduce the magnitude of loss. In order to isolate the effects of nonlinear probability weighting, we do not consider insurance here. To establish a baseline, the following proposition describes the main results from [9] regarding the properties of Nash equilibria in the three security games defined above with homogeneous risk neutral players (without self-insurance). We will compare these results with the equilibria under nonlinear probability weighting in subsequent sections.

Proposition 1. *Consider a set of N risk neutral players with homogeneous cost parameters (b and L).*

1. *Total Effort: There is a unique symmetric PNE except for the special case where $\frac{Nb}{L} = 1$. If $\frac{Nb}{L} < 1$, then each player invests to fully protect herself, i.e., $s_i^* = 1, i \in \{1, 2, \ldots, N\}$. Otherwise if $\frac{Nb}{L} > 1$, $s_i^* = 0$.*
2. *Weakest Link: At any PNE, all players have identical security investment. If $\frac{b}{L} > 1$, then $s_i^* = 0$ for every player i. Otherwise, any investment $s_i^* \in [0, 1]$ can constitute a PNE.*
3. *Best Shot: If $\frac{b}{L} > 1$, then $s_i^* = 0$ for every player i at the PNE. Otherwise, there is no symmetric PNE, and at most one player has a nonzero investment of 1, while all other players free ride without making any security investment.*

We make two preliminary observations regarding the above equilibria in games with risk neutral players. First, in the *Total Effort* game, the best response of a player is independent of the decisions of other players and only depends on her cost parameters, i.e., the interdependence has no impact on her strategy. Secondly, in both *Total Effort* and *Best Shot* games, the PNE causes the system to be either fully secure from attack, or to be fully unprotected. We will show that under behavioral probability weighting, both the best response and the equilibria have much richer structural properties and vary more smoothly with the parameters of the game.

4   Total Effort Game with Probability Weighting: Homogeneous Players

First we characterize the pure Nash equilibria in a *Total Effort* security game when the number of players is sufficiently large and players are homogeneous in their weighting functions and cost parameters. Unlike the risk neutral case described in Proposition 1, the best response in this case will be potentially discontinuous in the strategies of other players. Furthermore, for a sufficiently large number of players, we show that there always exists an interior equilibrium. This equilibrium is not necessarily unique, and in fact, can coexist with an equilibrium where all players invest 1.

With probability weighting, the expected utility of player i under investment $s_i \in [0, 1]$ is given by

$$\mathbb{E}u_i(s_i, s_{-i}) = -Lw\left(1 - \frac{s_i + \bar{s}_{-i}}{N}\right) - bs_i,$$

where $\bar{s}_{-i} = \sum_{j \neq i} s_j$ is the total investment in security by all players other than i. The function w is the Prelec weighting function defined in (1). The marginal utility is given by

$$\frac{\partial \mathbb{E}u_i}{\partial s_i} = \frac{L}{N}w'\left(1 - \frac{s_i + \bar{s}_{-i}}{N}\right) - b. \tag{3}$$

The solutions of $\frac{\partial \mathbb{E}u_i}{\partial s_i} = 0$ satisfy the first order condition of optimality, and are therefore candidate solutions for players' best responses and the PNE. Note that $\left(1 - \frac{s_i + \bar{s}_{-i}}{N}\right)$ is the objective attack probability faced by the players (without probability weighting). For a given strategy profile of other players, player i's strategy can change the objective attack probability in the interval $\mathcal{X}(\bar{s}_{-i}) := \left[1 - \frac{1 + \bar{s}_{-i}}{N}, 1 - \frac{\bar{s}_{-i}}{N}\right]$. In other words, when the number of players is N, each player can directly change the probability of successful attack by at most $\frac{1}{N}$.

Recall from Remark 1 that the minimum value of $w'(x)$ for $x \in [0, 1]$ is α. Therefore, if $\alpha > \frac{Nb}{L}$, from (3) we have $\frac{\partial \mathbb{E}u_i}{\partial s_i} > 0$ for $s_i \in [0, 1]$, and investing 1 is the only best response of player i irrespective of the strategies of the other players. Therefore, the only PNE strategy profile for $\alpha > \frac{Nb}{L}$ is when each player invests 1 in security. Note that in the special case where $\alpha = 1$ (i.e., $w(x) = x$), this reduces to the risk-neutral strategy profile given in Proposition 1.

Now suppose $\alpha < \frac{Nb}{L}$. In this case, the first order condition $w'(x) = \frac{Nb}{L}$ has two distinct interior solutions corresponding to objective attack probabilities $X_1 < \frac{1}{e}$ and $X_2 > \frac{1}{e}$, as illustrated in Fig. 2. It is easy to see that as the number of players N increases, $X_2 - X_1$ increases while $\frac{1}{N}$ decreases.

In the following proposition, we characterize the PNE for sufficiently large N such that $X_2 - X_1 > \frac{1}{N}$. This condition implies that at a given strategy profile of other players, $\mathcal{X}(\bar{s}_{-i})$ does not simultaneously contain both X_1 and X_2. This makes the analysis more tractable, and is a reasonable assumption for networked environments where the number of players is large.

Proposition 2. *Consider a Total Effort security game with homogeneous players with probability weighting parameter $\alpha < \frac{Nb}{L}$. Let N be sufficiently large so that $X_2 > \frac{1}{N} + X_1$, where $X_j, j = 1, 2$ are solutions to the equation $w'(x) = \frac{Nb}{L}$. Then,*

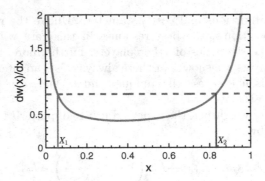

Fig. 2. Interior solutions of $w'(x) = \frac{Nb}{L}$ are denoted by X_1 and X_2. In this example, $\frac{Nb}{L} = 0.8$ and is shown by the horizontal line. The weighting parameter is $\alpha = 0.4$.

1. *any strategy profile with* $\left(1 - \frac{s_i + \bar{s}_{-i}}{N}\right) = X_2$ *is a PNE,*
2. *any strategy profile with* $\left(1 - \frac{s_i + \bar{s}_{-i}}{N}\right) = X_1$ *is not a PNE, and*
3. *there exists a PNE with all players investing 1 if and only if*
   (a) $X_1 \geq \frac{1}{N}$, *or*
   (b) $X_1 < \frac{1}{N}$ *and* $w(\frac{1}{N}) > \frac{b}{L}$.

Proof. **Part 1.** Consider any strategy profile $\mathbf{s}^* = \{s_1^*, \ldots, s_N^*\}$ with $\left(1 - \frac{s_i^* + \bar{s}_{-i}^*}{N}\right) = X_2$. The best response of each player i is obtained by solving the following optimization problem:

$$\max_{x \in [0,1]} \quad -Lw\left(1 - \frac{x + \bar{s}_{-i}^*}{N}\right) - bx. \tag{4}$$

At $x = s_i^*$, the player satisfies the first order condition of optimality $\frac{\partial \mathbb{E}u_i}{\partial s_i} = \frac{L}{N}w'(X_2) - b = 0$. For any $x < s_i^*$, we have $X = 1 - \frac{x + \bar{s}_{-i}}{N} > X_2$ and $\frac{\partial \mathbb{E}u_i}{\partial s_i} = \frac{L}{N}w'(X) - b > 0$ since $X_2 > \frac{1}{e}$. As a result, no $x < s_i^*$ would satisfy the first order necessary condition of optimality. On the other hand, for any $x > s_i^*$, we have $X = 1 - \frac{x + \bar{s}_{-i}}{N} < X_2$. However, from our assumption that $X_2 > \frac{1}{N} + X_1$, we would have $X = 1 - \frac{x + \bar{s}_{-i}}{N} > X_1$. Therefore, $\frac{\partial \mathbb{E}u_i}{\partial s_i} < 0$ for any $x > s_i^*$. As a result, $x = s_i^*$ is the only candidate for optimal investment, and it also satisfies the second order sufficient condition since $w''(X_2) > 0$ as $X_2 > \frac{1}{e}$. This concludes the proof.

Part 2. Consider any strategy profile $\mathbf{s}^* = \{s_1^*, \ldots, s_N^*\}$ with $\left(1 - \frac{s_i^* + \bar{s}_{-i}^*}{N}\right) = X_1$. For any $x > s_i^*$, we have $X = 1 - \frac{x + \bar{s}_{-i}}{N} < X_1$. Since $X_1 < \frac{1}{e}$, $w'(X) > \frac{Nb}{L}$ (see Fig. 2). Thus $\frac{\partial \mathbb{E}u_i}{\partial s_i} > 0$ for any $s_i > s_i^*$. As a result, $s_i = 1$ is also a candidate solution for the utility maximization problem of player i, along with s_i^*. However we show that for Prelec weighting functions, a player would always prefer to invest $s_i = 1$ over $s_i = s_i^*$. To simplify the notation, define $Y_1 = 1 - \frac{1+s_{-i}^*}{N}$. Note that $X_1 - Y_1 = \frac{1}{N}(1 - s_1^*)$. Now we compute

$$
\begin{aligned}
\mathbb{E}u_i(1, \mathbf{s}_{-i}^*) - \mathbb{E}u_i(s_i^*, \mathbf{s}_{-i}^*) &= -Lw(Y_1) - b + Lw(X_1) + bs_i^* \\
&= L(w(X_1) - w(Y_1)) - b(1 - s_i^*) \\
&= L(w(X_1) - w(Y_1)) - bN(X_1 - Y_1) \\
&= L(X_1 - Y_1)\left[\frac{w(X_1) - w(Y_1)}{(X_1 - Y_1)} - \frac{Nb}{L}\right] \\
&\geq L(X_1 - Y_1)\left[w'(X_1) - \frac{Nb}{L}\right] = 0,
\end{aligned}
$$

where the inequality is due to the fact that $w(x)$ is concave for $x \in [0, \frac{1}{e}]$, and $Y_1 < X_1 < \frac{1}{e}$, with equality at $\alpha = 1$. Therefore, between the potential interior solution s_i^* that satisfies the first order condition and the boundary solution $s_i = 1$, the player will always prefer the boundary solution. Since $X_1 > 0$, there always exists a player with $s_i^* < 1$ which would prefer to invest 1, and therefore the strategy profile is not a PNE.

Part 3. Suppose $X_1 \geq \frac{1}{N}$ and all players other than player i are investing 1. Then player i's investment can vary the objective probability of successful attack in the range $[0, \frac{1}{N}]$, and in this region, $\frac{\partial \mathbb{E}u_i}{\partial s_i} > 0$. As a result, investing $s_i = 1$ is the only best response for player 1. Thus $\mathbf{s}^* = \{1, \ldots, 1\}$ is a PNE.

On the other hand, suppose $X_1 < \frac{1}{N}$. Consider a strategy profile where all players other than i are investing 1, i.e., $\mathbf{s}_{-i}^* = \{s_1^* = 1, \ldots, s_{i-1}^* = 1, s_{i+1}^* = 1, \ldots, s_N^* = 1\}$. The following three strategies satisfy the first order necessary condition of optimality of the utility maximization problem (4), and thus are candidates for best responses: (i) $s_i^* = 1$ as $\frac{\partial \mathbb{E}u_i}{\partial s_i}|_{s_i^*=1} > 0$, (ii) $s_i^* = 1 - NX_1$ as $\frac{\partial \mathbb{E}u_i}{\partial s_i}|_{s_i^*=1-NX_1} = \frac{L}{N}w'(X_1) - b = 0$, and (iii) $s_i^* = 0$ as $\frac{\partial \mathbb{E}u_i}{\partial s_i}|_{s_i^*=0} = \frac{L}{N}w'(\frac{1}{N}) - b < 0$ (from Fig. 2 and the fact that $X_1 < \frac{1}{N}$).

From our analysis in Case 2, we know that the player would always prefer to invest 1 over investing $s_i = 1 - NX_1$. Therefore, a necessary and sufficient condition for all players investing 1 to be a PNE is $\mathbb{E}u_i(1, \mathbf{s}_{-i}^*) > \mathbb{E}u_i(0, \mathbf{s}_{-i}^*)$. Since

$$
\mathbb{E}u_i(1, \mathbf{s}_{-i}^*) - \mathbb{E}u_i(0, \mathbf{s}_{-i}^*) = -b + Lw(\frac{1}{N}),
$$

we have the equivalent condition that $\mathbf{s}^* = \{1, \ldots, 1\}$ is a PNE if and only if $w(\frac{1}{N}) > \frac{b}{L}$. Otherwise, player i would achieve greater utility by investing 0, and therefore, all players investing 1 is not a PNE.   □

Discussion: The above proposition completely characterizes the pure Nash equilibria in *Total Effort* security games with homogeneous nonlinear probability weighting for a sufficiently large number of players. It is instructive to compare this set of equilibria with the ones for risk neutral players given in Proposition 1. When $\frac{Nb}{L} > \alpha$, there always exists an interior PNE corresponding to a (true) probability of successful attack equal to X_2. As the number of players N increases, the probability of successful attack X_2 gradually increases to 1 (by the definition of X_2 and Fig. 2). This is in contrast to the situation with risk neutral players ($\alpha = 1$), where the probability of successful attack jumps suddenly from 0 to 1 once the number of players increases beyond a certain point.

Secondly, under behavioral probability weighting, there can *also* be an equilibrium where all players invest 1 (under the conditions described in the third part of the above result). This is a consequence of overweighting of small probabilities; individual players with $\alpha < 1$ do not find it profitable to reduce their investments ("free-ride"), since the resulting small increase of attack probability is perceived to be much larger.

In the next subsection, we discuss how the attack probabilities at PNE are affected by the weighting parameter α, and then analyze the social optimum in the *Total Effort* game.

4.1   Comparative Statics

Consider two *Total Effort* games where the parameters N, b and L are the same across the two games. The first game has homogeneous players with weighting parameter α_1, and the second game has homogeneous players with weighting parameter α_2, where $\alpha_1 < \alpha_2 < \frac{Nb}{L}$. By Proposition 2, both games have an interior PNE, with corresponding true probability of successful attack equal to X_2^1 and X_2^2, respectively. Note that for $i \in \{1, 2\}$, $X_2^i > \frac{1}{e}$, and is the solution to the equation $w_i'(x) = \frac{Nb}{L}$, where $w_i(x)$ be the Prelec function (1) with weighting parameter α_i, for $i \in \{1, 2\}$. As we illustrate in Fig. 3 for $\alpha_1 = 0.4$ and $\alpha_2 = 0.8$, $w_1'(x)$ is initially smaller than $w_2'(x)$ as x starts to increase from $\frac{1}{e}$, until the quantity $x = \bar{X}$ (which depends on the values of α_1 and α_2) at which $w_1'(x) = w_2'(x)$. For $x > \bar{X}$, $w_1'(x) > w_2'(x)$. We first formally prove this observation via the following lemma and proposition.

Lemma 1. *The function*

$$g(x) = (-\ln(x))^{\alpha_2 - \alpha_1} \exp[(-\ln(x))^{\alpha_1} - (-\ln(x))^{\alpha_2}], \quad \alpha_1 < \alpha_2 < 1,$$

is strictly decreasing in $x \in [\frac{1}{e}, 1]$.

Proof. We compute

$$g'(x) = \exp((-\ln(x))^{\alpha_1} - (-\ln(x))^{\alpha_2})$$
$$\times \left[(\alpha_1(-\ln(x))^{\alpha_1 - 1} - \alpha_2(-\ln(x))^{\alpha_2 - 1})\frac{-1}{x}(-\ln(x))^{\alpha_2 - \alpha_1}\right.$$
$$\left. -(\alpha_2 - \alpha_1)(-\ln(x))^{\alpha_2 - \alpha_1 - 1}\frac{1}{x}\right]$$
$$= -\frac{1}{x}\exp((-\ln(x))^{\alpha_1} - (-\ln(x))^{\alpha_2})(-\ln(x))^{\alpha_2 - \alpha_1 - 1}$$
$$\times [(\alpha_1(-\ln(x))^{\alpha_1} - \alpha_1 - \alpha_2(-\ln(x))^{\alpha_2} + \alpha_2].$$

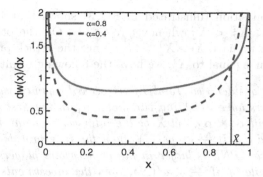

Fig. 3. $w'(x)$ for two different weighting parameters, $\alpha = 0.4$ and $\alpha = 0.8$. At $x = \bar{X}$, $w'_1(x) = w'_2(x)$.

When $x > \frac{1}{e}$, $(-\ln(x)) < 1$ and thus $1 > (-\ln(x))^{\alpha_1} > (-\ln(x))^{\alpha_2}$. As a result, $1 - (-\ln(x))^{\alpha_1} < 1 - (-\ln(x))^{\alpha_2}$. Since $\alpha_1 < \alpha_2$, this implies $\alpha_1(-\ln(x))^{\alpha_1} - \alpha_1 > \alpha_2(-\ln(x))^{\alpha_2} - \alpha_2$. Therefore, $g'(x) < 0$ over $x \in [\frac{1}{e}, 1]$. □

Proposition 3. *Consider two Prelec weighting functions w_1 and w_2 with parameters α_1 and α_2, respectively and let $\alpha_1 < \alpha_2$. Then there exists a unique $\bar{X} > \frac{1}{e}$ such that (i) $w'_1(\bar{X}) = w'_2(\bar{X})$, ii) for $x \in [\frac{1}{e}, \bar{X}]$, $w'_1(x) < w'_2(x)$, and iii) for $x \in [\bar{X}, 1]$, $w'_1(x) > w'_2(x)$.*

Proof. The first derivative of the Prelec weighting function is given by $w'(x) = w(x)\frac{\alpha}{x}(-\ln(x))^{\alpha-1}$. Therefore, if at a given x, $w'_1(x) = w'_2(x)$, we have

$$w_1(x)\alpha_1(-\ln(x))^{\alpha_1} = w_2(x)\alpha_2(-\ln(x))^{\alpha_2}$$

$$\implies \frac{\alpha_1}{\alpha_2} = \frac{w_2(x)(-\ln(x))^{\alpha_2}}{w_1(x)(-\ln(x))^{\alpha_1}}$$

$$\implies \frac{\alpha_1}{\alpha_2} = (-\ln(x))^{\alpha_2-\alpha_1}\exp((-\ln(x))^{\alpha_1} - (-\ln(x))^{\alpha_2}) = g(x).$$

From the definition of $g(x)$, $g(\frac{1}{e}) = 1 > \frac{\alpha_1}{\alpha_2}$. Furthermore, as $x \to 1$, $g'(x) \to -\infty$. As a result, $g(x)$ becomes smaller than $\frac{\alpha_1}{\alpha_2}$ for some $x < 1$. Thus there exists $\bar{X}$ at which $w'_1(x) = w'_2(x)$. The uniqueness of $\bar{X}$ follows from the strict monotonicity of $g(x)$ as proved in Lemma 1.

In order to prove the second and third parts of the lemma, it suffices to show that $w''_1(\bar{X}) > w''_2(\bar{X})$. Therefore, we compute $w''(x)$, which after some calculations yields

$$w''(x) = \frac{w'(x)}{-x\ln(x)}[1 + \ln(x) + \alpha((-\ln(x))^{\alpha} - 1)].$$

From the previous discussion, $\alpha_1(-\ln(x))^{\alpha_1} - \alpha_1 > \alpha_2(-\ln(x))^{\alpha_2} - \alpha_2$, and $w''(x) > 0$ for $x > \frac{1}{e}$. Therefore at $\bar{X}$, $w''_1(\bar{X}) > w''_2(\bar{X})$. This concludes the proof. □

The above proposition (illustrated in Fig. 3) shows that if $\frac{Nb}{L} > w'(\bar{X})$, then we have $\bar{X} < X_2^1 < X_2^2$ whenever $\alpha_1 < \alpha_2$. On the other hand, when $\alpha_2 < \frac{Nb}{L} < w'(\bar{X})$, we have $\bar{X} > X_2^1 > X_2^2$. Since the attack probability at the interior equilibrium is equal to X_2^i, we have the following result.

Proposition 4. *Consider two Total Effort security games with homogeneous players. The players have weighting parameter α_1 in the first game and α_2 in the second game with $\alpha_1 < \alpha_2$. Let $\bar{X}$ be the intersection point defined in Proposition 3 for α_1 and α_2. If $\frac{Nb}{L} > w'(\bar{X})$, then the true probability of successful attack at the interior PNE is larger in the game where players have weighting parameter α_2. Similarly, if $\frac{Nb}{L} < w'(\bar{X})$, then the investments by players with weighting parameter α_1 results in higher probability of successful attack at the interior PNE. If $\frac{Nb}{L} = w'(\bar{X})$, both games have identical attack probability at the interior PNE.*

Discussion: The above result shows that when the attack probability is close to 1, the players view security investments to be highly beneficial in terms of reducing the perceived attack probabilities (due to significant underweighting of probabilities closer to 1). This effect is more pronounced in players with a smaller α (as shown in Fig. 1), and as a result, the attack probability at the PNE is smaller compared to a game where players have higher α. On the other hand, when the quantity $\frac{Nb}{L}$ is smaller, the attack probability at the interior solution is more moderate and the players with smaller α do not find the perceived reduction of attack probability beneficial enough to make a high investment compared to players with larger α. Therefore, the nature of probability weighting plays a key role in shaping the attack probability at the PNE.

4.2   Social Optimum

We define the social welfare of the *Total Effort* game as the sum of the utilities of the individual players, as is commonly defined in the game theory literature [24]. Formally, for a given strategy profile $\mathbf{s}$, the social welfare function is defined as

$$\Psi(\mathbf{s}) = -NLw\left(1 - \frac{\sum_{i=1}^{N} s_i}{N}\right) - b\sum_{i=1}^{N} s_i. \tag{5}$$

Noting that the social welfare function only depends on the aggregate investment, we denote with some abuse of notation

$$\Psi(\bar{s}) = -N[Lw(1 - \bar{s}) + b\bar{s}], \tag{6}$$

where $\bar{s}$ is the average security investment by the players. As a result, the social welfare optimization performed by the social planner is independent of the number of players in the system. In fact, the optimal solution to the problem of the central planner is the same as the optimal investment when there is a single player in the game. In the following result, we discuss how the optimal investment under a central planner depends on the weighting parameter α and the cost parameters b and L. Subsequently, the following result will also be helpful in characterizing the nature of Nash equilibria in *Weakest Link* and *Best Shot* games.

The marginal of the social welfare function (6) is

$$\frac{1}{N}\frac{\partial \Psi}{\partial s} = Lw'(1-s) - b, \tag{7}$$

where we use s as the average security investment instead of $\bar{s}$ for notational convenience. Similar to the discussion prior to Proposition 2, if $\frac{b}{L} < \alpha$, then $\frac{\partial \Psi}{\partial s} > 0$ for $s \in [0,1]$, and as a result, the optimal investment is 1. Therefore we focus on the case where $\frac{b}{L} > \alpha$; in this case, there are two solutions to $w'(x) = \frac{b}{L}$, denoted as $X_1 < \frac{1}{e}$ and $X_2 > \frac{1}{e}$ as before. We will need the following lemma for our analysis.

Lemma 2. *Let z be such that $w'(z) = \frac{w(z)}{z}$. Then (i) z is unique, (ii) $z > \frac{1}{e}$ and (iii) for $x > z$, $w'(x) > \frac{w(x)}{x}$.*

Proof. For the Prelec weighting function,

$$w'(x) = w(x)\frac{\alpha}{x}(-\ln(x))^{\alpha-1}.$$

At any z with $w'(z) = \frac{w(z)}{z}$, we must have $\alpha(-\ln(z))^{\alpha-1} = 1$. Since $\alpha < 1$, we must have $-\ln(z) < 1$ or equivalently $z > \frac{1}{e}$. Furthermore, $(-\ln(x))^{\alpha-1}$ is strictly increasing in x for $\alpha < 1$. As a result, there is a unique $x = z$ at which $\alpha(-\ln(z))^{\alpha-1} = 1$, and for $x > z$, $w'(x) > \frac{w(x)}{x}$. Similarly, $w'(x) < \frac{w(x)}{x}$ for $x \in [\frac{1}{e}, z]$.   □

Proposition 5. *Let z be as defined in Lemma 2.*

1. *If $\frac{b}{L} < w'(z)$, the socially optimal average investment is 1.*
2. *Otherwise, the socially optimal average investment is $1 - X_2$, where $w'(X_2) = \frac{b}{L}$.*

Proof. Since $\frac{b}{L}$ is finite, from (7) we have $\frac{\partial \Psi}{\partial s} > 0$ at $s = 0$ and therefore investing 0 in security is not a utility maximizer. So we have three candidate solutions for utility maximization, $s_1 = 1 - X_1$, $s_2 = 1 - X_2$ or $s_3 = 1$.

From the analysis in Part 2 of Proposition 2 with $\bar{s}^*_{-i} = 0$ and $N = 1$, we have $\mathbb{E}u(1) > \mathbb{E}u(1 - X_1)$. Therefore, between the potential interior solution $s_1 = 1 - X_1$ that satisfies the first order condition and the boundary solution $s_3 = 1$, the player will always prefer the boundary solution.

Now, to compare the utilities at the solutions s_2 and 1, we compute

$$\frac{1}{N}(\Psi(1) - \Psi(s_2)) = Lw(1-s_2) - b(1-s_2)$$

$$= L(1-s_2)\left[\frac{w(1-s_2)}{1-s_2} - \frac{b}{L}\right]$$

$$= L(1-s_2)\left[\frac{w(1-s_2)}{1-s_2} - w'(1-s_2)\right].$$

From our discussion in Lemma 2, if $\frac{b}{L} = w'(1 - s_2) < w'(z)$, the player would prefer to invest 1, and otherwise will prefer to invest s_2.   □

5   Weakest Link and Best Shot Games

The Nash equilibrium strategies in *Weakest Link* and *Best Shot* games have very special properties, as the security level of the entire system is determined by the investment of a single player (the one with the smallest and largest investment, respectively). As a result, the levels of investment at the equilibria often depend on the optimal investment by a single user (i.e., investment in a game with $N = 1$), which we analyzed in the previous subsection. We first characterize the PNE in *Weakest Link* games.

Proposition 6. *Consider a Weakest Link game with homogeneous players having probability weighting parameter $\alpha \in (0, 1]$. Then at any PNE, all players have identical investment. If $\frac{b}{L} > w'(z)$, where z is as defined in Lemma 2, then there is a continuum of pure Nash equilibria with attack probability greater than or equal to X_2. When $\frac{b}{L} < w'(z)$, then there are additional equilibria (including the ones in the previous case) with attack probabilities less than $\frac{1}{e}$.*

Proof. The first part of the statement is easy to see, as no player prefers to invest more than the current minimum level of investment. In fact, if at a given strategy profile, no player can improve her utility by unilaterally deviating to a lower investment level, then the strategy profile is a PNE.

When $\frac{b}{L} > w'(z)$, our result in Proposition 5 states that a single player investing in isolation would prefer to invest $s^* = 1 - X_2$ where $X_2 > \frac{1}{e}$ is the interior solution to the first order condition $w'(x) = \frac{b}{L}$. Now suppose all players have identical security investment $s \leq s^*$, i.e., the objective attack probability $X \geq X_2$. Since for each player $w'(x) > \frac{b}{L}$ for $x > X_2$, no player would unilaterally deviate to make a lower investment. Therefore, any investment less than s^* by all the players would result in a PNE.

When $\frac{b}{L} < w'(z)$, the optimal investment when $N = 1$ is 1 since $\mathbb{E}u(1) > \mathbb{E}u(1 - X_2)$. Therefore, some very low attack probabilities (such as with investment $s = 1 - \epsilon, \epsilon \rightarrow 0$ where $\mathbb{E}u(s) > \mathbb{E}u(1 - X_2)$) can be supported at a PNE, in addition to the set of equilibria with attack probabilities greater than X_2.   □

Discussion: The main differences of the above result compared to the equilibria in *Weakest Link* games without probability weighting (Proposition 1) are twofold. First, for large enough $\frac{b}{L}$, the only possible equilibrium in the risk neutral case is when all players invest 0, while under probability weighting, there is a range of possible equilibrium investments. Secondly, for smaller values of $\frac{b}{L}(< 1)$, any investment level by the players can give rise to a PNE for risk neutral players, while that is no longer the case with behavioral probability weighting. However, at the social optimum of *Weakest Link* games, the investment level chosen by the central planner will coincide with the optimal investment stated in Proposition 5.

Proposition 7. *Consider a Best Shot game with players having identical cost and weighting parameters. Then at any PNE, at most one player can have a nonzero investment with investment level according to Proposition 5, and the rest of the players invest zero.*

Discussion: Recall that the structure of the PNE was similar in homogeneous *Best Shot* games without probability weighting, with at most one player having nonzero investment. However, the nonzero investment level was at one of the boundary points, either 0 or 1, and as a result, the equilibrium was either entirely protected or vulnerable (Proposition 1). With probability weighting, the nonzero equilibrium investment is one of the interior solutions when $\frac{b}{L} > w'(z)$, and the investment level gradually decreases as $\frac{b}{L}$ increases. Finally, note that the social optimum solution coincides with the PNE in *Best Shot* games.

6   Total Effort Game with Heterogeneous Players

In this section, we consider *Total Effort* games with player-specific weighting parameters α_i and cost parameters $\frac{b_i}{L_i}$. We first prove the existence of a PNE in games with a sufficiently large number of players, and subsequently give a constructive proof of uniqueness of PNE.

In the rest of the analysis, we denote the weighting function of player i as $w_i(x)$, and denote the solutions of $w_i'(x) = \frac{N b_i}{L_i}$, if any, as $X_1^i < \frac{1}{e}$ and $X_2^i > \frac{1}{e}$, respectively. Now we are ready to state our assumptions on the number of players, which are sufficient conditions for our results to hold.

Assumption 1. *Let the number of players N be large enough such that for every player i,*

1. $\alpha_i < \frac{N b_i}{L_i}$,
2. $X_2^i - X_1^i > \frac{1}{N}$,
3. $X_1^i < \frac{1}{N}$ and
4. $w_i(\frac{1}{N}) < \frac{b_i}{L_i}$.

The above conditions are guaranteed to be simultaneously satisfied for a sufficiently large number of players due to the properties of the weighting functions. Recall from our discussion in Proposition 2 that for homogeneous players, the last two of the above assumptions are necessary and sufficient to ensure that all players investing 1 is not an equilibrium.

Proposition 8. *Consider a Total Effort game where player i has player-specific weighting parameter α_i and cost ratio $\frac{b_i}{L_i}$. Let the number of players satisfy Assumption 1. Then there exists a PNE of the game.*

Proof. We show that the best response of a player is unique, continuous and is an affine decreasing function of the total security investment by other players. The result will then follow from Brouwer's fixed point theorem. Our analysis of the best response of a player holds for every player i, and therefore we drop the subscript in the rest of proof.

Let $y \in [0, N-1]$ be the total security investment by all other players. For a given y, the investment of the player can only change the true probability of successful attack within the interval $\mathcal{X}(y) := \left[1 - \frac{1+y}{N}, 1 - \frac{y}{N}\right]$, the length of which is $\frac{1}{N}$. Under Assumption 1, this interval might fall into one of four different cases. As y increases from 0, we gradually go through each of the cases, starting from either Case 1 or Case 2.

Case 1: $X_2 < 1 - \frac{1+y}{N}$

In this case, the interval $\mathcal{X}(y)$ lies to the right of X_2. Therefore, for any attack probability $x \in \mathcal{X}(y)$, $\frac{L}{N}w'(x) > b$ (from Fig. 2). Thus, $\frac{\partial \mathbb{E}u}{\partial s}\big|_{s=x} > 0$ and consequently, $b(y) = 1$ this case.

Case 2: $1 - \frac{1+y}{N} \le X_2 \le 1 - \frac{y}{N}$

In this case, $X_2 \in \mathcal{X}(y)$, and therefore, the player has a feasible investment strategy $s^* = N(1 - X_2) - y$ at which the first order condition is satisfied with equality. By identical arguments as in Part 1 of Proposition 2, the player is only going to invest at the interior solution s^*. The second requirement of Assumption 1 ensures that $X_1 \notin \mathcal{X}(y)$, and as result, the utility function remains concave, and therefore the best response is unique for any given y.

Since the optimal solution s^* must have the property that $1 - \frac{s^* + y}{N} = X_2$, it is continuous and linearly decreasing in y, with boundary values at 1 and 0 respectively for $y = N(1 - X_2) - 1$ and $y = N(1 - X_2)$.

Case 3: $X_1 < 1 - \frac{1+y}{N}$ and $X_2 > 1 - \frac{y}{N}$

In this case, the interval $\mathcal{X}(y)$ lies in the region between X_1 and X_2. Therefore, for any objective failure probability $x \in \mathcal{X}(y)$, $\frac{\partial \mathbb{E}u}{\partial s} < 0$. As a result, $b(y) = 0$.

Case 4: $1 - \frac{1+y}{N} \le X_1 \le 1 - \frac{y}{N}$

In this case, there are three candidate solutions for utility maximization, $s = 1$, $s = N(1 - X_1) - y$ and $s = 0$, analogous to the candidate solutions in Part 3 of Proposition 2. We have $X = 1 - \frac{s+y}{N}$ as the objective failure probability resulting from the strategies of the players. From an identical analysis as in Part 2 of Proposition 2 with $\bar{s}_{-i} = y$, we conclude that the player would always prefer to invest 1 over investing $s^* = N(1 - X_1) - y$. This leads to the possibility that the best response might have a discontinuous jump from 0 to 1 at some value of y in this region. However, we show that under the third and fourth conditions of Assumption 1, the player would always prefer to invest 0 over investing 1.

We compute

$$\mathbb{E}u(1, y) - \mathbb{E}u(0, y) = L\left[w(1 - \frac{y}{N}) - w(1 - \frac{1+y}{N})\right] - b$$

$$= L\left[w(\lambda + \frac{1}{N}) - w(\lambda)\right] - b$$

$$\le L\left[w(\frac{1}{N}) - w(0)\right] - b,$$

where $\lambda = 1 - \frac{1+y}{N}$. The last inequality follows because the function $h(\lambda) \triangleq w(\lambda + \frac{1}{N}) - w(\lambda)$ is a strictly decreasing function of λ for $\lambda \in [0, X_1]$. Indeed, $h'(\lambda) = w'(\lambda + \frac{1}{N}) - w'(\lambda) < 0$, as $w'(\lambda) \geq w'(X_1) = \frac{Nb}{L}$, and $w'(\lambda + \frac{1}{N}) < \frac{Nb}{L}$, as $X_1 < \lambda + \frac{1}{N} < X_2$. Therefore, if $w(\frac{1}{N}) < \frac{b}{L}$, the player would always prefer to invest 0 over investing 1, regardless of the value of y. Furthermore, since $X_1 < \frac{1}{N}$, and $1 - \frac{y}{N} = \frac{1}{N}$ at $y = N - 1$, the best response remains at 0 in this region of y.

Combining the analysis in all four cases together, the best response of any player is unique and continuous in the strategies of the other players, regardless of the value of $\alpha_i \in (0, 1]$ and $\frac{b_i}{L_i}$. In addition, the strategy space of each player is $[0, 1]$, which is compact and convex. Therefore, a PNE always exists by Brouwer's fixed point theorem.   □

In the following result, we establish the uniqueness of the PNE.

Proposition 9. *Consider a Total Effort game where each player i has player-specific weighting parameter α_i and cost ratio $\frac{b_i}{L_i}$. Let the number of players N satisfy Assumption 1. Then all PNE have the same objective probability of successful attack.*

Proof. Without loss of generality, let players be ordered such that $X_2^1 \leq X_2^2 \leq X_2^3 \leq \ldots \leq X_2^N$, where X_2^i is the largest solution to $w_i'(x) = \frac{Nb_i}{L_i}$; note that such a solution is guaranteed by the first requirement of Assumption 1. We present a numerical illustration in Fig. 4. Note that this ordering does not necessarily mean that the corresponding α_i's or cost ratios form a monotonic sequence. Under Assumption 1, no objective attack probability $X < X_2^1$ would be a PNE, since there would always exist a player with positive investment who would prefer to reduce her investment.

Now suppose there are two PNEs with different corresponding probabilities of successful attack. Consider the strategy profile with the smaller attack probability, denoted X^*. Note that we ruled out the possibility of $X^* < X_2^1$ above. There are two exhaustive cases: either $X_2^l < X^* < X_2^{l+1}$ for some player l, or $X_2^l = X^*$ for some player l.

Let $X_2^l < X^* < X_2^{l+1}$ for some player l. By the definition of the quantities X_2^i, we have $w_i'(X^*) < \frac{Nb_i}{L_i}$ for $i \in \{l+1, \ldots, N\}$, and therefore, $s_i^* = 0$ for $i \in \{l+1, \ldots, N\}$. Similarly, $w_i'(X^*) > \frac{Nb_i}{L_i}$ and $s_i^* = 1$ for $i \in \{1, \ldots, l\}$. In this case, $X^* = 1 - \frac{l}{N}$. Now at the second PNE with objective attack probability $Y^* > X^*$, the players in $\{1, \ldots, l\}$ would continue to invest 1, with the possibility of more players investing nonzero amounts if $Y^* \geq X_2^{l+1}$. But then the objective attack probability $X = 1 - \frac{\sum_i s_i}{N}$ would decrease from X^*, contradicting the assumption that $Y^* > X^*$.

The proof of the case where $X^* = X_2^l$ for some player l follows identical arguments, and therefore we omit its presentation.   □

7   Discussion and Conclusion

In this paper, we studied a class of interdependent security games where the players exhibit certain behavioral attributes vis-a-vis their perception of attack

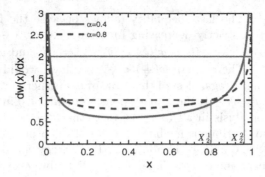

Fig. 4. First order conditions for heterogeneous players. The horizontal lines represent the quantities $\frac{Nb_i}{L_i}$, while the curved lines represent $w_i'(x)$, which depends on the player specific weighting parameter α_i. The quantity X_2^i represents the largest solution to the equation $w_i'(x) = \frac{Nb_i}{L_i}$.

probabilities while making security investment decisions. In particular, we considered the parametric form of probability weighting function proposed by Prelec [26], which is commonly used in behavioral decision theory settings, including prospect theory. We analyzed the properties of pure Nash equilibria (PNE) in three canonical interdependent security game models, (i) *Total Effort*, (ii) *Weakest Link*, and (iii) *Best Shot* games.

We first considered the *Total Effort* game with players having homogeneous weighting functions and cost parameters, and characterized the PNE strategies under a sufficiently large number of players. The equilibria with nonlinear probability weightings have much richer structural properties than the corresponding equilibria for risk-neutral players. There are only two types of equilibria with risk neutral players; one where the probability of successful attack is 1 (completely undefended), and the other where the probability is 0 (completely defended) with an abrupt transition to the latter as the number of players increase. However under behavioral probability weighting, there exist interior equilibria where the attack probability lies between 0 and 1. Furthermore, the equilibrium attack probability gradually increases to 1 with respect to certain cost parameters and the number of players. In addition to the interior equilibrium, there might coexist equilibria where players invest to fully secure themselves. In these equilibria, overweighting of low probabilities disincentivizes individuals from reducing their investments, since the perceived increase in attack probability due to reduced investment is much larger.

We also obtained interesting comparative statics results on the effect of the weighting parameter on the magnitude of the attack probability at equilibrium. If the probability of successful attack is sufficiently high, then players whose weighting functions are closer to linear prefer to invest relatively less in security, while players who exhibit a large underweighting of probabilities closer to 1 (certainty effect) prefer to invest more. This is due to the fact that the perceived reduction in attack probability is larger for the latter players. However, if the attack probability is only moderately high, we observe the opposite behavior; the attack probability at the equilibrium with highly nonlinear weighting functions is larger compared to the attack probability at equilibrium with players who more accurately perceive the true probabilities.

We subsequently analyzed the social welfare maximizing investment profiles in *Total Effort* games, which also had implications for the equilibria in *Weakest Link* and *Best Shot* games. In *Weakest Link* games, there often arise a multitude of equilibria with a continuum of attack probabilities, while in *Best Shot* games, at most one player makes a nonzero investment at any PNE. The investment levels at the equilibria in both these games have a more smooth variation in the game parameters compared to the investments by risk neutral players.

Finally, we analyzed *Total Effort* games where players have heterogeneous cost and weighting parameters, and established the existence of PNE and uniqueness of the corresponding attack probability when the number of players is sufficiently large. We leave a more comprehensive discussion on the effects of heterogeneity in weighting parameters for future work, in addition to several other future directions that we discuss below.

Future Work: There are several directions in which this line of investigation can be extended.

Cyber Insurance: The dichotomy between investing in security (to potentially reduce likelihood of attack) and purchasing cyber insurance (to decrease the magnitude of loss) has received considerable attention among information security researchers [3,23,25]. However, in practice, the market for cyber insurance has seen limited growth despite growing awareness in the industry about various security risks. Further analysis of behavioral risk preferences (such as probability weighting, loss aversion and reference dependence) of decision makers in the context of cyber insurance could potentially uncover important phenomena which is not easily captured in models that only consider the classical expected utility maximization framework. The work in [15] is a step in this direction, where the authors investigate the strategies of risk averse players (with concave utility functions) in *Weakest Link* security games in the presence of market insurance.

Network Structure: In this paper, we have only considered extreme forms of network effects between players, as only the average, the highest, or the lowest investment levels decide the overall failure probabilities. Analyzing the effects of probability weighting and other forms of deviations from classical expected utility maximization behavior in models that consider richer networked environments [19,21,22,28] is a challenging future direction.

Inefficiency of Equilibria: Selfish behavior by users often leads to reduced welfare and increased attack probability at equilibrium in interdependent security games with positive externalities. While there is prior work in the literature on price of anarchy [14] and price of uncertainty [11] of the current class of security games, investigating the effects of (behavioral) risk preferences of users, including probability weighting, on the inefficiency of equilibria remains an important avenue for future research.

Experimental Investigations: Finally, the results obtained in this paper compliments and further motivates experimental/empirical investigations of human decision making in the context of information security and privacy.

References

1. Baddeley, M.: Information security: lessons from behavioural economics. In: Security and Human Behavior (2011)
2. Barberis, N.C.: Thirty years of prospect theory in economics: a review and assessment. J. Econ. Perspect. **27**(1), 173–196 (2013)
3. Böhme, R., Schwartz, G.: Modeling cyber-insurance: towards a unifying framework. In: Workshop on the Economics of Information Security (WEIS) (2010)
4. Brown, M., Haskell, W.B., Tambe, M.: Addressing scalability and robustness in security games with multiple boundedly rational adversaries. In: Saad, W., Poovendran, R. (eds.) GameSec 2014. LNCS, vol. 8840, pp. 23–42. Springer, Heidelberg (2014)
5. Camerer, C.F., Loewenstein, G., Rabin, M.: Advances in Behavioral Economics. Princeton University Press, Princeton (2011)
6. Christin, N.: Network security games: combining game theory, behavioral economics, and network measurements. In: Katz, J., Baras, J.S., Altman, E. (eds.) GameSec 2011. LNCS, vol. 7037, pp. 4–6. Springer, Heidelberg (2011)
7. Christin, N., Egelman, S., Vidas, T., Grossklags, J.: It's all about the benjamins: an empirical study on incentivizing users to ignore security advice. In: Danezis, G. (ed.) FC 2011. LNCS, vol. 7035, pp. 16–30. Springer, Heidelberg (2012)
8. Gonzalez, R., Wu, G.: On the shape of the probability weighting function. Cogn. Psychol. **38**(1), 129–166 (1999)
9. Grossklags, J., Christin, N., Chuang, J.: Secure or insure?: a game-theoretic analysis of information security games. In: Proceedings of the 17th International Conference on World Wide Web, pp. 209–218. ACM (2008)
10. Grossklags, J., Christin, N., Chuang, J.: Security and insurance management in networks with heterogeneous agents. In: Proceedings of the 9th ACM Conference on Electronic Commerce, pp. 160–169. ACM (2008)
11. Grossklags, J., Johnson, B., Christin, N.: The price of uncertainty in security games. In: Moore, T., Pym, D., Ioannidis, C. (eds.) Economics of Information Security and Privacy, pp. 9–32. Springer, Heidelberg (2010)
12. Hota, A.R., Garg, S., Sundaram, S.: Fragility of the commons under prospect-theoretic risk attitudes. (2014, arXiv preprint). arXiv:1408.5951
13. Jiang, A.X., Nguyen, T.H., Tambe, M., Procaccia, A.D.: Monotonic maximin: a robust stackelberg solution against boundedly rational followers. In: Das, S.K., Nita-Rotaru, C., Kantarcioglu, M. (eds.) GameSec 2013. LNCS, vol. 8252, pp. 119–139. Springer, Heidelberg (2013)
14. Jiang, L., Anantharam, V., Walrand, J.: How bad are selfish investments in network security? IEEE/ACM Trans. Netw. **19**(2), 549–560 (2011)
15. Johnson, B., Böhme, R., Grossklags, J.: Security games with market insurance. In: Altman, E., Baras, J.S., Katz, J. (eds.) GameSec 2011. LNCS, vol. 7037, pp. 117–130. Springer, Heidelberg (2011)

16. Kahneman, D., Tversky, A.: Prospect theory: an analysis of decision under risk. Econom. J. Econom. Soc. **47**(2), 263–291 (1979)
17. Kar, D., Fang, F., Delle Fave, F., Sintov, N., Tambe, M.: A game of thrones: when human behavior models compete in repeated Stackelberg security games. In: Proceedings of the 2015 International Conference on Autonomous Agents and Multiagent Systems, pp. 1381–1390 (2015)
18. Kunreuther, H., Heal, G.: Interdependent security. J. Risk Uncertain. **26**(2–3), 231–249 (2003)
19. La, R.J.: Interdependent security with strategic agents and cascades of infection. IEEE Trans. Netw. (2015, To appear)
20. Laszka, A., Felegyhazi, M., Buttyan, L.: A survey of interdependent information security games. ACM Comput. Surveys (CSUR) **47**(2), 23:1–23:38 (2014)
21. Lelarge, M., Bolot, J.: A local mean field analysis of security investments in networks. In: Proceedings of the 3rd International Workshop on Economics of Networked Systems, pp. 25–30. ACM (2008)
22. Lelarge, M., Bolot, J.: Network externalities and the deployment of security features and protocols in the Internet. ACM SIGMETRICS Perform. Eval. Rev. **36**(1), 37–48 (2008)
23. Naghizadeh, P., Liu, M.: Voluntary participation in cyber-insurance markets. In: Workshop on the Economics of Information Security (WEIS) (2014)
24. Nisan, N., Roughgarden, T., Tardos, E., Vazirani, V.V.: Algorithmic Game Theory. Cambridge University Press, Cambridge (2007)
25. Pal, R., Golubchik, L., Psounis, K., Hui, P.: Will cyber-insurance improve network security? A market analysis. In: 2014 Proceedings IEEE INFOCOM, pp. 235–243. IEEE (2014)
26. Prelec, D.: The probability weighting function. Econometrica **66**(3), 497–527 (1998)
27. Rosoff, H., Cui, J., John, R.S.: Heuristics and biases in cyber security dilemmas. Environ. Syst. Decis. **33**(4), 517–529 (2013)
28. Schwartz, G.A., Sastry, S.S.: Cyber-insurance framework for large scale interdependent networks. In: Proceedings of the 3rd International Conference on High Confidence Networked Systems, pp. 145–154. ACM (2014)
29. Tversky, A., Kahneman, D.: Advances in prospect theory: cumulative representation of uncertainty. J. Risk Uncertain. **5**(4), 297–323 (1992)
30. Varian, H.: System reliability and free riding. In: Camp, L.J., Lewis, S. (eds.) Economics of Information Security. AIS, pp. 1–15. Springer, Heidelberg (2004)
31. Yang, R., Kiekintveld, C., Ordóñez, F., Tambe, M., John, R.: Improving resource allocation strategies against human adversaries in security games: an extended study. Artif. Intell. **195**, 440–469 (2013)
32. Zhuang, J.: Modeling attacker-defender games with risk preference. Current Research Project Synopses. Paper 69 (2014). http://research.create.usc.edu/current_synopses/69

Making the Most of Our Regrets: Regret-Based Solutions to Handle Payoff Uncertainty and Elicitation in Green Security Games

Thanh H. Nguyen[1]([✉]), Francesco M. Delle Fave[1], Debarun Kar[1],
Aravind S. Lakshminarayanan[2], Amulya Yadav[1], Milind Tambe[1],
Noa Agmon[3], Andrew J. Plumptre[4], Margaret Driciru[5],
Fred Wanyama[5], and Aggrey Rwetsiba[5]

[1] University of Southern California, Los Angeles, USA
{thanhhng,dellefav,dkar,amulyaya,tambe}@usc.edu
[2] Indian Institute of Technology Madras, Chennai, India
aravindsrinivas@gmail.com
[3] Bar-Ilan University, Ramat Gan, Israel
agmon@cs.biu.ac.il
[4] Wildlife Conservation Society, New York, USA
aplumptre@wcs.org
[5] Uganda Wildlife Authority, Kampala, Uganda
{margaret.driciru,fred.wanyama,aggrey.rwetsiba}@ugandawildlife.org

Abstract. Recent research on Green Security Games (GSG), i.e., security games for the protection of wildlife, forest and fisheries, relies on the promise of an abundance of available data in these domains to learn adversary behavioral models and determine game payoffs. This research suggests that adversary behavior models (capturing bounded rationality) can be learned from real-world data on where adversaries have attacked, and that game payoffs can be determined precisely from data on animal densities. However, previous work has, as yet, failed to demonstrate the usefulness of these behavioral models in capturing adversary behaviors based on real-world data in GSGs. Previous work has also been unable to address situations where available data is insufficient to accurately estimate behavioral models or to obtain the required precision in the payoff values.

In addressing these limitations, as our first contribution, this paper, for the first time, provides validation of the aforementioned adversary behavioral models based on real-world data from a wildlife park in Uganda. Our second contribution addresses situations where real-world data is not precise enough to determine exact payoffs in GSG, by providing the first algorithm to handle payoff uncertainty in the presence of adversary behavioral models. This algorithm is based on the notion of minimax regret. Furthermore, in scenarios where the data is not even sufficient to learn adversary behaviors, our third contribution is to provide a novel algorithm to address payoff uncertainty assuming a perfectly rational attacker (instead of relying on a behavioral model); this algorithm allows for a significant scaleup for large security games. Finally, to

MHR Khouzani et al. (Eds.): GameSec 2015, LNCS 9406, pp. 170–191, 2015.
DOI: 10.1007/978-3-319-25594-1_10

reduce the problems due to paucity of data, given mobile sensors such as Unmanned Aerial Vehicles (UAV), we introduce new payoff elicitation strategies to strategically reduce uncertainty.

1 Introduction

Following the successful deployments of Stackelberg Security Games (SSG) for infrastructure protection [1,13,24], recent research on security games has focused on Green Security Games (GSG) [4,7,21,27]. Generally, this research attempts to optimally allocate limited security resources in a vast geographical area against environmental crime, e.g., improving the effectiveness of protection of wildlife or fisheries [4,27].

Research in GSGs has differentiated itself from work in SSGs (which often focused on counter-terrorism), not only in terms of the domains of application but also in terms of the amounts of data available. In particular, prior research on SSGs could not claim the presence of large amounts of adversary data [24]. In contrast, GSGs are founded on the promise of an abundance of adversary data (about where the adversaries attacked in the past) that can be used to accurately learn adversary behavior models which capture their bounded rationality [4,7,27]. Furthermore, GSG research assumes that available domain data such as animal/fish density is sufficient to help determine payoff values precisely. However, there remain four key shortcomings in GSGs related to these assumptions about data. First, despite proposing different adversary behavioral models (e.g., Quantal Response [28]), GSG research has yet to evaluate these models on any real-world data. Second, the amount of real-world data available is not always present in abundance, introducing different types of uncertainties in GSGs. In particular, in some GSG domains, there is a significant need to handle uncertainty in both the defender and the adversary's payoffs since information on key domain features, e.g., animal density, terrain, etc. that contribute to the payoffs is not precisely known. Third, in some GSG domains, we may even lack sufficient attack data to learn an adversary behavior model, and simultaneously must handle the aforementioned payoff uncertainty. Finally, defenders have access to mobile sensors such as UAVs to elicit information over multiple targets at once to reduce payoff uncertainty, yet previous work has not provided efficient techniques to exploit such sensors for payoff elicitation [17].

In this paper, we address these challenges by proposing four key contributions. As our first contribution, we provide the first results demonstrating the usefulness of behavioral models in SSGs using real-world data from a wildlife park. To address the second limitation of uncertainty over payoff values, our second contribution is ARROW (i.e., **A**lgorithm for **R**educing **R**egret to **O**ppose **W**ildlife crime), a novel security game algorithm that can solve the *behavioral minimax regret problem*. MiniMax Regret (MMR) is a robust approach for handling uncertainty that finds the solution which minimizes the maximum regret (i.e., solution quality loss) with respect to a given uncertainty set [8]. A key advantage of using MMR is that it produces less conservative solutions than the standard maximin approach [17]. ARROW is the first algorithm to compute MMR in the presence of an adversary behavioral model; it is also the first to handle payoff uncertainty in both players' payoffs in SSGs. However, jointly handling of adversary bounded rationality and payoff uncertainty creates the challenge of solving a non-convex optimization problem; ARROW provides an efficient solution to this problem. (Note that we primarily assume a zero-sum game as done in some prior GSG research; however as discussed our key techniques generalize to non-zero sum games as well.)

Our third contribution addresses situations where we do not even have data to learn a behavior model. Specifically, we propose ARROW-Perfect, a novel MMR-based algorithm to handle uncertainty in *both* players' payoffs, assuming a perfectly rational adversary without any requirement of data for learning. ARROW-Perfect exploits the adversary's perfect rationality as well as extreme points of payoff uncertainty sets to gain significant additional efficiency over ARROW.

Another significant advantage of MMR is that it is very useful in guiding the preference elicitation process for learning information about the payoffs [3]. We exploit this advantage by presenting two new elicitation heuristics which select *multiple* targets at a time for reducing payoff uncertainty, leveraging the multi-target-elicitation capability of sensors (e.g., UAVs) available in green security domains. Lastly, we conduct extensive experiments, including evaluations of ARROW based on data from a wildlife park.

2   Background and Related Work

Stackelberg Security Games: In SSGs, the defender attempts to protect a set of T targets from an attack by an adversary by optimally allocating a set of R resources $(R < T)$ [24]. The key assumption here is that the defender commits to a (*mixed*) strategy first and the adversary can observe that strategy and then attacks a target. Denote by $\mathbf{x} = \{x_t\}$ the defender's strategy where x_t is the *coverage probability* at target t, the set of feasible strategies is $\mathbf{X} = \{\mathbf{x} : 0 \leq x_t \leq 1, \sum_t x_t \leq R\}$.[1] If the adversary attacks t when the defender is not protecting it, the adversary receives a reward R_t^a, otherwise, the adversary gets a penalty P_t^a. Conversely, the defender receives a penalty P_t^d in the former case and a reward R_t^d in the latter case. Let $(\mathbf{R^a}, \mathbf{P^a})$ and $(\mathbf{R^d}, \mathbf{P^d})$ be the payoff vectors. The players' expected utilities at t is computed as:

$$U_t^a(\mathbf{x}, \mathbf{R^a}, \mathbf{P^a}) = x_t P_t^a + (1 - x_t) R_t^a \tag{1}$$

$$U_t^d(\mathbf{x}, \mathbf{R^d}, \mathbf{P^d}) = x_t R_t^d + (1 - x_t) P_t^d \tag{2}$$

Boundedly Rational Attacker: In SSGs, attacker bounded rationality is often modeled via behavior models such as Quantal Response (QR) [14,15]. QR predicts the adversary's probability of attacking t, denoted by $q_t(\mathbf{x}, \mathbf{R^a}, \mathbf{P^a})$ (as

---

[1] The true mixed strategy would be a probability assignment to each pure strategy, where a pure strategy is an assignment of R resources to T targets. However, that is equivalent to the set $\mathbf{X}$ described here, which is a more compact representation [12].

shown in Eq. 3 where the parameter λ governs the adversary's rationality). Intuitively, the higher the expected utility at a target is, the more likely that the adversary will attack that target.

$$q_t(\mathbf{x}, \mathbf{R^a}, \mathbf{P^a}) = \frac{e^{\lambda U_t^a(\mathbf{x}, \mathbf{R^a}, \mathbf{P^a})}}{\sum_{t'} e^{\lambda U_{t'}^a(\mathbf{x}, \mathbf{R^a}, \mathbf{P^a})}} \tag{3}$$

The recent SUQR model (Subjective Utility Quantal Response) is shown to provide the best performance among behavior models in security games [18]. SUQR builds on the QR model by integrating the following subjective utility function into QR instead of the expected utility:

$$\hat{U}_t^a(\mathbf{x}, \mathbf{R^a}, \mathbf{P^a}) = w_1 x_t + w_2 R_t^a + w_3 P_t^a \tag{4}$$

where (w_1, w_2, w_3) are parameters indicating the importance of the three target features for the adversary. The adversary's probability of attacking t is then predicted as:

$$\hat{q}_t(\mathbf{x}, \mathbf{R^a}, \mathbf{P^a}) = \frac{e^{\hat{U}_t^a(\mathbf{x}, \mathbf{R^a}, \mathbf{P^a})}}{\sum_{t'} e^{\hat{U}_t^a(\mathbf{x}, \mathbf{R^a}, \mathbf{P^a})}} \tag{5}$$

In fact, SUQR is motivated by the lens model which suggested that evaluation of adversaries over targets is based on a linear combination of multiple observable features [5]. One key advantage of these behavioral models is that they can be used to predict attack frequency for multiple attacks by the adversary, wherein the attacking probability is a normalization of attack frequency.

Payoff Uncertainty: One key approach to modeling payoff uncertainty is to express the adversary's payoffs as lying within specific intervals [10]: for each target t, we have $R_t^a \in [R_{min}^a(t), R_{max}^a(t)]$ and $P_t^a \in [P_{min}^a(t), P_{max}^a(t)]$. Let $\mathbf{I}$ denote the set of payoff intervals at all targets. An MMR-based solution was introduced in previous work to address payoff uncertainty in SSGs; yet it had two weaknesses: (i) this MMR-based solution is unable to handle uncertainty in both players' payoffs since it assumes that the defender's payoffs are exactly known; and (ii) it has failed to address payoff uncertainty in the context of adversary behavioral models [17].

Green Security Games: This paper focuses on wildlife protection — many species such as rhinos and tigers are in danger of extinction from poaching [16,22]. To protect wildlife, game-theoretic approaches have been advocated to generate ranger patrols [27] wherein the forest area is divided into a grid where each cell is a target. These ranger patrols are designed to counter poachers (whose behaviors are modeled using SUQR) that attempt to capture animals by setting snares. A similar system has also been developed for protecting fisheries [4]. Unfortunately, this previous work in wildlife protection [27] has four weaknesses as discussed in Sect. 1.

3   Behavioral Modeling Validation

Our first contribution addresses the first limitation of previous work mentioned in Sect. 1: understanding the extent to which existing behavior models capture real-world behavior data from green security domains. We used a real-world patrol and poaching dataset from Uganda Wildlife Authority supported by Wildlife Conservation Society. This dataset was collected from 1-year patrols in the Queen Elizabeth national park.[2]

3.1   Dataset Description

Our dataset had different types of observations (poacher sighting, animal sighting, etc.) with $40,611$ observations in total recorded by rangers at various locations in the park. The latitude and longitude of the location corresponding to each observation was recorded using a GPS device, thus providing reliable data. Each observation has a feature that specified the total count of the category of observation recorded, for example, number and type of animals sighted or poaching attacks identified, at a particular location. The date and time for a particular patrol was also present in the dataset. We discretized the park area into 2423 grid cells, with each grid cell corresponding to a 1 km × 1 km area within the park. After the discretization, each observation fell within one of the 2423 target cells and we therefore aggregated the animal densities and the number of poaching attacks within each target cell. We considered attack data from the year 2012 in our analysis, which has 2352 attacks in total.

Gaussian Smoothing of Animal Densities: Animal density at each target is computed based on the patrols conducted by the rangers and are thus observations at a particular instant of time. Animal density also has a spatial component, meaning that it is unlikely to change abruptly between grid cells. Therefore, to account for movement of animals over a few kilometers in general, we do a blurring of the current recording of animal densities over the cells. To obtain the spatial spread based on recordings of animal sightings, we use Gaussian smoothing; more specifically we use a Gaussian Kernel of size 5 × 5 with $\sigma = 2.5$ to smoothen out the animal densities over all the grid cells.

Distance as a Feature: In addition to animal density, the poachers' payoffs should take into account the distance (or effort) the poacher takes in reaching the grid cell. Therefore, we also use distance as a feature of our SUQR models. Here, the subjective utility function (Eq. 4) is extended to include the distance feature: $\hat{U}_t^a(\mathbf{x}, \mathbf{R}^a, \mathbf{P}^a) = w_1 x_t + w_2 R_t^a + w_3 P_t^a + w_4 \Phi_t$ where Φ_t is the distance from the attacker current position to target t. For calculating distance, we took a set of 10 entry points based on geographical considerations. The distance to each target location is computed as the minimum over the distances to this target from the 10 entry points.

---

[2] This is the preliminary work on modeling poachers' behaviors. Further study on building more complex behavioral models would be a new interesting research topic for future work.

3.2   Learning Results

We compare the performance of 13 behavioral models[3] as follows (Fig. 1):
(i) SUQR-3, which corresponds to SUQR with three features (coverage probability as discussed in Sect. 2, poacher reward which is considered to be same as
the animal density and poacher penalty which is kept uniform over all targets);
(ii) SUQR-4, which corresponds to SUQR with four features (coverage probability, animal density, poacher penalty and distance to the target location);
(iii) QR; (iv) eight versions of the ϵ-optimal model, a bounded rationality
model [20] where the adversary chooses to attack any one of the targets with an
utility value which is within ϵ of the optimal target's utility, with equal probability; (v) a random adversary model; and (vi) a perfectly rational model.

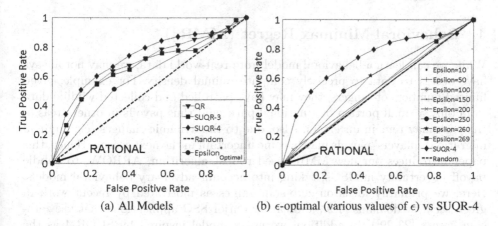

(a) All Models   (b) ϵ-optimal (various values of ϵ) vs SUQR-4

Fig. 1. ROC plots on Uganda dataset

From the 2352 total attacks in our dataset, we randomly sampled (10 times)
20 % of the attack data for testing and trained the three models: SUQR-3, SUQR-
4 and QR on the remaining 80 % data. For each train-test split, we trained our
behavioral models to learn their parameters, which are used to get probabilities
of attack on each grid cell in the test set. Thus, for each grid cell, we get the
actual label (whether the target was attacked or not) along with our predicted
probability of attack on the cell. Using these labels and the predicted probabilities, we plotted a Receiver Operating Characteristic (ROC) curve (in Fig. 1) to
analyze the performance of the various models.

The result shows that the perfectly rational model, that deterministically
classifies which target gets attacked (unlike SUQR/QR which give probabilities
of attack on all targets), achieves an extremely poor prediction accuracy. We also
observe that the ϵ^*-optimal model performs worse than QR and SUQR models
(Fig. 1(a)). Here, by ϵ^*-optimal model, we mean the model corresponding to

---

[3] Models involving cognitive hierarchies [26] are not applicable in Stackelberg games
given that attacker plays knowing the defender's actual strategy.

the ϵ that generates the best prediction (Fig. 1(b)). In our case, the best value of ϵ is 250. For the ϵ-optimal model, no matter what ϵ we choose, the curves from the ϵ-optimal method never gets above the SUQR-4 curve, demonstrating that SUQR-4 is a better model than ϵ-optimal. Furthermore, SUQR-4 (Area Under the Curve (AUC) = 0.73) performs better than both QR (AUC = 0.67) and SUQR-3 (AUC = 0.67), thus highlighting the importance of distance as a feature in the adversary's utility. Thus, SUQR-4 provides the highest prediction accuracy and thus will be our model of choice in the rest of the paper.

In summary, comparing many different models shows for the first time support for SUQR from real-world data in the context of GSGs. The SUQR-4 model convincingly beats QR, ϵ-optimal, perfect-rationality and the random model, thus showing the validity of using SUQR in predicting adversary behaviors in GSGs.

4   Behavioral Minimax Regret (MMR$_b$)

While we can learn a behavioral model from real-world data, we may not always have access to data to precisely compute animal density. For example, given limited numbers of rangers, they may have patrolled and collected wildlife data from only a small portion of a national park, and thus payoffs in other areas of the park may remain uncertain. Also, due to the dynamic changes (e.g., animal migration), players' payoffs may become uncertain in the next season. Hence, this paper introduces our new MMR-based robust algorithm, ARROW, to handle payoff uncertainty in GSGs, taking into account adversary behavioral models. Here, we primarily focus on zero-sum games as motivated by recent work in green security domains [4,9], and earlier major SSG applications that use zero-sum games [23,29]). In addition, we use a model inspired by SUQR-4 as the adversary's behavioral model, given its high prediction accuracy presented in Sect. 3. More specifically, the subjective utility function in Eq. (4) is extended to: $\hat{U}^a_t(\mathbf{x}, \mathbf{R}^a, \mathbf{P}^a) = w_1 x_t + w_2 R^a_t + w_3 P^a_t + w_4 \Phi_t$ where Φ_t is some other feature (e.g., distance) of target t. In fact, our methods generalize to non-zero-sum games with a general class of QR (see Online Appendix A).[4]

We now formulate MMR$_b$ with uncertain payoffs for both players in zero-sum SSG with a boundedly rational attacker.

Definition 1. *Given* $(\mathbf{R}^a, \mathbf{P}^a)$, *the defender's **behavioral regret** is the loss in her utility for playing a strategy* $\mathbf{x}$ *instead of the optimal strategy, which is represented as follows:*

$$R_b(\mathbf{x}, \mathbf{R}^a, \mathbf{P}^a) = \max_{\mathbf{x}' \in \mathbf{X}} F(\mathbf{x}', \mathbf{R}^a, \mathbf{P}^a) - F(\mathbf{x}, \mathbf{R}^a, \mathbf{P}^a) \tag{6}$$

$$\text{where } F(\mathbf{x}, \mathbf{R}^a, \mathbf{P}^a) = \sum_t \hat{q}_t(\mathbf{x}, \mathbf{R}^a, \mathbf{P}^a) U^d_t(\mathbf{x}, \mathbf{R}^d, \mathbf{P}^d) \tag{7}$$

---

[4] Online Appendix: https://www.dropbox.com/s/620aqtinqsul8ys/Appendix.pdf? dl=0.

Table 1. A 2-target, 1-resource game.

Targets	Attacker reward	Attacker penalty
1	[2, 3]	[−2, 0]
2	[5, 7]	[−10, −9]

Behavioral regret measures the distance in terms of utility loss from the defender strategy $\mathbf{x}$ to the optimal strategy given the attacker payoffs. Here, $F(\mathbf{x}, \mathbf{R}^a, \mathbf{P}^a)$ is the defender's utility (which is non-convex fractional in $\mathbf{x}$) for playing $\mathbf{x}$ where the attacker payoffs, whose response follows SUQR, are $(\mathbf{R}^a, \mathbf{P}^a)$. The defender's payoffs in zero-sum games are $\mathbf{R}^d = -\mathbf{P}^a$ and $\mathbf{P}^d = -\mathbf{R}^a$. In addition, the attacking probability, $\hat{q}_t(\mathbf{x}, \mathbf{R}^a, \mathbf{P}^a)$, is given by Eq. 5. When the payoffs are uncertain, if the defender plays a strategy $\mathbf{x}$, she receives different behavioral regrets w.r.t to different payoff instances within the uncertainty intervals. Thus, she could receive a **behavioral max regret** which is defined as follows:

Definition 2. *Given payoff intervals* $\mathbf{I}$, *the* ***behavioral max regret*** *for the defender to play a strategy* $\mathbf{x}$ *is the maximum behavioral regret over all payoff instances:*

$$MR_b(\mathbf{x}, \mathbf{I}) = \max_{(\mathbf{R}^a, \mathbf{P}^a) \in \mathbf{I}} R_b(\mathbf{x}, \mathbf{R}^a, \mathbf{P}^a) \qquad (8)$$

Definition 3. *Given payoff intervals* $\mathbf{I}$, *the* ***behavioral minimax regret*** *problem attempts to find the defender optimal strategy that minimizes the* MR_b *she receives:*

$$MMR_b(\mathbf{I}) = \min_{\mathbf{x} \in \mathbf{X}} MR_b(\mathbf{x}, \mathbf{I}) \qquad (9)$$

Intuitively, behavorial minimax regret ensures that the defender's strategy minimizes the loss in the solution quality over the uncertainty of all possible payoff realizations.

Example 1. In the 2-target zero-sum game as shown in Table 1, each target is associated with uncertainty intervals of the attacker's reward and penalty. For example, if the adversary successfully attacks Target 1, he obtains a reward which belongs to the interval [2, 3]. Otherwise, he receives a penalty which lies within the interval [−2, 0]. The attacker's response, assumed to follow SUQR, is defined by the parameters $(w_1 = -10.0, w_2 = 2.0, w_3 = 0.2, w_4 = 0.0)$. Then the defender's optimal mixed strategy generated by behavioral MMR (Eq. 9) corresponding to this SUQR model is $\mathbf{x} = \{0.35, 0.65\}$. The attacker payoff values which give the defender the maximum regret w.r.t this behavioral MMR strategy are $(3.0, 0.0)$ and $(5.0, -10.0)$ at Target 1 and 2 respectively. In particular, the defender obtains an expected utility of -0.14 for playing $\mathbf{x}$ against this payoff instance. On the other hand, she would receive a utility of 2.06 if playing the optimal strategy $\mathbf{x}' = \{0.48, 0.52\}$ against this payoff instance. As a result, the defender gets a maximum regret of 2.20.

5   ARROW Algorithm: Boundedly Rational Attacker

Algorithm 1 presents the outline of ARROW to solve the MMR_b problem in Eq. 9. Essentially, ARROW's two novelties compared to previous work [17] — addressing uncertainty in both players' payoffs and a boundedly rational attacker — lead to two new computational challenges: (1) uncertainty in defender payoffs makes the defender's expected utility at every target t non-convex in $\mathbf{x}$ and $(\mathbf{R^d}, \mathbf{P^d})$ (Eq. 2); and (2) the SUQR model is in the form of a logit function which is non-convex. These two non-convex functions are combined when calculating the defender's utility (Eq. 7) — which is then used in computing MMR_b (Eq. 9), making it computationally expensive. Overall, MMR_b can be reformulated as minimizing the max regret r such that r is no less than the behavioral regrets over all payoff instances within the intervals:

$$\min_{\mathbf{x} \in \mathbf{X}, r \in \mathbb{R}} r \tag{10}$$
$$\text{s.t. } r \geq F(\mathbf{x'}, \mathbf{R^a}, \mathbf{P^a}) - F(\mathbf{x}, \mathbf{R^a}, \mathbf{P^a}), \forall (\mathbf{R^a}, \mathbf{P^a}) \in \mathbf{I}, \mathbf{x'} \in \mathbf{X}$$

In (10), the set of (non-convex) constraints is infinite since $\mathbf{X}$ and $\mathbf{I}$ are continuous. One practical approach to optimization with large constraint sets is *constraint sampling* [6], coupled with *constraint generation* [2]. Following this approach, ARROW samples a subset of constraints in Problem (10) and gradually expands this set by adding violated constraints to the relaxed problem until convergence to the optimal MMR_b solution.

Specifically, ARROW begins by sampling pairs $(\mathbf{R^a}, \mathbf{P^a})$ of the adversary payoffs uniformly from $\mathbf{I}$. The corresponding optimal strategies for the defender given these payoff samples, denoted $\mathbf{x'}$, are then computed using the PASAQ algorithm [28] to obtain a finite set S of sampled constraints (Line 2). These sampled constraints are then used to solve the corresponding *relaxed* MMR_b program (line 4) using the R.ARROW algorithm (described in Sect. 5.1) — we call this problem *relaxed* MMR_b as it only has samples of constraints in (10). We thus obtain the optimal solution $(lb, \mathbf{x^*})$ which provides a lower bound (lb) on the true MMR_b. Then constraint generation is applied to determine violated constraints

---

Algorithm 1. ARROW Outline

---
1 Initialize $S = \phi, ub = \infty, lb = 0$;
2 Randomly generate sample $(\mathbf{x'}, \mathbf{R^a}, \mathbf{P^a})$, $S = S \cup \{\mathbf{x'}, (\mathbf{R^a}, \mathbf{P^a})\}$;
3 **while** $ub > lb$ **do**
4   Call R.ARROW to compute relaxed MMR_b w.r.t S. Let $\mathbf{x^*}$ be its optimal solution with objective value lb;
5   Call M.ARROW to compute $MR_b(\mathbf{x^*}, \mathbf{I})$. Let the optimal solution be $(\mathbf{x'^{,*}}, \mathbf{R^{a,*}}, \mathbf{P^{a,*}})$ with objective value ub;
6      $S = S \cup \{\mathbf{x'^{,*}}, \mathbf{R^{a,*}}, \mathbf{P^{a,*}}\}$;
7 return $(lb, \mathbf{x^*})$;

(if any). This uses the M.ARROW algorithm (described in Sect. 5.2) which computes $MR_b(\mathbf{x}^*, \mathbf{I})$ — the optimal regret of $\mathbf{x}^*$ which is an upper bound (ub) on the true MMR_b. If $ub > lb$, the optimal solution of M.ARROW, $\{\mathbf{x}'^{,*}, \mathbf{R}^{a,*}, \mathbf{P}^{a,*}\}$, provides the maximally violated constraint (line 5), which is added to S. Otherwise, $\mathbf{x}^*$ is the minimax optimal strategy and $lb = ub = MMR_b(\mathbf{I})$.

5.1 R.ARROW: Compute Relaxed MMR$_b$

The first step of ARROW is to solve the relaxed MMR_b problem using R.ARROW. This relaxed MMR_b problem is non-convex. Thus, R.ARROW presents two key ideas for efficiency: (1) binary search (which iteratively searches the defender's utility space to find the optimal solution) to remove the fractional terms (i.e., the attacking probabilities in Eq. 5) in relaxed MMR_b; and (2) it then applies piecewise-linear approximation to linearize the non-convex terms of the resulting decision problem at each binary search step (as explained below). Overall, relaxed MMR_b can be represented as follows:

$$\min_{\mathbf{x} \in \mathbf{X}, r \in \mathbb{R}} r \tag{11}$$
$$\text{s.t. } r \geq F(\mathbf{x}'^{,k}, \mathbf{R}^{a,k}, \mathbf{P}^{a,k}) - F(\mathbf{x}, \mathbf{R}^{a,k}, \mathbf{P}^{a,k}), \forall k = 1 \ldots K$$

where $(\mathbf{x}'^{,k}, \mathbf{R}^{a,k}, \mathbf{P}^{a,k})$ is the kth sample in S (i.e., the payoff sample set as described in Algorithm 1) where $k = 1 \ldots K$ and K is the total number of samples in S. In addition, r is the defender's max regret for playing $\mathbf{x}$ against sample set S. Finally, $F(\mathbf{x}'^{,k}, \mathbf{R}^{a,k}, \mathbf{P}^{a,k})$ is the defender's optimal utility for every sample of attacker payoffs $(\mathbf{R}^{a,k}, \mathbf{P}^{a,k})$ where $\mathbf{x}'^{,k}$ is the corresponding defender's optimal strategy (which can be obtained via PASAQ [28]). The term $F(\mathbf{x}, \mathbf{R}^{a,k}, \mathbf{P}^{a,k})$, which is included in relaxed MMR_b's constraints, is non-convex and fractional in $\mathbf{x}$ (Eq. 7), making (11) non-convex and fractional. We now detail the two key ideas of R.ARROW.

Binary Search. In each binary search step, given a value of r, R.ARROW tries to solve the decision problem (**P1**) that determines if there exists a defender strategy $\mathbf{x}$ such that the defender's regret for playing $\mathbf{x}$ against any payoff sample in S is no greater than r.

$$\boxed{(\mathbf{P1}) : \exists \mathbf{x} \text{ s.t. } r \geq F(\mathbf{x}'^{,k}, \mathbf{R}^{a,k}, \mathbf{P}^{a,k}) - F(\mathbf{x}, \mathbf{R}^{a,k}, \mathbf{P}^{a,k}), \forall k = 1 \ldots K?}$$

We present the following Proposition 1 showing that (**P1**) can be converted into the *non-fractional* optimization problem (**P2**) (as shown below) of which the optimal solution is used to determine the feasibility of (**P1**):

$$\boxed{\begin{aligned} (\mathbf{P2}): &\min_{\mathbf{x} \in \mathbf{X}, v \in \mathbb{R}} v \\ &\text{s.t. } v \geq \sum_t \left[F(\mathbf{x}'^{,k}, \mathbf{R}^{a,k}, \mathbf{P}^{a,k}) - r - U_t^{d,k}(\mathbf{x}) \right] e^{\hat{U}_t^a(\mathbf{x}, \mathbf{R}^{a,k}, \mathbf{P}^{a,k})}, \forall k = 1 \ldots K \end{aligned}}$$

where $U_t^{d,k}(\mathbf{x}) = - \left[x_t P_t^{a,k} + (1 - x_t) R_t^{a,k} \right]$ is the defender's expected utility at target t given $\mathbf{x}$ and the kth payoff sample.

Proposition 1. *Suppose that* $(v^*, \mathbf{x}^*)$ *is the optimal solution of (P2). If* $v^* \leq 0$, *then* $\mathbf{x}^*$ *is a feasible solution of the decision problem (P1). Otherwise, (P1) is infeasible.*

The proof of Proposition 1 is in Online Appendix B. Given that the decision problem **(P1)** is now converted into the optimization problem **(P2)**, as the next step, we attempt to solve **(P2)** using piecewise linear approximation.

Piecewise Linear Approximation. Although (P2) is non-fractional, its constraints are non-convex. We use a piecewise linear approximation for the RHS of the constraints in (P2) which is in the form of $\sum_t f_t^k(x_t)$ where the term $f_t^k(x_t)$ is a non-convex function of x_t (recall that x_t is the defender's coverage probability at target t). The feasible region of the defender's coverage x_t for all t, $[0,1]$, is then divided into M equal segments $\{[0, \frac{1}{M}], [\frac{1}{M}, \frac{2}{M}], \ldots, [\frac{M-1}{M}, 1]\}$ where M is given. The values of $f_t^k(x_t)$ are then approximated by using the segments connecting pairs of consecutive points $\left(\frac{i-1}{M}, f_t^k\left(\frac{i-1}{M}\right)\right)$ and $\left(\frac{i}{M}, f_t^k\left(\frac{i}{M}\right)\right)$ for $i = 1 \ldots M$ as follows:

$$f_t^k(x_t) \approx f_t^k(0) + \sum_{i=1}^{M} \alpha_{t,i}^k x_{t,i} \tag{12}$$

where $\alpha_{t,i}^k$ is the slope of the ith segment which can be determined based on the two extreme points of the segment. Also, $x_{t,i}$ refers to the portion of the defender's coverage at target t belonging to the ith segment, i.e., $x_t = \sum_i x_{t,i}$.

Example 2. When the number of segments $M = 5$, it means that we divide $[0, 1]$ into 5 segments $\{[0, \frac{1}{5}], [\frac{1}{5}, \frac{2}{5}], [\frac{2}{5}, \frac{3}{5}], [\frac{3}{5}, \frac{4}{5}], [\frac{4}{5}, 1]\}$. Suppose that the defender's coverage at target t is $x_t = 0.3$, since $\frac{1}{5} < x_t < \frac{2}{5}$, we obtain the portions of x_t that belongs to each segment is $x_{t,1} = \frac{1}{5}$, $x_{t,2} = 0.1$, and $x_{t,3} = x_{t,4} = x_{t,5} = 0$ respectively. Then each non-linear term $f_t^k(x_t)$ is approximated as $f_t^k(x_t) \approx f_t^k(0) + \frac{1}{5}\alpha_{t,1}^k + 0.1\alpha_{t,2}^k$ where the slopes of the 1st and 2nd segments are $\alpha_{t,1}^k = 5\left[f_t^k\left(\frac{1}{5}\right) - f_t^k(0)\right]$ and $\alpha_{t,2}^k = 5\left[f_t^k\left(\frac{2}{5}\right) - f_t^k\left(\frac{1}{5}\right)\right]$ respectively.

By using the approximations of $f_t^k(x_t)$ for all k and t, we can reformulate **(P2)** as the MILP **(P2')** which can be solved by the solver CPLEX:

$$\textbf{(P2'):} \quad \min_{x_{t,i}, z_{t,i}, v} \quad v \tag{13}$$

$$\text{s.t. } v \geq \sum_t f_t^k(0) + \sum_t \sum_i \alpha_{t,i}^k x_{t,i}, \forall k = 1 \ldots K \tag{14}$$

$$\sum_{t,i} x_{t,i} \leq R, 0 \leq x_{t,i} \leq \frac{1}{M}, \forall t = 1 \ldots T, i = 1 \ldots M \tag{15}$$

$$z_{t,i}\frac{1}{M} \leq x_{t,i}, \forall t = 1 \ldots T, i = 1 \ldots M - 1 \tag{16}$$

$$x_{t,i+1} \leq z_{t,i}, \forall t = 1 \ldots T, i = 1 \ldots M - 1 \tag{17}$$

$$z_{t,i} \in \{0, 1\}, \forall t = 1 \ldots T, i = 1 \ldots M - 1 \tag{18}$$

where $z_{t,i}$ is an auxiliary integer variable which ensures that the portions of x_t satisfies $x_{t,i} = \frac{1}{M}$ if $x_t \geq \frac{i}{M}$ ($z_{t,i} = 1$) or $x_{t,i+1} = 0$ if $x_t < \frac{i}{M}$ ($z_{t,i} = 0$)

(constraints (15–18)). Constraints (14) are piecewise linear approximations of constraints in (P2). In addition, constraint (15) guarantees that the resource allocation condition, $\sum_t x_t \leq R$, holds true and the piecewise segments $0 \leq x_{t,i} \leq \frac{1}{M}$.

Finally, we provide Theorem 1 showing that R.ARROW guarantees a solution bound on computing relaxed MMR_b. The proof of Theorem 1 is in the Online Appendix C.

Theorem 1. *R.ARROW provides an $O\left(\epsilon + \frac{1}{M}\right)$-optimal solution of relaxed MMR_b where ϵ is the tolerance of binary search and M is the number of piecewise segments.*

5.2 M.ARROW: Compute MR_b

Given the optimal solution $\mathbf{x}^*$ returned by R.ARROW, the second step of ARROW is to compute MR_b of $\mathbf{x}^*$ using M.ARROW (line 5 in Algorithm 1). The problem of computing MR_b can be represented as the following non-convex maximization problem:

$$\max_{\mathbf{x}' \in \mathbf{X},(\mathbf{R}^a,\mathbf{P}^a) \in I} F(\mathbf{x}',\mathbf{R}^a,\mathbf{P}^a) - F(\mathbf{x}^*,\mathbf{R}^a,\mathbf{P}^a) \qquad (19)$$

Overall, it is difficult to apply the same techniques used in R.ARROW for M.ARROW since it is a subtraction of two non-convex fractional functions, $F(\mathbf{x}',\mathbf{R}^a,\mathbf{P}^a)$ and $F(\mathbf{x}^*,\mathbf{R}^a,\mathbf{P}^a)$. Therefore, we use local search with multiple starting points which allows us to reach different local optima.

6 ARROW-Perfect Algorithm: Perfectly Rational Attacker

While ARROW incorporates an adversary behavioral model, it may not be applicable for green security domains where there may be a further paucity of data in which not only payoffs are uncertain but also parameters of the behavioral model are difficult to learn accurately. Therefore, we introduce a novel MMR-based algorithm, ARROW-Perfect, to handle uncertainty in both players' payoffs assuming a perfectly rational attacker. In general, ARROW-Perfect follows the same *constraint sampling* and *constraint generation* methodology as ARROW. Yet, by leveraging the property that the attacker's optimal response is a pure strategy (given a perfectly rational attacker) and the game is zero-sum, we obtain the *exact optimal solutions* for computing both relaxed MMR and max regret in *polynomial time* (while we cannot provide such guarantees for a boundedly rational attacker). In this case, we call the new algorithms for computing relaxed MMR and max regret: R.ARROW-Perfect and M.ARROW-Perfect respectively.

6.1 R.ARROW-Perfect: Compute Relaxed MMR

In zero-sum games, when the attacker is perfectly rational, the defender's utility for playing a strategy $\mathbf{x}$ w.r.t the payoff sample $(\mathbf{R}^{a,k}, \mathbf{P}^{a,k})$ is equal to $F(\mathbf{x}, \mathbf{R}^{a,k}, \mathbf{P}^{a,k}) = -U_t^a(\mathbf{x}, \mathbf{R}^{a,k}, \mathbf{P}^{a,k})$ if the attacker attacks target t. Since the adversary is perfectly rational, therefore, $F(\mathbf{x}, \mathbf{R}^{a,k}, \mathbf{P}^{a,k}) = -\max_t U_t^a(\mathbf{x}, \mathbf{R}^{a,k}, \mathbf{P}^{a,k})$, we can reformulate the relaxed MMR in (11) as the following linear minimization problem:

$$\min_{\mathbf{x} \in \mathbf{X}, r \in \mathbb{R}} r \tag{20}$$

$$\text{s.t. } r \geq F(\mathbf{x}'^{,k}, \mathbf{R}^{a,k}, \mathbf{P}^{a,k}) + U_t^a(\mathbf{x}, \mathbf{R}^{a,k}, \mathbf{P}^{a,k}), \forall k = 1 \ldots K, \forall t = 1 \ldots T \tag{21}$$

where $F(\mathbf{x}'^{,k}, \mathbf{R}^{a,k}, \mathbf{P}^{a,k})$ is the defender's optimal utility against a perfectly rational attacker w.r.t payoff sample $(\mathbf{R}^{a,k}, \mathbf{P}^{a,k})$ and $\mathbf{x}'^{,k}$ is the corresponding optimal strategy which is the Maximin solution. In addition, constraint (21) ensures that the regret $r \geq F(\mathbf{x}'^{,k}, \mathbf{R}^{a,k}, \mathbf{P}^{a,k}) + \max_t U_t^a(\mathbf{x}, \mathbf{R}^{a,k}, \mathbf{P}^{a,k})$ for all payoff samples. This linear program can be solved exactly in polynomial time using any linear solver, e.g., CPLEX.

6.2 M.ARROW-Perfect: Compute Max Regret

Computing max regret (MR) in zero-sum games presents challenges that previous work [17] can not handle since the defender's payoffs are uncertain while [17] assumes these payoff values are known. In this work, we propose a new exact algorithm, M.ARROW-Perfect, to compute MR in polynomial time by exploiting insights of zero-sum games.

In zero-sum games with a perfectly rational adversary, Strong Stackelberg Equilibrium is equivalent to Maximin solution [30]. Thus, given the strategy $\mathbf{x}^*$ returned by relaxed MMR, max regret in (19) can be reformulated as follows:

$$\max_{\mathbf{x}' \in \mathbf{X}, (\mathbf{R}^a, \mathbf{P}^a) \in \mathbf{I}, v} v - F(\mathbf{x}^*, \mathbf{R}^a, \mathbf{P}^a) \tag{22}$$

$$\text{s.t. } v \leq -[x_t' P_t^a + (1 - x_t') R_t^a], \forall t \tag{23}$$

where v is the Maximin/SSE utility for the defender against the attacker payoff $(\mathbf{R}^a, \mathbf{P}^a)$. Moreover, the defender's utility for playing $\mathbf{x}^*$ can be computed as $F(\mathbf{x}^*, \mathbf{R}^a, \mathbf{P}^a) = -[x_j^* P_j^a + (1 - x_j^*) R_j^a]$ if the adversary attacks target j. Thus, we divide the attacker payoff space into T subspaces such that within the jth subspace, the adversary always attacks target j against the defender strategy $\mathbf{x}^*$, for all $j = 1 \ldots T$. By solving these T sub-max regret problems corresponding to this division, our final global optimal solution of max regret will be the maximum of all T sub-optimal solutions.

Next, we will explain how to solve these sub-max regret problems. Given the jth attacker payoff sub-space, we obtain the jth sub-max regret problem as:

$$\max_{\mathbf{x}' \in \mathbf{X}, (\mathbf{R}^a, \mathbf{P}^a) \in \mathbf{I}, v} v + (x_j^* P_j^a + (1 - x_j^*) R_j^a) \tag{24}$$

$$\text{s.t. } v \leq -[x_t' P_t^a + (1 - x_t') R_t^a], \forall t \tag{25}$$

$$x_j^* P_j^a + (1 - x_j^*) R_j^a \geq x_t^* P_t^a + (1 - x_t^*) R_t^a, \forall t \tag{26}$$

where constraints (26) ensures that the adversary attacks target j against the defender strategy $\mathbf{x}^*$. Here, constraints (25) are non-convex for all targets. We provide the following proposition which allows us to linearize constraints (25) for all targets but j.

Proposition 2. *Given target j, the lower bounds of the attacker's payoffs at all targets except j, $\{R^a_{min}(t), P^a_{min}(t)\}_{t \neq j}$, are optimal solutions of $\{R^a_j, P^a_j\}_{t \neq j}$ for the jth sub-max regret problem.*

The proof of Proposition 2 is in Online Appendix D. Now, only constraint (25) w.r.t target j remains non-convex for which we provide further steps to simplify it. Given the defender strategy $\mathbf{x}'$, we define the attack set as including all targets with the attacker's highest expected utility: $\Gamma(\mathbf{x}') = \{t : U^a_t(\mathbf{x}', \mathbf{R^a}, \mathbf{P^a}) = \max_{t'} U^a_{t'}(\mathbf{x}', \mathbf{R^a}, \mathbf{P^a})\}$. We provide the following observations based on which we can determine the optimal value of the attacker's reward at target j, R^a_j, for the sub-max regret problem (24–26) (according to the Proposition 3 below):

Observation 1. *If $\mathbf{x}'$ is the optimal solution of computing the jth sub-max regret in (24–26), target j belongs to the attack set $\Gamma(\mathbf{x}')$.*

Since $\mathbf{x}'$ is the Maximin or SSE solution w.r.t attacker payoffs $(\mathbf{R^a}, \mathbf{P^a})$, the corresponding attack set $\Gamma(\mathbf{x}')$ has the maximal size [11]. In other words, if a target t belongs to the attack set of any defender strategy w.r.t $(\mathbf{R^a}, \mathbf{P^a})$, then $t \in \Gamma(\mathbf{x}')$. In (24–26), because target j belongs to the attack set $\Gamma(\mathbf{x}^*)$, we obtain $j \in \Gamma(\mathbf{x}')$.

Observation 2. *If $\mathbf{x}'$ is the optimal solution of computing the jth sub-max regret in (24–26), the defender's coverage at target j: $x'_j \geq x^*_j$.*

Since $j \in \Gamma(\mathbf{x}')$ according to Observation 1, the defender utility for playing $\mathbf{x}'$ is equal to $v = -[x'_j P^a_j + (1 - x'_j) R^a_j]$. Furthermore, the max regret in (24) is always not less than zero, meaning that $v \geq -[x^*_j P^a_j + (1 - x^*_j) R^a_j]$. Thus, we obtain $x'_j \geq x^*_j$.

Proposition 3. *Given target j, the upper bound of the attacker's reward at j, $R^a_{max}(j)$, is an optimal solution of the attacker reward R^a_j for the jth sub-max regret problem.*

Proof. Suppose that $R^a_j < R^a_{max}(j)$ is optimal in (24–26) and $\mathbf{x}'$ is the corresponding defender optimal strategy, then $v = -[x'_j P^a_j + (1 - x'_j) R^a_j]$ according to the Observation 1. We then replace R^a_j with $R^a_{max}(j)$ while other rewards/penalties and $\mathbf{x}'$ remain the same. Since $R^a_j < R^a_{max}(j)$, this new solution is also feasible for (24–26) and target j still belongs to $\Gamma(\mathbf{x}')$. Therefore, the corresponding utility of the defender for playing $\mathbf{x}'$ will be equal to $-[x'_j P^a_j + (1 - x'_j) R^a_{max}(j)]$. Since $R^a_j < R^a_{max}(j)$ and $x'_j \geq x^*_j$ (Observation 2), the following inequality holds true:

$$- [x'_j P^a_j + (1 - x'_j) R^a_{max}(j)] + [(x^*_j P^a_j + (1 - x^*_j) R^a_{max}(j)] \tag{27}$$

$$= -[x'_j P^a_j + (1 - x'_j) R^a_j] + [(x^*_j P^a_j + (1 - x^*_j) R^a_j] + [x'_j - x^*_j][R^a_{max}(j) - R^a_j] \tag{28}$$

$$\geq -[x'_j P^a_j + (1 - x'_j) R^a_j] + [(x^*_j P^a_j + (1 - x^*_j) R^a_j]. \tag{29}$$

This inequality indicates that the defender's regret w.r.t $R^a_{max}(j)$ (the LHS of the inequality) is no less than w.r.t R^a_j (the RHS of the inequality). Therefore, $R^a_{max}(j)$ is an optimal solution of the attacker's reward at target j for (24–26).   ∎

Based on the Proposition 2 & 3 and the Observation 1, the jth sub-max regret (24–26) is simplified to the following optimization problem:

$$\max_{\mathbf{x}' \in \mathbf{X}, P^a_j, v} \ v + (x^*_j P^a_j + (1 - x^*_j) R^a_{max}(j)) \tag{30}$$

$$\text{s.t. } v = -\left[x'_j P^a_j + (1 - x'_j) R^a_{max}(j) \right] \tag{31}$$

$$v \leq -\left[x'_t P^a_{min}(t) + (1 - x'_t) R^a_{min}(t) \right], \forall t \neq j \tag{32}$$

$$P^a_{max}(j) \geq P^a_j \geq \max \left\{ P^a_{min}(j), \frac{C - (1 - x^*_j) R^a_{max}(j)}{x^*_j} \right\} \tag{33}$$

where $C = \max_{t \neq j} x^*_t P^a_{min}(t) + (1 - x^*_t) R^a_{min}(t)$ is a constant. In addition, constraints (31–32) refer to constraint (25) (where constraint (31) is a result of Observation 1) and constraints (33) is equivalent to constraint (26). The only remaining non-convex term is $x'_j P^a_j$ in constraint (31). We then alleviate the computational cost incurred based on Theorem 2 which shows that if the attack set $\Gamma(\mathbf{x}')$ is known beforehand, we can convert (30–33) into a simple optimization problem which is straightforward to solve.

Theorem 2. *Given the attack set $\Gamma(\mathbf{x}')$, the jth sub-max regret problem (30–33) can be represented as the following optimization problem on the variable v only:*

$$\max_v v + \frac{av + b}{cv + d} \tag{34}$$

$$\text{s.t. } v \in [l_v, u_v]. \tag{35}$$

where v is the defender utility for playing $\mathbf{x}'$ in (30–33).

The proof of Theorem 2 is in Online Appendix E. The constants (a, b, c, d, l_v, u_v) are determined based on the attack set $\Gamma(\mathbf{x}')$, the attacker's payoffs $\{R^a_{min}(t), P^a_{min}(t)\}_{t \neq j}$ and $R^a_{max}(j)$, and the number of the defender resources R. Here, the total number of possible attack sets $\Gamma(\mathbf{x}')$ is maximally T sets according to the property that $R^a_t > R^a_{t'}$ for all $t \in \Gamma(\mathbf{x}')$ and $t' \notin \Gamma(\mathbf{x}')$ [11]. Therefore, we can iterate over all these possible attack sets and solve the corresponding optimization problems in (34–35). The optimal solution of each sub-max regret problem (30–33) will be the maximum over optimal solutions of (34–35). The final optimal solution of the max regret problem (22) will be the maximum over optimal solutions of all these sub-max regret problems.

In summary, we provide the M.ARROW-Perfect algorithm to exactly compute max regret of playing the strategy $\mathbf{x}^*$ against a perfectly rational attacker in zero-sum games by exploiting the insight of extreme points of the uncertainty intervals as well as attack sets. Furthermore, we provide Theorem 3 (its proof is in the Online Appendix F) showing that the computational complexity of solving max regret is polynomial.

Algorithm 2. Elicitation process

---

1 Input: budget: B, regret barrier: δ, uncertainty intervals: $\mathbf{I}$;
2 Initialize regret $r = +\infty$, cost $c = 0$;
3 **while** $c < B$ *and* $r > \delta$ **do**
4 $\quad$ $(r, \mathbf{x}^*, (\mathbf{x}'^{,*}, \mathbf{R}^{\mathbf{a},*}, \mathbf{P}^{\mathbf{a},*})) = ARROW(\mathbf{I})$;
5 $\quad$ $\mathbf{P} = calculatePath(\mathbf{x}^*, (\mathbf{x}'^{,*}, \mathbf{R}^{\mathbf{a},*}, \mathbf{P}^{\mathbf{a},*}))$;
6 $\quad$ $\mathbf{I} = collectInformationUAV(\mathbf{P})$; $c = updateCost(\mathbf{P})$;
7 return $(r, \mathbf{x}^*)$;

Theorem 3. *M.ARROW-Perfect provides an optimal solution for computing max regret against a perfectly rational attacker in $O(T^3)$ time.*

7 UAV Planning for Payoff Elicitation (PE)

Our final contribution is to provide PE heuristics to select the best UAV path to reduce uncertainty in payoffs, given any adversary behavioral model. Despite the limited availability of mobile sensors in conservation areas (many of them being in developing countries), these UAVs may still be used to collect accurate imagery of these areas periodically, e.g., every six months to reduce payoff uncertainty. Since the UAV availability is limited, it is important to determine the best UAV paths such that reducing payoff uncertainty at targets on these paths could help reducing the defender's regret the most. While a UAV visits multiple targets to collect data, planning an optimal path (which considers all possible outcomes of reducing uncertainty) is computationally expensive. Thus, we introduce the *current solution*-based algorithm which evaluates a UAV path based solely on the MMR$_b$ solution given current intervals.[5]

We first present a general elicitation process for UAV planning (Algorithm 2). The input includes the defender's initial budget B (e.g., limited time availability of UAVs), the regret barrier δ which indicates how much regret (utility loss) the defender is willing to sacrifice, and the uncertainty intervals $\mathbf{I}$. The elicitation process consists of multiple rounds of flying a UAV and stops when the UAV cost exceeds B or the defender's regret is less than δ. At each round, ARROW is applied to compute the optimal MMR$_b$ solution given current $\mathbf{I}$; ARROW then outputs the regret r, the optimal strategy $\mathbf{x}^*$, and the corresponding most unfavorable strategy and payoffs $(\mathbf{x}'^{,*}, \mathbf{R}^{\mathbf{a},*}, \mathbf{P}^{\mathbf{a},*})$ which provide the defender's max regret (line 4). Then the best UAV path is selected based on these outputs (line 5). Finally, the defender controls the UAV to collect data at targets on that path to obtain new intervals and then updates the UAV flying cost (line 6).

The key aspects of Algorithm 2 are in lines 4 and 5 where the MMR$_b$ solution is computed by ARROW and the *current solution* heuristic is used to determine the best UAV path. In this heuristic, the *preference value* of a target t, denoted

---

[5] A similar idea was introduced in [2] although in a very different domain without UAV paths.

$pr(t)$, is measured as the distance in the defender utility between $\mathbf{x}^*$ and the most unfavorable strategy $\mathbf{x}'^{,*}$ against attacker payoffs $(\mathbf{R}^{a,*}, \mathbf{P}^{a,*})$ at that target, which can be computed as follows: $pr(t) = \hat{q}_t(\mathbf{x}^*, \mathbf{R}^{a,*}, \mathbf{P}^{a,*}) U_t^d(\mathbf{x}^*, \mathbf{R}^d, \mathbf{P}^d) - \hat{q}_t(\mathbf{x}'^{,*}, \mathbf{R}^{a,*}, \mathbf{P}^{a,*}) U_t^d(\mathbf{x}'^{,*}, \mathbf{R}^d, \mathbf{P}^d)$ where $\mathbf{R}^d = -\mathbf{P}^{a,*}$ and $\mathbf{P}^d = -\mathbf{R}^{a,*}$. Intuitively, targets with higher preference values play a more important role in reducing the defender's regret. We use the sum of preference values of targets to determine the best UAV path based on the two heuristics: **Greedy heuristic:** The chosen path consists of targets which are iteratively selected with the maximum pr value and then the best neighboring target. **MCNF Heuristic:** We represent this problem as a Min Cost Network Flow (MCNF) where the cost of choosing a target t is $-pr(t)$. For example, there is a grid of four cells (t_1, t_2, t_3, t_4) (Fig. 2(a)) where each cell is associated with its preference value, namely $(pr(1), pr(2), pr(3), pr(4))$. Suppose that a UAV covers a path of two cells every time it flies and its entry locations (where the UAV takes off or land) can be at any cell. The MCNF for UAV planning is shown in Fig. 2(b) which has two layers where each cell t_i has four copies $(t_i^1, t_i^2, t_i^3, t_i^4)$ with edge costs $c(t_i^1, t_i^2) = c(t_i^3, t_i^4) = -pr(i)$. The connectivity between these two layers corresponds to the grid connectivity. There are *Source* and *Sink* nodes which determine the UAV entry locations. The edge costs between the layers and between the *Source* or *Sink* to the layers are set to zero.

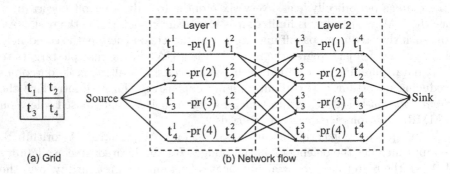

(a) Grid   (b) Network flow

Fig. 2. Min cost network flow

8   Experimental Results

We use CPLEX for our algorithms and Fmincon of MATLAB on a 2.3 GHz/4 GB RAM machine. *Key comparison results are statistically significant under bootstrap-t ($\alpha = 0.05$)* [25]. More results are in the Online Appendix G.

8.1   Synthetic Data

We first conduct experiments using synthetic data to simulate a wildlife protection area. The area is divided into a grid where each cell is a target, and we create different payoff structures for these cells. Each data point in our results is averaged over *40 payoff structures* randomly generated by GAMUT [19]. The attacker reward/defender penalty refers to the animal density while the attacker penalty/defender reward refers to, for example, the amount of snares that are confiscated by the defender [27]. Here, the defender's regret indicates the animal loss and thus can be used as a measure for the defender's patrolling effectiveness. Upper and lower bounds for payoff intervals are generated randomly from [−14, −1] for penalties and [1, 14] for rewards with an interval size of 4.0.

Solution Quality of ARROW. The results are shown in Fig. 3 where the x-axis is the grid size (number of targets) and the y-axis is the defender's max regret. First, we demonstrate the importance of handling the attacker's bounded rationality in ARROW by comparing solution quality (in terms of the defender's regret) of ARROW with ARROW-Perfect and Maximin. Figure 3(a) shows that the defender's regret significantly increases when playing ARROW-Perfect and Maximin strategies compared to playing ARROW strategies, which demonstrates the importance of behavioral MMR.

Second, we examine how ARROW's parameters influence the MMR_b solution quality; which we show later affects its runtime-solution quality tradeoff. We examine if the defender's regret significantly increases if (i) the number of starting points in M.ARROW decreases (i.e., ARROW with 20 (ARROW-20), 5 (ARROW-5) and 1 (ARROW-1) starting points for M.ARROW and 40 iterations to iteratively add 40 payoff samples into the set S), or (ii) when ARROW only uses R.ARROW (without M.ARROW) to solve relaxed MMR_b (i.e., R.ARROW with 50 (R.ARROW-50) and 100 (R.ARROW-100) uniformly random payoff samples). Figure 3(b) shows that the number of starting points in M.ARROW does not have a significant impact on solution quality. In particular, ARROW-1's solution quality is approximately the same as ARROW-20 after 40 iterations. This result shows that the shortcoming of local search in M.ARROW (where solution quality depends on the number of starting points) is compensated by a sufficient number (e.g., 40) of iterations in ARROW. Furthermore, as R.ARROW-50 and R.ARROW-100 only solve relaxed MMR_b, they both lead to much higher regret. Thus, it is important to include M.ARROW in ARROW.

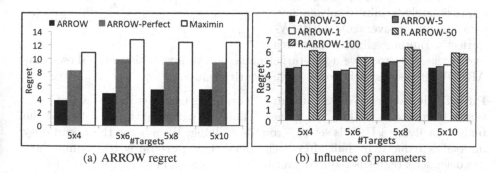

(a) ARROW regret   (b) Influence of parameters

Fig. 3. Solution quality of ARROW

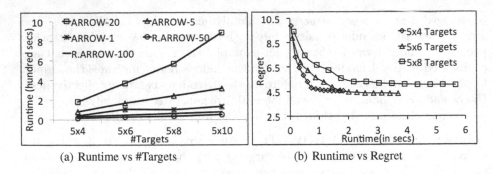

(a) Runtime vs #Targets   (b) Runtime vs Regret

Fig. 4. Runtime performance of ARROW

Runtime Performance of ARROW. Figure 4(a) shows the runtime of ARROW with different parameter settings. In all settings, ARROW's runtime linearly increases in the number of targets. Further, Fig. 3(a) shows that ARROW-1 obtains approximately the same solution quality as ARROW-20 while running significantly faster (Fig. 4(a)). This result shows that one starting point of M.ARROW might be adequate for solving MMR_b in considering the trade-off between runtime performance and solution quality. Figure 4(b) plots the trade-off between runtime and the defender's regret in 40 iterations of ARROW-20 for 20–40 targets which shows a useful anytime profile.

Runtime Performance of ARROW-Perfect. Figure 5 shows the runtime performance of ARROW-Perfect compared to ARROW and a non-linear solver (i.e., fmincon of Matlab) to compute MMR of the perfectly rational attacker case. While the runtime of both ARROW and non-linear solver increase quickly w.r.t the number of targets (e.g., it takes them approximately 20 min to solve a 200-target game on average), ARROW-Perfect's runtime slightly increases

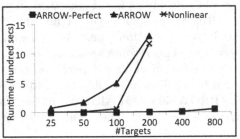

Fig. 5. Runtime performance of ARROW-Perfect

and reaches 53 s to solve a 800-target game on average. This result shows that ARROW-Perfect is scalable for large security games.

Payoff Elicitation. We evaluate our PE strategies using synthetic data of 5×5-target (target = 2×2 km cell) games. The UAV path length is 3 cells and the budget for flying a UAV is set to 5 rounds of flying. We assume the uncertainty interval is reduced by half after each round. Our purpose is to examine how the defender's regret is reduced over different rounds. The empirical results are shown in Fig. 6 where the x-axis is the number of rounds and the y-axis is the regret obtained after each round (Fig. 6(a)) or the accumulative runtime of the

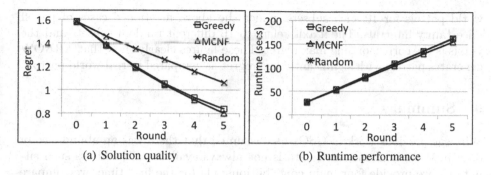

(a) Solution quality

(b) Runtime performance

Fig. 6. UAV planning: uncertainty reduction over rounds

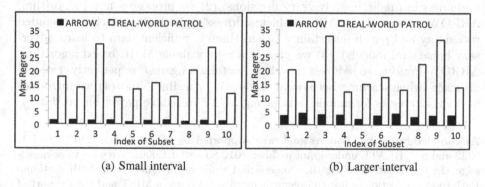

(a) Small interval

(b) Larger interval

Fig. 7. Real world max regret comparison

elicitation process over rounds (Fig. 6(b)). We compare three heuristics: (1) randomly choosing a path (Random) (2) Greedy, and (3) MCNF. Figure 6 shows that the defender's regret is reduced significantly by using Greedy and MCNF in comparison with Random. As mentioned, the difference are statistically significant ($\alpha = 0.05$). Also, both Greedy and MCNF run as quickly as Random.

8.2 Real-World Data

Lastly, we use our wildlife dataset from Uganda (Sect. 3) to analyze the effectiveness of past patrols conducted by rangers (in the wildlife park) compared with the patrol strategies generated by ARROW. We choose multiple subsets of 50 grid cells each, randomly sampled from the 2423 grid cells for our analysis. Before these wildlife areas were patrolled, there was uncertainty in the features values in those areas. We simulate these conditions faced by real world patrollers by introducing uncertainty intervals in the real-world payoffs. In our experiments, we impose uncertainty intervals on the animal density for each target, though two cases: a small and a large interval of sizes 5 and 10 respectively. Figure 7(a) and (b) show the comparison of the max regret achieved by ARROW and real world patrols for 10 such subsets, under the above mentioned cases of payoff uncertainty intervals. The x-axis refers to 10 different random subsets and the y-axis is the corresponding max regret. These figures clearly show that ARROW generates patrols with significantly less regret as compared to real-world patrols.

9 Summary

Whereas previous work in GSGs had assumed that there was an abundance of data in these domains, such data is not always available. To address such situations, we provide four main contributions: (1) for the first time, we compare key behavioral models, e.g., SUQR/QR on real-world data and show SUQR's usefulness in predicting adversary decisions; (2) we propose a novel algorithm, ARROW, to solve the MMR_b problem addressing both the attacker's bounded rationality and payoff uncertainty (when there is sufficient data to learn adversary behavioral models); (3) we present a new scalable MMR-based algorithm, ARROW-Perfect, to address payoff uncertainty against a perfectly rational attacker (when learning behavioral models is infeasible), and (4) we introduce new PE strategies for mobile sensors, e.g., UAV to reduce payoff uncertainty.

Acknowledgements. This research was supported by MURI Grant W911NF-11-1-0332 and by CREATE under grant number 2010-ST-061-RE0001. We wish to acknowledge the contribution of all the rangers and wardens in Queen Elizabeth National Park to the collection of law enforcement monitoring data in MIST and the support of Uganda Wildlife Authority, Wildlife Conservation Society and MacArthur Foundation, US State Department and USAID in supporting these data collection financially.

References

1. Basilico, N., Gatti, N., Amigoni, F.: Leader-follower strategies for robotic patrolling in environments with arbitrary topologies. In: AAMAS (2009)
2. Boutilier, C., Patrascu, R., Poupart, P., Schuurmans, D.: Constraint-based optimization and utility elicitation using the minimax decision criterion. Artif. Intell. **170**, 686–713 (2006)
3. Braziunas, D., Boutilier, C.: Assessing regret-based preference elicitation with the utpref recommendation system. In: EC (2010)
4. Brown, M., Haskell, W.B., Tambe, M.: Addressing scalability and robustness in security games with multiple boundedly rational adversaries. In: GameSec (2014)
5. Brunswik, E.: The Conceptual Framework of Psychology. University of Chicago Press, New York (1952)
6. De Farias, D.P., Van Roy, B.: On constraint sampling in the linear programming approach to approximate dynamic programming. Math. Oper. Res. **29**, 462–478 (2004)
7. Fang, F., Stone, P., Tambe, M.: When security games go green: designing defender strategies to prevent poaching and illegal fishing. In: IJCAI (2015)
8. French, S.: Decision Theory: An Introduction to the Mathematics of Rationality. Halsted Press, New York (1986)

9. Haskell, W.B., Kar, D., Fang, F., Tambe, M., Cheung, S., Denicola, L.E.: Robust protection of fisheries with compass. In: IAAI (2014)
10. Kiekintveld, C., Islam, T., Kreinovich, V.: Security games with interval uncertainty. In: AAMAS (2013)
11. Kiekintveld, C., Jain, M., Tsai, J., Pita, J., Ordez, F., Tambe, M.: Computing optimal randomized resource allocations for massive security games. In: AAMAS (2009)
12. Korzhyk, D., Conitzer, V., Parr, R.: Complexity of computing optimal stackelberg strategies in security resource allocation games. In: AAAI (2010)
13. Letchford, J., Vorobeychik, Y.: Computing randomized security strategies in networked domains. In: AARM (2011)
14. McFadden, D.: Conditional logit analysis of qualitative choice behavior. Technical report (1972)
15. McKelvey, R., Palfrey, T.: Quantal response equilibria for normal form games. Game Econ. Behav. **10**(1), 6–38 (1995)
16. Montesh, M.: Rhino poaching: a new form of organised crime1. University of South Africa, Technical report (2013)
17. Nguyen, T.H., Yadav, A., An, B., Tambe, M., Boutilier, C.: Regret-based optimization and preference elicitation for stackelberg security games with uncertainty. In: AAAI (2014)
18. Nguyen, T.H., Yang, R., Azaria, A., Kraus, S., Tambe, M.: Analyzing the effectiveness of adversary modeling in security games. In: AAAI (2013)
19. Nudelman, E., Wortman, J., Shoham, Y., Leyton-Brown, K.: Run the gamut: a comprehensive approach to evaluating game-theoretic algorithms. In: AAMAS (2004)
20. Pita, J., Jain, M., Tambe, O.M., Kraus, S., Magori-cohen, R.: Effective solutions for real-world stackelberg games: when agents must deal with human uncertainties. In: AAMAS (2009)
21. Qian, Y., Haskell, W.B., Jiang, A.X., Tambe, M.: Online planning for optimal protector strategies in resource conservation games. In: AAMAS (2014)
22. Secretariat, G.: Global tiger recovery program implementation plan: 2013–14. Report, The World Bank, Washington, DC (2013)
23. Shieh, E., An, B., Yang, R., Tambe, M., Baldwin, C., DiRenzo, J., Maule, B., Meyer, G.: Protect: a deployed game theoretic system to protect the ports of the united states. In: AAMAS (2012)
24. Tambe, M.: Security and Game Theory: Algorithms, Deployed Systems. Cambridge University Press, Lessons Learned (2011)
25. Wilcox, R.: Applying Contemporary Statistical Techniques. Academic Press, New York (2002)
26. Wright, J.R., Leyton-Brown, K.: Level-0 meta-models for predicting human behavior in games. In: ACM-EC, pp. 857–874 (2014)
27. Yang, R., Ford, B., Tambe, M., Lemieux, A.: Adaptive resource allocation for wildlife protection against illegal poachers. In: AAMAS (2014)
28. Yang, R., Ordonez, F., Tambe, M.: Computing optimal strategy against quantal response in security games. In: AAMAS (2012)
29. Yin, Z., Jiang, A.X., Tambe, M., Kiekintveld, C., Leyton-Brown, K., Sandholm, T., Sullivan, J.P.: Trusts: scheduling randomized patrols for fare inspection in transit systems using game theory. AI Mag. **33**, 59 (2012)
30. Yin, Z., Korzhyk, D., Kiekintveld, C., Conitzer, V., Tambe, M.: Stackelberg vs. nash in security games: Interchangeability, equivalence, and uniqueness. In: AAMAS (2010)

A Security Game Model for Environment Protection in the Presence of an Alarm System

Nicola Basilico[1]([✉]), Giuseppe De Nittis[2], and Nicola Gatti[2]

[1] Department of Computer Science, University of Milan, Milan, Italy
nicola.basilico@unimi.it
[2] Dipartimento di Elettronica, Informazione e Bioingegneria,
Politecnico di Milano, Milan, Italy

Abstract. We propose, to the best of our knowledge, the first Security Game where a *Defender* is supported by a spatially uncertain *alarm system* which non–deterministically generates *signals* once a target is under attack. Spatial uncertainty is common when securing large environments, e.g., for wildlife protection. We show that finding the equilibrium for this game is $\mathcal{FNP}$–hard even in the zero–sum case and we provide both an exact algorithm and a heuristic algorithm to deal with it. Without false positives and missed detections, the best patrolling strategy reduces to stay in a place, wait for a signal, and respond to it at best. This strategy is optimal even with non–negligible missed detection rates.

1 Introduction

Security Games model the task of protecting physical environments as a non–cooperative game between a *Defender* and an *Attacker* [12]. Usually taking place under a Stackelberg (a.k.a. leader–follower) paradigm [18], they have been shown to outperform other approaches such as, e.g., MDPs [16] and they have been employed in a number of on–the–field systems [9,13]. Recent research lines aim at refining the models by incorporating features from real–world applications, e.g., in [1,20] the Attacker may have different observation models and limited planning capabilities, in [6] realistic aspects of infrastructures to be protected are taken into account. Patrolling is one of the recently studied applications where the Defender controls mobile resources (such as patrolling robots) and the Attacker aims at compromising some locations denoted as targets [2]. Equilibrium strategies prescribe how the Defender should schedule resources in time to maximize its expected utility.

Infrastructures and environments that need to be surveilled are usually characterized by the presence of locally installed sensory systems. *Detection sensors* are able to gather measurements about suspicious events that an *alarm system* can process to generate alarm *signals*. These physical devices often present some degree of inaccuracy, such as false positives rates or missed detections rates. Alarm signals are spatially uncertain, meaning that they do not precisely localize the detected event, but provide a probabilistic belief over the locations

© Springer International Publishing Switzerland 2015
MHR Khouzani et al. (Eds.): GameSec 2015, LNCS 9406, pp. 192–207, 2015.
DOI: 10.1007/978-3-319-25594-1_11

potentially under attack. Spatial uncertainty is common when dealing with complex infrastructures or large open environments, where a *broad area surveillance* activity, in which an attack is detected but only approximately localized, triggers a *local investigation* activity, where guards have to find and clear the attack. A similar approach is adopted in a number of real–world problems where cheap and spatially uncertain sensors cover the targets to be protected. In [10], the problem of poaching of endangered species is studied and a device to help rangers against this threat is proposed. The introduction of cheap wide–range sensors, affordable by the conservation agencies, could significantly improve the behavior of the rangers, giving them information about the areas in which a potential attack is occurring. Other applications include UAVs surveillance [4], wildfires detection with CCD cameras [14] and monitoring agricultural fields [11]. In [21] a system for surveillance based on wireless sensor networks is proposed.

To the best of our knowledge, [8] is the only work integrating sensors in Security Games. It assumes sensors with no spatial uncertainty in detecting attacks on single targets. When no false positives are possible, an easy variation of the algorithm for the case without sensors [2] can be used, while, when false positives are present, the problem is computationally intractable.

Contributions. In this paper, we propose the first Security Game model that integrates a spatially uncertain alarm system in game–theoretic settings for patrolling. Each alarm signal carries the information about the set of targets that can be under attack and is described by a probability of being generated when each target is attacked. Moreover, the Defender can control only one patroller. We show that finding the equilibrium is $\mathcal{FNP}$–hard even in the zero–sum case and we give an exact exponential–time algorithm and a heuristic algorithm to deal with it. When no false positives and no missed detections are present, the optimal Defender strategy is to stay in a fixed location, wait for a signal, and respond to it at best. This strategy keeps being optimal even when non–negligible missed detection rates are allowed. Finally, we experimentally evaluate the scalability of our exact algorithm and we compare it with respect to the heuristic one in terms of solution quality.

2   Problem Formulation

Basic patrolling security game models [2,19] are turn–based extensive–form games with infinite horizon and imperfect information between two agents: an *Attacker* $\mathcal{A}$ and a *Defender* $\mathcal{D}$. The environment to be patrolled is formally described by an undirected connected graph $G = (V, E)$. Each edge $(i, j) \in E$ requires one turn to be traversed, while we denote with $\omega_{i,j}^*$ the temporal cost (in turns) of the shortest path between any i and $j \in V$. We denote by $T \subseteq V$ the subset of vertices called *targets* that have some value for $\mathcal{D}$ and $\mathcal{A}$. Each target $t \in T$ is characterized by a value $\pi(t) \in (0, 1]$ and a penetration time $d(t) \in \mathbb{N}$ measuring the number of turns needed to complete an attack over t. At each turn of the game, agents $\mathcal{A}$ and $\mathcal{D}$ play simultaneously: if $\mathcal{A}$ has not attacked in the previous turns, it can observe the position of $\mathcal{D}$ in the graph and decides whether to attack a target[1] or to wait for a turn, while $\mathcal{D}$ has no information about the actions undertaken by $\mathcal{A}$ in previous turns and decides the next vertex to patrol among all those adjacent to the current one. If $\mathcal{D}$ patrols a target t that is under attack of $\mathcal{A}$ before $d(t)$, $\mathcal{A}$ is captured. The game is constant sum (then equivalent to a zero sum game): if $\mathcal{A}$ is captured, $\mathcal{D}$ receives a utility of 1 and $\mathcal{A}$ receives 0, while, if an attack over t has success, $\mathcal{D}$ receives $1 - \pi(t)$ and $\mathcal{A}$ receives $\pi(t)$; finally, if $\mathcal{A}$ waits forever, $\mathcal{D}$ receives 1 and $\mathcal{A}$ receives 0. The appropriate solution concept is the leader–follower equilibrium. The game being constant sum, the best leader's strategy is its maxmin/minmax strategy.

Our *Patrolling Game* (PG) extends the above model introducing a *spatial uncertain alarm system* available to $\mathcal{D}$. The system is defined as a tuple (S, p) where $S = \{s_1, \cdots, s_m\}$ is a set of $m \geq 1$ *signals* and $p : S \times T \to [0, 1]$ is a function that specifies the probability of having the system generating a signal s given that target t has been attacked. With a slight abuse of notation, for a signal s we define $T(s) = \{t \in T \mid p(s \mid t) > 0\}$ and, similarly, for a target t we have $S(t) = \{s \in S \mid p(s \mid t) > 0\}$. In this work, we initially assume that the alarm system is not affected by false positives, i.e. a signal is generated but no attack has occurred, or missed detections, i.e. the signal is not generated even though an attack has occurred. In our model, at each turn, before deciding its next move, agent $\mathcal{D}$ can observe whether or not a signal has been generated by the alarm system.

We observe that, since no false positive and no missed detection are present, $\mathcal{D}$ will always receive a signal as soon as $\mathcal{A}$ starts an attack. This allows us to identify, in our game model, a number of subgames, each in which $\mathcal{D}$ is in a given vertex v and an attack is started. The solution to our PG can be safely found by, at first, finding the best strategies of $\mathcal{D}$ in responding to a signal from any $v \in V$ and, subsequently, on the basis of such signal–response strategies, by finding the best patrolling strategy over G. In Sect. 3 , we present algorithms to find the best signal–response strategies, while, in Sect. 4, we focus on the best patrolling strategies.

3   Finding the Best Signal–Response Strategy

We study the subgame in which $\mathcal{D}$ is in a vertex v and $\mathcal{A}$ decides to attack. We call it *Signal–Response Game given v* (SRG–v). The actions available to $\mathcal{A}$ are given by T and its strategy $\sigma_v^{\mathcal{A}}$ is defined as a probability distribution over T. We denote with $\sigma_{v,s}^{\mathcal{D}}$ the generic strategy of $\mathcal{D}$ when it is at v and receives a signal s and we discuss below the problem of defining the space of actions available to $\mathcal{D}$. We denote with g_v the expected utility of $\mathcal{A}$, the expected utility of $\mathcal{D}$ is $1 - g_v$. We show that, independently of how we define the space of actions of $\mathcal{D}$, the problem of finding the best $\sigma_v^{\mathcal{D}} = (\sigma_{v,s_1}^{\mathcal{D}}, \ldots, \sigma_{v,s_m}^{\mathcal{D}})$ is $\mathcal{FNP}$–hard [5]. We do this by assessing the complexity of its decision version.

---

[1] As is customary, we assume that $\mathcal{A}$ can instantly reach the target of its attack. This assumption can be easily relaxed as shown in [3].

Definition 1. k-**SRG**-v
INSTANCE: an instance of SRG-v *as defined above;*
QUESTION: is there $\sigma^{\mathcal{D}}$ *such that* $g_v \leq k$?

Theorem 1. k-**SRG**-v *is* $\mathcal{NP}$-*hard.*

Proof. Let us consider the following reduction from HAMILTONIAN–PATH. Given an instance of HAMILTONIAN–PATH $G_H = (V_H, E_H)$, we build an instance for k-**SRG**-v as:

- $V = V_H \cup \{v\}$;
- $E = E_H \cup \{(v, h), \forall h \in V_H\}$;
- $T = V_H$;
- $d(t) = |V_H|$;
- $\pi(t) = 1$, for all $t \in T$;
- $S = \{s\}$;
- $p(s \mid t) = 1$, for all $t \in T$;
- $k = 0$.

If $g_s \leq 0$, then there must exist a path starting from v and visiting all the targets in T by $d = |V_H|$. Given the edge costs and penetration times assigned in the above construction, the path must visit each target exactly once. Therefore, since $T = V_H$, the game's value is less or equal than zero if and only if G_H admits an Hamiltonian path. This concludes the proof. ☐

Given that an SRG–v is a subgame of the PG, it follows that finding the best strategy of $\mathcal{D}$ in PG is $\mathcal{FNP}$-hard. Since computing maxmin/minmax strategies can be done in polynomial time in the size of the payoffs matrix by means of linear programming, the difficulty of SRG–v resides in the generation of the payoffs matrix whose size is in the worst case exponential in the size of the graph (unless $\mathcal{P} = \mathcal{NP}$).

Now we focus on the problem of defining the set of actions available to $\mathcal{D}$ when it is in v and receives signal s. We define a generic *route* r as a sequence of vertices visited by $\mathcal{D}$. We denote with $r(i)$ the i–th vertex visited along r and with $A_r(r(i)) = \sum_{h=0}^{i-1} \omega^*_{r(h),r(h+1)}$ the time needed by $\mathcal{D}$ to visit $r(i)$ starting from $r(0)$. We restrict our attention on a subset of routes, that we call *covering routes*, with the following properties: $r(0) = v$ (i.e., the starting vertex is v), $\forall i \geq 1$ it holds $r(i) \in T(s)$, where s is the signal generated by the alarm system (i.e., only targets potentially under attack are visited) and $\forall i \geq 1$ it holds $A_r(r(i)) \leq d(r(i))$ (i.e., all the targets are visited within their penetration times) with $\mathcal{D}$ moving on the shortest paths between each pair of targets. Notice that a covering route r may visit a strict subset of $T(s)$. The set of actions available to $\mathcal{D}$ is given by all the covering routes. Given a covering route r, with a slight abuse of notation, we define the *covering set* $T(r)$ as the set of targets visited along r and we denote with $c(r)$ the temporal cost of the corresponding path, that is $c(r) = A_r(r(|T(r)|))$. Notice that in the worst case the number of covering routes is $O(|T(s)|^{|T(s)|})$, but using all of them may be unnecessary since some covering routes will never be played by $\mathcal{D}$ due to strategy domination and therefore they can be safely discarded [15]. We introduce two definitions of dominance that we use below.

Definition 2 (Intra–Set Dominance). Given two different covering routes r, r' for (v, s) such that $T(r) = T(r')$, if $c(r) \leq c(r')$ then r dominates r'.

Definition 3 (Inter–Set Dominance). Given two different covering routes r, r' for (v, s), if $T(r) \supset T(r')$ then r dominates r'.

Definition 2 suggests that we can safely use only one route per covering set. Covering sets suffice for computing the payoffs matrix of the game and in the worst case are $O(2^{|T(s)|})$, with a remarkable reduction of the search space w.r.t. $O(|T(s)|^{|T(s)|})$. However, any algorithm working directly with covering sets instead of covering routes should also decide whether or not a set of targets is a covering one: this problem is hard.

Definition 4. COV–SET
INSTANCE: a graph $G = (V, E)$, a target set T with penetration times d, and a starting vertex v;
QUESTION: is T a covering set for some covering route r?

By trivially adapting the same reduction for Theorem 1 we can state the following theorem.

Theorem 2. *COV–SET is $\mathcal{NP}$–complete.*

Computing a covering route for a given set of targets (or deciding that no covering route exists) is not doable in polynomial time unless $\mathcal{P} \neq \mathcal{NP}$. In addition, Theorem 2 suggests that no algorithm for COV–SET can have complexity better than $O(2^{|T(s)|})$ unless there is a better algorithm for HAMILTONIAN–PATH than the best algorithm known in the literature. This seems to suggest that enumerating all the possible subsets of targets and, for each of them, checking whether or not it is covering requires a complexity worse than $O(2^{|T(s)|})$. Surprisingly, we show in the next section that there is an algorithm with complexity $O(2^{|T(s)|})$ (neglecting polynomial terms) to enumerate all and only the covering sets and, for each of them, one covering route. Therefore, the complexity of our algorithm matches (neglecting polynomial terms) the complexity of the best known algorithm for HAMILTONIAN–PATH.

Definition 3 suggests that we can reduce further the set of actions available to $\mathcal{D}$. Given a covering set Q (where $Q = T(r)$ for some r), we say that Q is *maximal* if there is no route r' such that $Q \subset T(r')$. In the best case, when there is a route covering all the targets, the number of maximal covering sets is 1, while the number of covering sets is $2^{|T(s)|}$, thus considering only maximal covering sets allows an exponential reduction of the payoffs matrix. In the worst case, when all the possible subsets of $|T(s)|/2$ targets are maximal covering sets, the number of maximal covering sets is $O(2^{|T(s)|-2})$, while the number of covering sets is $O(2^{|T(s)|-1})$, allowing a reduction of the payoffs matrix by a factor of 2. Furthermore, if we knew *a priori* that Q is a maximal covering set we could avoid to search for covering routes for any set of targets that strictly contains Q. When designing an algorithm to solve this problem, Definition 3 could then be exploited to introduce some kind of pruning technique for saving average compute time. However, the following result shows that deciding whether a covering set is maximal is hard.

Definition 5. MAX–COV–SET
INSTANCE: a graph $G = (V, E)$, a target set (T, d), a starting vertex v, and a covering set $T' \subset T$;
QUESTION: is T' maximal?

Theorem 3. *MAX–COV–SET is in co–$\mathcal{NP}$ and no polynomial time for it exists unless $\mathcal{P} = \mathcal{NP}$.*

Proof. Any covering route r such that $T(r) \supset T'$ is a NO certificate for MAX–COV–SET, placing it in co–$\mathcal{NP}$. (Notice that, due to Theorem 2, having a covering set would not suffice given that we cannot verify in polynomial time whether it is actually covering unless $\mathcal{P} = \mathcal{NP}$.)

Let us suppose we have a polynomial–time algorithm for MAX–COV–SET, called A. Then (since $\mathcal{P} \subseteq \mathcal{NP} \cap$ co–$\mathcal{NP}$) we have a polynomial algorithm for the complement problem, i.e., deciding whether all the covering routes for T' are dominated. Let us consider the following algorithm: given an instance for COV–SET specified by graph $G = (V, E)$, a set of target T with penetration times d, and a starting vertex v:

1. assign to targets in T a lexicographic order $t_1, t_2, \ldots, t_{|T|}$;
2. for every $t \in T$, verify if $\{t\}$ is a covering set in $O(n)$ time by comparing $w^*_{v,t}$ and $d(t)$; if at least one is not a covering set, then output NO and terminate; otherwise set $\hat{T} = \{t_1\}$ and $k = 2$;
3. apply algorithm A on the following instance: graph $G = (V, E)$, target set $\{\hat{T} \cup \{t_k\}, \hat{d}\}$ (where $\hat{d}$ is d restricted to $\hat{T} \cup \{t_k\}$), start vertex v, and covering set $\hat{T}$;
4. if A's output is YES (that is, $\hat{T}$ is not maximal) then set $\hat{T} = \hat{T} \cup \{t_k\}$, $k = k + 1$ and restart from step 3; if A's output is NO and $k = |T|$ then output YES; if A's output is NO and $k < |T|$ then output NO;

Thus, the existence of A would imply the existence of a polynomial algorithm for COV–SET which (under $\mathcal{P} \neq \mathcal{NP}$) would contradict Theorem 2. This concludes the proof. $\square$

Nevertheless, we show in the following section that there is an algorithm enumerating all and only the maximal covering sets and one route for each of them (which potentially leads to an exponential reduction of the time needed for solving the linear program) with only an additional polynomial cost w.r.t. the enumeration of all the covering sets and therefore, neglecting polynomial terms, has a complexity $O(2^{|T(s)|})$.

3.1 Computing $\mathcal{D}$'s actions

Here we provide an algorithm to find the set of actions available to $\mathcal{D}$ when it is in v and receives signal s. Let us denote $C_{v,t}^k$ a collection of covering sets $Q_{v,t}^k$s such that $Q_{v,t}^k$ has cardinality k and admits a covering route r whose starting vertex is v and whose last vertex is t. Each $Q_{v,t}^k$ is associated with a cost $c(Q_{v,t}^k)$ representing the temporal cost of the shortest covering route for $Q_{v,t}^k$ that specifies t as the k–th target to visit. Upon this basic structure, our algorithm iteratively computes covering sets collections and costs for increasing cardinalities, that is from $k = 1$ possibly up to $k = |T|$ including one target at each iteration. Using a dynamic programming approach, we assume to have solved up to cardinality $k - 1$ and we specify how to complete the task for cardinality k. Detailed steps are reported in Algorithm 1, while in the following we provide an intuitive description. Given $Q_{v,t}^{k-1}$, we can compute a set of targets Q^+ (Line 6) such that for each target $t' \in Q^+$, $t' \notin Q_{v,t}^{k-1}$ and, if t' is appended to the shortest covering route for $Q_{v,t}^{k-1}$, it will be visited before $d(t')$. If Q^+ is not empty, for each $t' \in Q^+$, we extend $Q_{v,t}^{k-1}$ (Line 8) by including it and naming the resulting covering set as $Q_{v,t'}^k$ since it has cardinality k and we know it admits a covering route with last vertex t'. Such route is obtainable by appending t' to the covering route for $Q_{v,t}^{k-1}$ and has cost $c(Q_{v,t}^{k-1}) + \omega_{t,t'}^*$. This value is assumed to be the cost of the extended covering set. (In Line 9 we make use of a procedure $Search(Q, C)$ which outputs Q if $Q \in C$ and $\emptyset$ otherwise). If such extended covering set is not present in collection $C_{v,t'}^k$ or is already present with a higher cost (Line 10), then collection and cost are updated (Lines 11 and 12). After the iteration for cardinality k is completed, for each covering set Q in collection $C_{v,t}^k$, $c(Q)$ represents the temporal cost of the shortest covering route with t as last target.

Algorithm 1. ComputeCovSets_Basic(v, s)

1: $\forall t \in T(s), k \in \{2, \ldots, |T(s)|\}, C_{v,t}^1 = \{t\}, C_{v,t}^k = \emptyset$
2: $\forall t \in T(s), c(\{t\}) = \omega_{v,t}^*, c(\emptyset) = \infty$
3: **for all** $k \in \{2 \ldots |T(s)|\}$ **do**
4:     **for all** $t \in T(s)$ **do**
5:         **for all** $Q_{v,t}^{k-1} \in C_{v,t}^{k-1}$ **do**
6:             $Q^+ = \{t' \in T(s) \setminus Q_{v,t}^{k-1} \mid c(Q_{v,t}^{k-1}) + \omega_{t,t'}^* \leq d(t')\}$
7:             **for all** $t' \in Q^+$ **do**
8:                 $Q_{v,t'}^k = Q_{v,t}^{k-1} \cup \{t'\}$
9:                 $U = Search(Q_{v,t'}^k, C_{v,t'}^k)$
10:                 **if** $c(U) > c(Q_{v,t}^{k-1}) + \omega_{t,t'}^*$ **then**
11:                     $C_{v,t'}^k = C_{v,t'}^k \cup \{Q_{v,t'}^k\}$
12:                     $c(Q_{v,t'}^k) = c(Q_{v,t}^{k-1}) + \omega_{t,t'}^*$
13:                 **end if**
14:             **end for**
15:         **end for**
16:     **end for**
17: **end for**

After Algorithm 1 completed its execution, for any arbitrary $T' \subseteq T$ we can easily obtain the temporal cost of its shortest covering route as

$$c^*(T') = \min_{Q \in Y_{|T'|}} c(Q)$$

where $Y_{|T'|} = \cup_{t \in T} \{Search(T', C_{v,t}^{|T'|})\}$ (notice that if T' is not a covering set then $c^*(T') = \infty$). Algorithm 1 is dubbed "basic" because it does not specify how to carry out two sub–tasks we describe in the following.

The first one is the *annotation of dominated covering sets*. Each time Lines 11 and 12 are executed, a covering set is added to some collection. Let us call it Q and assume it has cardinality k. Each time a new Q has to be included at cardinality k, we mark all the covering sets at cardinality $k - 1$ that are dominated by Q (as per Definition 3). The sets that can be dominated are in the worst case $|Q|$, each of them has to be searched in collection $C_{v,t}^{k-1}$ for each feasible terminal t and, if found, marked as dominated. The number of terminal targets and the cardinality of Q are at most n and the *Search* procedure can be efficiently executed in $O(|T(s)|)$ using a binary tree approach. Therefore, dominated covering sets can be annotated with a $O(|T(s)|^3)$ extra cost at each iteration of Algorithm 1. We can only mark and not delete dominated covering sets since they can generate non–dominated ones.

The second task is the *generation of routes*. Algorithm 1 focuses on covering sets and does not maintain a list of corresponding routes. In fact, to build the payoffs matrix for SRG–v we do not strictly need covering routes since covering sets would suffice to determine payoffs. However, we do need them operatively since $\mathcal{D}$ should know in which order targets have to be covered to physically play an action. This task can be accomplished by maintaining an additional list of routes where each route is obtained by appending terminal vertex t' to the route stored for $Q_{v,t}^{k-1}$ when set $Q_{v,t}^{k-1} \cup \{t'\}$ is included in its corresponding collection. At the end of the algorithm only routes that correspond to non–dominated covering sets are filtered out. Maintaining such a list introduces a $O(1)$ cost.

Algorithm 1, in the worst case, has to compute covering sets up to cardinality $|T(s)|$. The number of operations is then bounded by $\sum_{i=1}^{|T(s)|} \binom{|T(s)|}{i-1} i(|T(s)| - 1)$ which is $O(|T(s)|^2 2^{|T(s)|})$. With annotations of dominances and routes generation the whole algorithm yields a worst case complexity of $O(|T(s)|^5 2^{|T(s)|})$.

3.2 A Heuristic Algorithm

We know that no polynomial–time algorithm solves exactly the COV–SET problem (unless $\mathcal{NP} = \mathcal{P}$) and therefore any exact algorithm of our problem cannot scale to tackle large settings. In this section, we focus on the design of a heuristic algorithm that can be used for very large instances of patrolling games with spatially uncertain alarms. We note that even if we had a polynomial–time approximation algorithm for COV–SET we would need to call the algorithm $O(2^{|T(s)|})$ times, one per set of targets, and therefore we would not have a polynomial–time approximation algorithm for our problem. This is why we do not focus on the design of approximation algorithms for COV–SET.

Our heuristic algorithm works as follows. Given v and s, for each target $t \in T(s)$ such that $w^*_{v,t} \leq d(t)$ we generate a covering route r with $r(0) = v$ and $r(1) = t$. Thus, $\mathcal{D}$ has at least one covering route per target (that can be covered in time from v). Each route r is expanded by inserting a target $t' \notin T(s) \setminus T(r)$ after position p and shifting each target that was at position $i > p$ in r at position $i + 1$. The pair (t', p) that determines the next expansion is chosen as the pair maximizing a heuristic function $h_r(t', p)$ among all the pairs leading to covering routes (i.e., insertions that make $A_r(t'') > d(t'')$ for some t'' are excluded). Route r is repeatedly expanded in greedy fashion until no insertion is possible. As a result, our algorithm generates at most $|T(s)|$ covering routes.

The heuristic function is defined as $h_r : \{T(s) \setminus T(r)\} \times \{1 \ldots |T(r)|\} \to \mathbb{Z}$, where $h_r(t', p)$ evaluates the cost of expanding r by inserting target t' after the p-th position of r. The basic idea (inspired by [17]) is to adopt a conservative approach, trying to preserve feasibility. Given a route r, let us define the *possible forward shift* of r as the minimum temporal margin in r between the arrival at a target t and $d(t)$: $PFS(r) = \min_{t \in T(r)}(d(t) - A_r(t))$. The *extra mileage* $e_r(t', p)$ for inserting target t' after position p is the additional traveling cost to be paid: $e_r(t', p) = (A_r(r(t')) + \omega^*_{r(p),t'} + \omega^*_{t',r(p+1)}) - A_r(r(p + 1))$. The *advance time* that such insertion gets with respect to $d(t')$ is defined as: $a_r(t', p) = d(t') - (A_r(r(p)) + \omega^*_{r(p),t'})$. Finally, $h_r(t', p)$ is defined as: $h_r(t', p) = \min\{a_r(t', p); (PFS(r) - e_r(t', p))\}$.

We partition the set $T(s)$ in two sets T_{tight} and T_{large} where $t \in T_{tight}$ if $d(t) < \delta \cdot w^*_{v,t}$ and $t \in T_{large}$ otherwise ($\delta \in \mathbb{R}$ is a parameter). The previous inequality is a non–binding choice we made to discriminate targets with a tight penetration time from those with a large one. Initially, we insert all the tight targets and only subsequently we insert the non–tight targets. It can be easily observed that our heuristic algorithm runs in $O(|T(s)|^3)$ given that heuristic h_r can be computed in $O(|T(s)|^2)$.

3.3   Solving SRG–v

Now we can formulate the problem of computing the equilibrium signal response strategy for $\mathcal{D}$. Let us denote with $\sigma^{\mathcal{D}}_{v,s}(r)$ the probability with which $\mathcal{D}$ plays route r under signal s and with $R_{v,s}$ the set of all the routes available to $\mathcal{D}$ generated by some algorithm. We introduce function $\mathcal{U}_\mathcal{A}(r, t)$ returning $\pi(t)$ if r is not a route covering t and 0 otherwise. The best $\mathcal{D}$ strategy (maxmin strategy) can be found by solving the following linear mathematical programming problem:

$$\min \quad g_v \quad \text{s.t.}$$

$$\sum_{s \in S(t)} p(s \mid t) \sum_{r \in R_{v,s}} \sigma^{\mathcal{D}}_{v,s}(r) \mathcal{U}_\mathcal{A}(r, t) \leq g_v \qquad \forall t \in T$$

$$\sum_{r \in R_{v,s}} \sigma^{\mathcal{D}}_{v,s}(r) = 1 \qquad \forall s \in S$$

$$\sigma^{\mathcal{D}}_{v,s}(r) \geq 0 \ \forall r \in R_{v,s}, s \in S$$

4   Finding the Best Patrolling Strategy

We now focus on the problem of finding the best patrolling strategy given that we know (from Sect. 3.3) the best signal–response strategy for each vertex v in which $\mathcal{D}$ can place. Given the current vertex of $\mathcal{D}$ and the sequence of the last, say n, vertices visited by $\mathcal{D}$ (where n is a tradeoff between effectiveness of the solution and computational effort), a patrolling strategy is usually defined as a randomization over the next adjacent vertices [2]. We define $v^* = \arg\min_{v \in V}\{g_v\}$, where g_v is the value returned by the optimization problem described in Sect. 3.3, as the vertex that guarantees the maximum expected utility to $\mathcal{D}$ over all the SRG–vs. We show that the maxmin equilibrium strategy in PG prescribes that $\mathcal{D}$ places at v^*, waits for a signal, and responds to it.

Theorem 4. *Without false positives and missed detections, if $\forall t \in T$ we have that $|S(t)| \geq 1$, then any patrolling strategy is dominated by the placement in v^*.*

Proof. Any patrolling strategy different from the placement in v^* should necessarily visit a vertex $v' \neq v^*$. Since the alarm system is not affected by missed detections, every attack will raise a signal which, in turn, will raise a response yielding a utility of g_x where x is the current position of $\mathcal{D}$ at the moment of the attack. Since $\mathcal{A}$ can observe the current position of $\mathcal{D}$ before attacking, $x = \arg\max_{v \in P}\{g_v\}$ where P is the set of the vertices patrolled by $\mathcal{D}$. Obviously, for any $P \supseteq \{v^*\}$ we would have that $g_x \geq g_{v^*}$ and therefore placing at v^* and waiting for signal is the best strategy for $\mathcal{D}$. $\qquad\square$

4.1   Computing the Best Placement

Under the absence of false positives and missed detections, Theorem 4 simplifies the computation of the patrolling strategy by reducing it to the problem of finding v^*. To such aim, we must solve SRG–v for each possible starting vertex v and select the one with maximum expected utility for $\mathcal{D}$. Since all the vertices are possible starting points, we should face this difficult problem (see Theorem 1) $|V|$ times, computing, for each signal, the covering routes from all the vertices. To avoid this issue, we ask whether there exists an algorithm that in the worst case allows us to consider a number of iterations, such that solving the problem for a given node v could help us finding the solution for another node v'. So, considering a specific set of targets, we wonder whether a solution for COV–SET with starting vertex v can be used to derive, in polynomial time, a solution to COV–SET for another starting vertex v'. To answer this question, we need to encode an instance of COV–SET in a different way, embedding the selection of the starting node in the structure of the graph. More precisely, we represent an instance of COV–SET $I = \langle G = (V, E), T, d, v \rangle$ with the equivalent instance $I' = \langle G' = (V', E'), c', T, d', \hat{v} \rangle$ defined in the following way:

- $V' = V \cup \{\hat{v}\}$, $E' = E \cup \{(\hat{v}, v_i), \forall v_i \in V\}$
- c' is a weight function such that $c(e) = 1$ if $e \in E \cup \{(\hat{v}, v)\}$ and $c(e) = M$ otherwise (M is a big constant);
- $d'(t) = d(t) + 1, \forall t \in T$.

With $\hat{v}$ we denote the dummy vertex that is always the starting node. We highlight the fact that, under this new encoding scheme, changing the starting vertex translates to rewriting the weights of c. Following the approach of [7], we can show that even the locally modified version of this problem, where a single weight is updated, is hard.

Definition 6. LM–COV–ROUTE *(Locally modified)*
INSTANCE: a graph $G = (V, E)$, a set of targets T with penetration times d, two weight functions c_1 and c_2 that coincide except for one edge, and a covering route r_1 such that, under c_1, $T(r_1) = T$.
QUESTION: is T a covering set under c_2?

Theorem 5. LM–*COV–ROUTE is $\mathcal{NP}$–complete.*

Proof. Let us consider the Restricted Hamiltonian Circuit problem (RHC) which is known to be $\mathcal{NP}$-complete. RHC is defined as follows: given a graph $G_H = (V_H, E_H)$ and an Hamiltonian path $P = \{h_1, \ldots, h_n\}$ for G_H such that $h_i \in V_H$ and $(h_1, h_n) \notin E_H$, find an Hamiltonian circuit for G_H. From such instance of RHC, following the approach of [7], we build the following instance for LM–COV–ROUTE:

- $V = T = V_H \cup \{v_s, v_t\}$;
- $E = \{(v_s, h_1), (v_s, h_{n-1})\} \cup \{(v_t, u)|(u, h_{n-1}) \in E_H\} \cup \overline{E_H}$ where $\overline{E_H}$ is the complete set of edges obtained by augmenting E_H;
- $d(v_s) = 0$, $d(v_t) = n + 1$, $d(t) = n$ for any $t \in T$;
- $c_1(e) = 1$ if $e \in E \cup \{(v_s, h_1), (v_s, h_{n-1})\} \cup \{(v_t, u)|(u, h_{n-1}) \in E_H\}$, $c_1(e) = (1 + \epsilon)$ otherwise (for any $\epsilon > 0$);
- $c_1 = c_2$ except for $c_2(v_s, h_1) = 1 + \epsilon$;
- $r_1 = \langle v_s, h_1, \cdots, h_n, v_t \rangle$.

It is easy to verify that G_H admits a Hamiltonian circuit if and only if T admits a covering route under c_2.   □

This shows that iteratively applying Algorithm 1 to SRG–v for each starting vertex v is the best we can do.

4.2   Robustness to Missed Detections

A deeper analysis of Theorem 4 can show that its scope does include cases where missed detections are present up to a non–negligible extent. For such cases, placement–based strategies keep being optimal even in the case when the alarm systems fails in detecting an attack. We encode the occurrence of this robustness property in the following proposition, which we shall prove by a series of examples.

Proposition 1. *There exist Patrolling Games where staying in a vertex, waiting for a signal, and responding to it is the optimal patrolling strategy for $\mathcal{D}$ even with a missed detection rate $\alpha = 0.5$.*

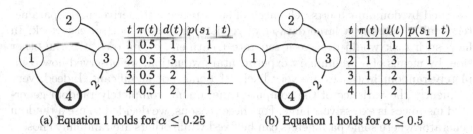

(a) Equation 1 holds for $\alpha \leq 0.25$ (b) Equation 1 holds for $\alpha \leq 0.5$

Fig. 1. Two examples proving Proposition 1.

Proof. The expected utility for $\mathcal{D}$ given by the placement in v^* is $(1-\alpha)(1-g_{v^*})$, where $(1-\alpha)$ is the probability with which the alarm system correctly generates a signal upon an attack and $(1-g_{v^*})$ denotes $\mathcal{D}$'s payoff when placed in v^*. A non–placement–based patrolling strategy will prescribe, by definition, to move between at least two vertices. From this simple consideration, we observe that an upper bound to the expected utility of any non–placement strategy is entailed by the case where $\mathcal{D}$ alternately patrols vertices v^* and v_2^*, where v_2^* is the second best vertex in which $\mathcal{D}$ can statically place. Such scenario give us an upper bound over the expected utility of non–placement strategies, namely $1-g_{v_2^*}$. It follows that a sufficient condition for the placement in v^* being optimal is given by the following inequality:

$$(1-\alpha)(1-g_{v^*}) > (1-g_{v_2^*}) \tag{1}$$

To prove Proposition 1, it then suffices to provide a Patrolling Game instance where Eq. 1 holds under some non–null missed detection rate α. In Fig. 1(a) and (b), we report two of such examples. The depicted settings have unitary edges except where explicitly indicated. For both, without missed detections, the best patrolling strategy is a placement $v^* = 4$. When allowing missed detections, in Fig. 1(a) it holds that $g_{v^*} = 0$ and $g_{v_2^*} = 0.75$, where $v^* = 4$ and $v_2^* = 1$. Thus, by Eq. 1, placement $v^* = 4$ is the optimal strategy for $\alpha \leq 0.25$. Under the same reasoning scheme, in Fig. 1(b) we have that $g_{v^*} = 0$ and $g_{v_2^*} = 0.5$, making the placement $v^* = 4$ optimal for any $\alpha \leq 0.5$.   □

5   Experimental Evaluation

We evaluate the scalability of Algorithm 1 and the quality of the solution returned by our heuristic algorithm for a set of instances of SRG–v. We do not include results on the evaluation of the algorithm to solve completely a PG, given that it trivially requires asymptotically $|V|$ times the effort required by the resolution of a single instance of SRG–v.

Testbed. In real deployment scenarios, the model parameters should be derived from the particular features that characterize the particular setting one must deal with. Besides the graph topology, which depends on the environment, target values and deadlines can be derived from available statistics or manually assigned by domain experts. The need of such process to derive model parameters makes building a large dataset of realistic instances not an easy task. In fact, such task would deserve a separate treatment by its own. On the other side, by means of a preliminary experimental evaluation, we observed how completely random instances are very likely of being not significant. Indeed, very frequently the variance of the compute time among completely random generated instances is excessively large. For these reasons, we decided to use a random generator where some parameters can be fixed while others are randomly chosen. We restricted our attention to basic, but significant, instances with all–targets graphs, arc costs set to 1, penetration times to $|T(s)|-1$, and the number of arcs is drawn from a normal distribution with mean ϵ, said edge density and defined as $\epsilon = |E|/\frac{|T(s)|(|T(s)|-1)}{2}$ (other parameters are randomly generated from uniform distributions, unless otherwise specified). Instances constructed with such mechanism include hard ones since the existence of a covering route over $T(s)$ would imply the existence of an Hamiltonian path on the graph. We explore two parameter dimensions: the number of targets $|T|$ and ϵ. Algorithms are developed in MATLAB and run on a 2.33 GHz LINUX machine.

Exact Algorithm Scalability. Table 1 shows the total compute time required to solve instances with a single signal, that can be generated by any target under attack. Table 2 refers to instances with multiple signals, where the targets covered by a signal and the probability that a target triggers a signal are randomly chosen according to a uniform distribution (in this second table $|T|$ is fixed to 16). Values are averages over 100 random instances and give insights on the computation effort along the considered dimensions. The results show that the problem is computationally challenging even for a small number of targets and signals.

Table 1. Compute times (in seconds) for single–signal instances.

ϵ \ $|T|$	6	8	10	12	14	16	18
.25	0.07	0.34	1.91	11.54	82.26	439.92	4068.8
.5	0.07	0.38	4.04	53.14	536.7	4545.4	≥ 5000
.75	0.09	0.96	11.99	114.3	935.74	7276.62	≥ 7000
1	0.14	1.86	17.46	143.05	1073.19	7964.49	≥ 8000

Table 2. Compute times (in seconds) for multi–signal instances.

m \ $|T(s)|$	5	10	15
2	-	17.83	510.61
3	-	33	769.3
4	0.55	35.35	1066.76
5	0.72	52.43	1373.32

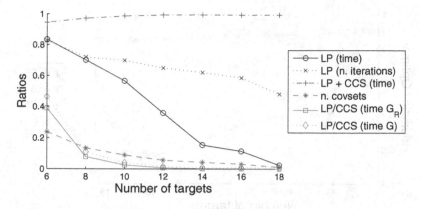

Fig. 2. Ratios evaluating dominances.

Figure 2 shows the impact of discarding dominated actions from the game. It depicts the trend of some performance ratios for different metrics. We shall call $\mathcal{G}$ the complete game including all $\mathcal{D}$'s dominated actions and $\mathcal{G}_R$ the reduced game; CCS will denote the full version of Algorithm 1 and LP will denote the linear program to solve SRG–v. Each instance has edge density $\epsilon = .25$ and is solved for a random starting vertex v; we report average ratios for 100 instances. "n. covsets" is the ratio between the number of covering sets in $\mathcal{G}_R$ and in $\mathcal{G}$. Dominated actions constitute a large percentage, increasing with the number of targets. This result indicates that the structure of the problem has some redundancy. LP times (iterations) report the ratio between $\mathcal{G}_R$ and $\mathcal{G}$ for the time (iterations) required to solve the minmax linear program. A relative gain directly proportional to the percentage of dominated covering sets is observable (LP has less variables and constraints). A similar trend is not visible when considering the same ratio for the total time which includes CCS. Indeed, the time needed by CCS largely exceed LP's and removal of dominated actions determines a polynomial additional cost which can be seen in the slightly increasing trend of the curve. The relative gap between LP and CCS compute times can be assessed by looking at the LP/CCS curve: when more targets are considered the time taken by LP is negligible w.r.t. CCS's. This shows that removing dominated actions is useful, allowing a small improvement in the average case, and assuring an exponential improvement in the worst case.

Heuristic Solution Quality. Figure 3 reports the performance of the heuristic algorithm (here we set $\delta = 2$) in terms of $\mathcal{D}$'s expected utility ratio $(1 - g_v)/(1 - \hat{g}_v)$, where g_v is the expected utility of $\mathcal{A}$ at the equilibrium considering all the covering sets and $\hat{g}_v$ is the expected utility of $\mathcal{A}$ at the equilibrium when covering sets are generated by our heuristic algorithm. The performance of our heuristic algorithm is well characterized by ϵ, providing fairly good approximations for $\epsilon > 0.25$, the ratio going to 1 as $|T|$ increases, because there are more edges and, consequently, there is a higher probability for the heuristics to find longer routes. The figure suggests that assessing the membership of our problem to the $\mathcal{APX}$ class could be an interesting problem.

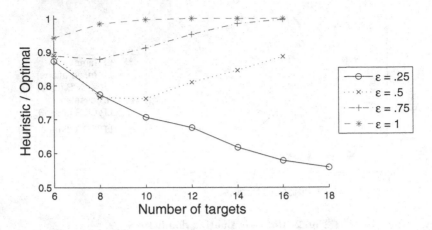

Fig. 3. Optimal vs heuristic algorithms.

6   Conclusions and Future Research

In this paper, to the best of our knowledge, we provide the first Security Game for large environments surveillance, e.g. for wildlife protection, that can exploit an alarm system with spatially uncertain signals. We propose a simple model of alarm systems that can be widely adopted with every specific technology and we include it in the state–of–art patrolling models obtaining a new security game model. We show that the problem of finding the best patrolling strategy to respond to a given alarm signal is $\mathcal{FNP}$–hard even when the game is zero sum. Then, we provide an exponential–time exact algorithm to find the best patrolling strategy to respond to a given alarm signal. We provide also a heuristic algorithm returning approximate solutions to deal with very large game instances. Furthermore, we show that if every target is alarmed and no missed detections are present, then the best patrolling strategy prescribes that the patroller stays in a given place waiting for a alarm signal. We show that such a strategy may be optimal even for missed detection rates up to 50 %. Finally, we experimentally evaluate our algorithms in terms of scalability (for the exact algorithm) and approximation ratio (for the heuristic algorithm).

In future works, we shall study the membership (or not) of our problem to $\mathcal{APX}$ class, design approximation algorithms with theoretical guarantees and investigate the impact of missed detections and false positives.

References

1. An, B., Brown, M., Vorobeychik, Y., Tambe, M.: Security games with surveillance cost and optimal timing of attack execution. In: AAMAS, pp. 223–230 (2013)
2. Basilico, N., Gatti, N., Amigoni, F.: Patrolling security games: definition and algorithms for solving large instances with single patroller and single intruder. Artif. Intell. **184–185**, 78–123 (2012)

3. Basilico, N., Gatti, N., Rossi, T.: Capturing augmented sensing capabilities and intrusion delay in patrolling-intrusion games. In: IEEE Symposium on Computational Intelligence and Games, CIG 2009, pp. 186–193, September 2009
4. Basilico, N., Carpin, S., Chung, T.: Distributed online patrolling with multi-agent teams of sentinels and searchers. In: DARS (2014)
5. Bellare, M., Goldwasser, S.: The complexity of decision versus search. SIAM J. Comput. 23(1), 97–119 (1994)
6. Blum, A., Haghtalab, N., Procaccia, A.D.: Lazy defenders are almost optimal against diligent attackers. In: AAAI, pp. 573–579 (2014)
7. Böckenhauer, H.J., Forlizzi, L., Hromkovič, J., Kneis, J., Kupke, J., Proietti, G., Widmayer, P.: Reusing optimal tsp solutions for locally modified input instances. In: IFIP TCS, pp. 251–270 (2006)
8. Munoz de Cote, E., Stranders, R., Basilico, N., Gatti, N., Jennings, N.: Introducing alarms in adversarial patrolling games. In: AAMAS, pp. 1275–1276 (2013)
9. Delle Fave, F.M., Jiang, A., Yin, Z., Zhang, C., Tambe, M., Kraus, S., Sullivan, J.P.: Game-theoretic patrolling with dynamic execution uncertainty and a case study on a real transit system. JAIR 50, 321–367 (2014)
10. Ford, B.J., Kar, D., Fave, F.M.D., Yang, R., Tambe, M.: PAWS: adaptive game-theoretic patrolling for wildlife protection. In: International conference on Autonomous Agents and Multi-Agent Systems, AAMAS 2014, Paris, France, 5–9 May 2014, pp. 1641–1642 (2014)
11. Garcia-Sanchez, A.J., Garcia-Sanchez, F., Garcia-Haro, J.: Wireless sensor network deployment for integrating video-surveillance and data-monitoring in precision agriculture over distributed crops. Comput. Electron. Agric. 75(2), 288–303 (2011)
12. Jain, M., An, B., Tambe, M.: An overview of recent application trends at the AAMAS conference: security, sustainability, and safety. AI Mag. 33(3), 14–28 (2012)
13. Jain, M., Tsai, J., Pita, J., Kiekintveld, C., Rathi, S., Tambe, M., Ordóñez, F.: Software assistants for randomized patrol planning for the lax airport police and the federal air marshal service. Interfaces 40(4), 267–290 (2010)
14. Ko, B.C., Park, J.O., Nam, J.Y.: Spatiotemporal bag-of-features for early wildfire smoke detection. Image Vis. Comput. 31(10), 786–795 (2013)
15. Osborne, M.J.: An Introduction to Game Theory, vol. 3. Oxford University Press, New York (2004)
16. Paruchuri, P., Tambe, M., Ordóñez, F., Kraus, S.: Security in multiagent systems by policy randomization. In: AAMAS, pp. 273–280 (2006)
17. Savelsbergh, M.W.: Local search in routing problems with time windows. Ann. Oper. Res. 4(1), 285–305 (1985)
18. Von Stengel, B., Zamir, S.: Leadership with commitment to mixed strategies (2004)
19. Vorobeychik, Y., An, B., Tambe, M., Singh, S.P.: Computing solutions in infinite-horizon discounted adversarial patrolling games. In: Proceedings of the Twenty-Fourth International Conference on Automated Planning and Scheduling, ICAPS 2014, Portsmouth, New Hampshire, USA, 21–26 June 2014 (2014)
20. Yang, R., Ford, B., Tambe, M., Lemieux, A.: Adaptive resource allocation for wildlife protection against illegal poachers. In: International Conference on Autonomous Agents and Multiagent Systems (AAMAS) (2014)
21. Yick, J., Mukherjee, B., Ghosal, D.: Wireless sensor network survey. Comput. Netw. 52(12), 2292–2330 (2008)

Determining a Discrete Set of Site-Constrained Privacy Options for Users in Social Networks Through Stackelberg Games

Sarah Rajtmajer[1]([✉]), Christopher Griffin[2], and Anna Squicciarini[3]

[1] Department of Mathematics, The Pennsylvania State University,
State College, USA
smr48@psu.edu
[2] Mathematics Department, United States Naval Academy, Annapolis, USA
griffinch@ieee.org
[3] College of Information Sciences and Technology,
The Pennsylvania State University, State College, USA
acs20@psu.edu

Abstract. The privacy policies of an online social network play an important role in determining user involvement and satisfaction, and in turn site profit and success. In this paper, we develop a game theoretic framework to model the relationship between the set of privacy options offered by a social network site and the sharing decisions of its users within these constraints. We model the site and the users in this scenario as the leader and followers, respectively, in a Stackelberg game. We formally establish the conditions under which this game reaches a Nash equilibrium in pure strategies and provide an approximation algorithm for the site to determine a discrete set of privacy options to maximize payoff. We validate hypotheses in our model on data collected from a mock-social network of users' privacy preferences both within and outside the context of peer influence, and demonstrate that the qualitative assumptions of our model are well-founded.

1 Introduction

At its core, an online social network (SN) is an infrastructure for user-generated shared content. Users have the ability to exercise control over their individual channels in the network, by deciding which content to share and with whom to share it. The SN site benefits from shared content in important ways. Shared content attracts new users, deepens the involvement of existing users, strengthens the community, and can be leveraged for monetization.

Individual behavior online, like individual behavior offline, is also subject to social norms and peer influence [12,15,24]. Notions of what is appropriate in content sharing online is defined comparatively, so that subtle shifts in local behavior may have much farther-reaching consequences for the network as a whole. In sum, unlike the SN site which is ultimately a business operating with

MHR Khouzani et al. (Eds.): GameSec 2015, LNCS 9406, pp. 208–227, 2015.
DOI: 10.1007/978-3-319-25594-1_12

a business model, users are individuals with more complex incentives, concerns and considerations operating voluntarily within the constraints of the SN.

Questions related to privacy in SNs have gained increasing interest over the last few years as the ubiquity of social media has become apparent and anecdotes of repercussions for over-disclosure more available. Many users are now aware of the risks associated with revelation online and concerned with protecting personal information from widespread dissemination. Advocates of fine-grained privacy policies argue that detailed user management of privacy settings for shared content can avert some of the potential risks users face in online SNs [20, 28]. Users can sort their data into categories to be shared with certain individuals in the network (i.e., friends, friends of friends, groups, everyone). SNs like Facebook and Google+ have implemented this model, allowing users to create narrower social circles from among their list of friends and to define which content is shared with whom. Unfortunately, studies have also shown that users often do not take advantage of finely-tuned options available to them. The majority of users on both Facebook and Twitter maintain the default privacy settings established by the site [12, 19], which tend to be more permissive than users would like [23].

In this work, we focus on the topic of privacy, from the perspectives of both the SN site and its users. We seek to determine an optimal discrete set of privacy options to be made available to users for content sharing. We define optimality here from the perspective of the site, taking into account user satisfaction. Intuitively, the site is to choose a set of options for users' shared content in order to maximize sharing. Yet, the site should allow users to maintain a level of control over their content without being overwhelmed by too many or too complex privacy settings from which to choose.

We model the conflicting yet complementary goals of the SN site and its users as a Stackelberg game whereby the leader (the site) moves first in setting the privacy options to be made available to user-members for shared content. Followers (users) respond by selecting privacy settings from among these options. Payoff to the site can be expressed in terms of amount of shared content and total user happiness. Payoff to each user depends on how closely the available options approximate his ideal sharing preferences, which is in turn a function of an inherent comfort and peer influences. We formally present this two-level game as well as a characterization of its convergence to a Nash equilibrium in pure strategies under certain simplifying assumptions. We develop an agent-based model to approximate optimal strategies on arbitrary network graphs and validate model assumptions with a study of 60 individuals, run over a mock-SN.

The remainder of this paper is organized as follows. The next section reviews related work, followed by our problem statement, succeeded by an overview of our model in Sect. 4. Section 5 presents approximation algorithms, and Sect. 6 discusses the experimental study we carried out. We conclude the paper in Sect. 7.

2   Related Work

The scale and gravity of privacy and security risks associated with online social networks have led to a rich body of work addressing a wide spectrum of these issues. By sharing their personal information, users in SNs become vulnerable to attacks from other users, the SN itself, third-party applications linked to their SN profiles, or other outside attackers able to de-anonymize user data published by the SN site. See [2,18] for recent reviews. These attacks may take the form of identity theft [12], scraping and harvesting [21], social phishing [17], or automated social engineering [3]. The risk of a breach of privacy in some form is particularly salient for users who are not closely monitoring their privacy settings or leaving privacy settings at their default values.

As a means of mediating some of these risks, there is a growing literature using machine learning to determine individual default privacy settings. *PriMa* [31] and *Privacy Wizard* [8] are examples of supervised learning algorithms which look at the behavior and preferences of a user, the behavior and preferences of his peer group or related users, and offer a classification of default settings for different types of shared content. We see this work as complementary to ours in that it does not suggest a method for the determining the privacy settings from which a user may choose, but rather once these options are in place, gives a method for selecting defaults amongst them which may most closely match a user's preferences.

This work is related in general to the body of work on game theory in social networks, both offline and online. Fundamental research efforts exploring cooperation in structured human populations include [23,26,38]. In the realm of online social networks, game theoretic models have been implemented for the study of the evolution of various social dilemmas and associated changes in network structure [9,16,25].

Most closely related to our work is the subset of this research concerning agent-based decision-making related to privacy and security in online social networks. Chen and colleagues model users' disclosure of personal attributes as a weighted evolutionary game and discuss the relationship between network topology and revelation in environments with varying level of risk [5].

In a series of papers considering the circumstances of deception in online SNs, Squicciarini et al. characterize a user's willingness to release, withhold or lie about information as a function of risk, reward and peer pressure within different game-theoretic frameworks [29,33]. They describe the relationship between a site and its users, determining that in the in the presence of a binding agreement to cooperate (strong guarantees on privacy), most users will agree to share real identifying information in return for registration in the system [34]. Authors also use a game theory to model of collective privacy management for photo sharing in SNs [32,35]. Their approach proposes automated privacy settings for shared images based on an extended notion of content co-ownership.

To the best of our knowledge, a game-theoretic approach to determining the privacy policy of an online SN has not been considered before in the literature.

In a previous work [11], we tackled the simpler question of determining a mandatory lower-bound on shared content. That is, we have addressed the SN site's decision of selecting the minimum amount of shared personal information which should be required of user with an active account in the network. For example, Facebook requires all users with a personal account to give a first name, last name, valid email address, password, gender and birth date. In fact, Facebook institutes further sharing requirements on various elements of a user's profile, e.g., a user's cover photo is always public [6].

3   Problem Statement

We assume a captive social network site, wherein users share pieces of personal content freely within the network and possibly with selected subgroups of network users, according to a set of privacy options for shared content made available by the site to its users.

We assume the site benefits when users share as freely as possible and it is of course incentivized to create options that promote the widest distribution of posted content. The site, however, must also be wary to consider users who are inherently more cautious about public sharing. A site requiring all shared content to be public, for example, may lure some users to post publicly who might otherwise have only shared with a narrower group, i.e., "friends only". But in other cases, this policy might have a detrimental effect for the site, as users may choose not to post at all. In any case, if the privacy setting a user would prefer for a piece of content is not presented the user will experience some degree of dissatisfaction in having to select an alternative. Figure 1 illustrates the problem space.

Users react to the options offered by choosing what to disclose and with whom. Examples of these settings in practice may include "visible to only me", "share with specific individuals", "share with friends", "share with my network" and "public". We abstract away from the details of how privacy options

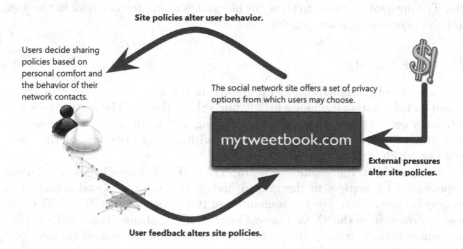

Site policies alter user behavior.

Users decide sharing policies based on personal comfort and the behavior of their network contacts.

The social network site offers a set of privacy options from which users may choose.

mytweetbook.com

External pressures alter site policies.

User feedback alters site policies.

Fig. 1. There is a natural push and pull between a SN site and its users with regard to sharing policies.

are presented to users, and map them to real values on the interval $[0, 1]$. The granularity of these options should be fine enough to meet users' needs, but coarse enough to be manageable in implementation for both the users and the SN site.

We formulate the site's utility as a function of user happiness and shared content, so that minimally the site would like to make sure that no user is unable to share content as freely as he would like due to a lack of available sharing options. In fact, the site would stand to profit by pushing users toward the upper boundary of their sharing comfort, and having a carefully chosen set of options may enable this to happen.

We model each user's utility function as a weighted sum of discomfort and peer pressure. Specifically, each user will act to minimize the difference between his selected privacy setting and his personal comfort level, and the difference between his selected privacy setting and the average privacy settings of his peers. The intuition is that users have an inherent degree of disclosure they feel most comfortable with, but are also influenced by their peers when making sharing decisions [7,14]. Since these two dimensions may not be considered equally for all users, we introduce weights to capture interpersonal differences in susceptibility to peer pressure. Precisely, we offer the option of including weights on either the peer pressure or personal comfort components of the user's utility function allowing customization of the model for non-homogeneous users and an opportunity to strengthen the model in the presence of additional information on user behavior, which the site may learn through observation.

4   Model Overview

We define two optimization problems: one for the SN user and one for the SN site. The optimal solutions to these problems determine the behavior of the user and site regarding privacy policies.

4.1   User Model

Our user model extends the model presented in [11] for the modeling of a lower-bound on information disclosure for membership in the SN. The motivations and actions of users with respect to content sharing in this framework are consistent with this prior work, but will be enacted within the constraints of the site's problem which is significantly different.

Assume a SN is represented by a graph $G = (V, E)$, where V is a set of users (represented by vertices in the graph) and E is the set of social connections (edges) between them. For the remainder of this paper, assume $|V| = N$. Users post information to the SN for reasons known only to themselves. Unlike in [30], we assume users who are perfectly honest, but may choose to omit (or keep private) a certain amount of information. Previous work has observed [10,30] that users have distinct sharing behaviors for different types of information, depending on the "social" value of such information (e.g., users are more willing to share their gender than their phone number). Assume there are M types of information. Since it is nontrivial to specify what a piece of information corresponds to in a SN, we will abstract away from any specific characterization of information, and assume User $i \in V$ accumulates *postable* information of type j at a rate of $\beta_i^j(t)$ (given in bits per second). Each user chooses a proportion (probability) of information of type j to share, denoted by $x_i^j(t) \in [0,1]$.

In general, users do not change their privacy policy frequently [22], and thus we can consider a simplified problem in which we attempt to find optimal values for (fixed) x_i^j ($i \in \{1,\ldots,N\}$, $j \in \{1,\ldots,M\}$). To do this, we define optimality in terms of:

1. Peer Pressure (and reputation),
2. Comfort level

Comfort level in the context of privacy and information disclosure refers to the degree of disclosure users feel comfortable with. This notion, often used to characterize information sharing in online sites (e.g. [1,7]), is also adopted in our model. Users reaching their optimal comfort level wish not to change any of their information sharing practices. Reputation and peer pressure are self-explanatory, and are combined in a single dimension as they are highly correlated [30].

Without loss of generality, focus on *one* information type, $x_i \in [0,1]$. To model peer pressure, we assume that individuals are encouraged to behave in accordance with the norms of their social group. Thus for User i, we define:

$$\bar{x}_{-i} = \frac{\sum_{j \in N_G(i)} v_{ij} x_j}{V_G(i)}$$

where $v_{ij} \geq 0$ and

$$V_G(i) = \sum_j v_{ij} \qquad (1)$$

is the weighted neighborhood size of i in G. If $v_{ij} = 1$ for all j, then $V_G(i) = |N_G(i)|$, the size of the neighborhood of i in G. The neighborhood may be defined in terms of the social graph of the user, or it may be a more restrictive subset of peers with whom the user actively interacts. Let the peer pressure function for User i be given by:

$$P_i(x) = v_i f_P(x - \bar{x}_{-i}) \qquad (2)$$

where f_P is a concave function with maximum at 0 and $v_i \geq 0$ is the subjective weight User i places on the peer pressure function. Thus, the payoff $P_i(x)$ is maximized as x_i approaches $\bar{x}_{-i}$.

We note that an alternate and equally reasonable approach to defining $P_i(x)$ is as:

$$\tilde{P}_i(x) = \sum_{j \in N_G(i)} v_{ij} f_P(x - x_j) \qquad (3)$$

where $v_{ij} \geq 0$. In this case, User i attempts to minimize a weighted function of the difference in privacy levels from all of his neighbors simultaneously.

Estimated weights on the link between User i and User j might be obtained, for example, as a function of the frequency and type of online interactions between them. This formulation increases the complexity of the problem and ultimately makes computation more cumbersome, but allows a richer model when more detailed information about users' relationships and peer influence is present.

By similar argument, assume that User i has a sharing level x_i^+ at which he is happiest. The comfort function $f_C(z)$ for User i is given by:

$$C_i(x) = w_i f_C(x - x_i^+)$$

for $w_i \geq 0$, which can be thought of as a user's tendency to act in preference to his own comfort rather than in response peer pressure. Here again, f_C is concave with maximum at 0, so that the comfort of User i is maximized as x_i approaches x_i^+.

In practice x_i^+ may be difficult to determine for an unknown User i. However, we assume that based on user demographics, as well as observed overall user behavior for a mass of users, either at the individual or group level, it is possible to infer of x_i^+, or at least an expected value $E[x_i^+]$ within a tolerated window of error.

Thus, the total objective function for User i is:

$$J_i(x_i; x_{-i}) = P_i(x_i) + C_i(x_i) = v_i f_P \left(x_i - \frac{\sum_{j \in N_G(i)} x_j}{|N_G(i)|} \right) + w_i f_C(x_i - x_i^+) \quad (4)$$

or, the weighted variant:

$$\tilde{J}_i(x_i; x_{-i}) = \tilde{P}_i(x_i) + C_i(x_i) = \sum_{j \in N_G(i)} v_{ij} f_P(x_i - x_j) + w_i f_C(x_i - x_i^+). \quad (5)$$

Here, x_{-i} indicates the privacy choices of all other users besides i and we write $J_i(x_i; x_{-i})$ to indicate that User i's utility is a function not only of his own decisions, but also of the decisions of the other users.

When f_P and f_C are concave, the following proposition holds [27]:

Proposition 1. *Assume that each x_i is constrained to lie in a convex set $X_i \subseteq [0,1]$ for $i = 1, \ldots, N$. There is at least one value x_i^* for each User i so that every user's objective function is simultaneously maximized and $(x_1^*, \ldots, x_N^*)$ is a Nash Equilibrium for the multi-player game defined by any combination of objective functions $J_1, \ldots, J_N$ or $\tilde{J}_1, \ldots, \tilde{J}_N$.* □

By similar reasoning, the preceding proposition can be extended to the case of multiple independent information types. In this case for each $j = 1, \ldots, M$ there is an equilibrium solution $x_i^{j^*}$ $i = 1, \ldots, N$. Correlated payoffs for information sharing among information types are beyond the scope of the current work.

In general, in this case, each user would have an information sharing strategy $\mathbf{x}_i \in [0,1]^M$ and a corresponding multi-dimensional payoff function. The existence of a Nash equilibrium would be guaranteed for convex functions with convex constraints.

4.2 Site Model for the Determination of a Discrete Set of Privacy Options for Shared Content

For the remainder of this paper, we will assume a user objective function of the form $\tilde{J}_i$ and fix $f_C(z) = f_P(z) = -z^2$, which is concave with maximum at zero. Furthermore, and for notational simplicity, we will consider the minimizing form of the problem in which User i minimizes $-\tilde{J}_i$.

Assume the site offers a discrete set of privacy settings $l_1, \ldots, l_K \in [0,1]$. Each user must choose from among these options for each piece of shared content. This is equivalent to choosing a generic privacy policy within a social network. Let $\mathbf{l}$ be the vector of these options. Define:

$$y_{ij} = \begin{cases} 1 & \text{Player } i \text{ chooses privacy level } j \\ 0 & \text{otherwise} \end{cases} \tag{6}$$

these binary variables indicate the privacy levels of each player. Naturally we require:

$$\sum_j y_{ij} = 1 \tag{7}$$

Let $\mathbf{y}$ be the matrix of y_{ij} values. Furthermore:

$$x_i(\mathbf{y}; \mathbf{l}) = \sum_{j=1}^{K} y_{ij} l_j$$

For given values y_{ij} ($i = 1, \ldots, N$ and $j = 1, \ldots, K$), the payoff to Player i is:

$$H_i(\mathbf{y}; \mathbf{l}) = \sum_{j \in N(i)} v_{ij}(x_i - x_j)^2 + w_i(x_i - x_i^+)^2 \tag{8}$$

Note, this is simply $-\tilde{J}_i$. Then the net payoff to the site is:

$$J(\mathbf{y}; \mathbf{l}) = \sum_i \left(\sum_j \pi_j y_{ij} - \lambda H_i \right), \tag{9}$$

where π_j is the benefit the site receives for a piece of content shared with privacy setting j and λ is the weight applied to the payoff of the users; i.e., the weight the site places on user happiness. When $\mathbf{y}$ is determined endogenously by the players, then the site's bi-level combinatorial optimization problem is:

$$
\begin{cases}
\max_{l} \ J(\mathbf{y}; \mathbf{l}) = \sum_i \left(\sum_j \pi_j y_{ij} - \lambda H_i \right) \\[2mm]
s.t. \ l_1, \ldots, l_K \in [0, 1] \\
\quad l_j \le l_{j+1} \quad j = 1, \ldots K - 1 \\[2mm]
\forall i
\begin{cases}
H_i(\mathbf{y}; \mathbf{l}) = \min_{\mathbf{y}_i} \ \sum_{j \in N(i)} v_{ij}(x_i - x_j)^2 + w_i(x_i - x_i^+)^2 \\[2mm]
\quad s.t. \ x_i = \sum_{j=1}^K y_{ij} l_j \\[2mm]
\quad \sum_j y_{ij} = 1 \\[2mm]
\quad y_{ij} \in \{0, 1\}
\end{cases}
\end{cases}
\tag{10}
$$

In this problem, each User i must decide the value of y_{ij} independently of all other users, while being simultaneously affected by her choice. It is clear that the sub-game has a solution in mixed strategies from Proposition 1, but what is less clear is whether it has a solution in pure strategies.

Consider the user game-theoretic sub-problem:

$$
\forall i
\begin{cases}
H_i(\mathbf{y}; \mathbf{l}) = \min_{\mathbf{y}_i} \ \sum_{j \in N(i)} v_{ij}(x_i - x_j)^2 + w_i(x_i - x_i^+)^2 \\[2mm]
\quad s.t. \ x_i = \sum_{j=1}^K y_{ij} l_j \\[2mm]
\quad \sum_j y_{ij} = 1 \\[2mm]
\quad y_{ij} \in \{0, 1\}
\end{cases}
$$

Define the energy function:

$$
H_0(\mathbf{y}; \mathbf{l}) = \sum_{i \in V} \sum_{j \in N(i)} v_{ij}(x_i - x_j)^2 + w_i(x_i - x_i^+)^2
\tag{11}
$$

It is straightforward to see there is a $\mathbf{y}^*$ that minimizes $H_0(\mathbf{y}; \mathbf{l})$. We characterize the conditions under which this $\mathbf{y}^*$ is a Nash Equilibrium in pure strategies for the players. Suppose the optimal solution $\mathbf{y}^*$ yields x_i^* with $x_i^* = l_j$ for some $j \in \{1, \ldots, K\}$. If User i chooses to deviate from this strategy, then her change in payoff is:

$$
\Delta H_i = H_i^* - H_i = \sum_{j \in N(i)} v_{ij} \left[(x_i^* - x_j^*)^2 - (x_i - x_j^*)^2 \right] + w_i \left[(x_i^* - x_i^+)^2 - (x_i - x_i^+)^2 \right]
\tag{12}
$$

while:

$$\Delta H_j = II_j^* - H_j = v_{ji} \left[(x_j^* - x_i^*)^2 - (x_j^* - x_i)^2 \right] = v_{ji} \left[(x_i^* - x_j^*)^2 - (x_i - x_j^*)^2 \right] \tag{13}$$

for each $j \in N(i)$. Under a symmetric weight assumption (i.e., $v_{ij} = v_{ji}$), we have:

$$\Delta H_0 = \sum_{i \in V} \Delta H_i = 2 \sum_{j \in N(i)} [v_{ij}(x_i^* - x_j^*)^2 - (x_i - x_j^*)^2] +$$
$$w_i[(x_i^* - x_i^+)^2 - (x_i - x_i^+)^2] \tag{14}$$

Let:

$$A = \sum_{j \in N(i)} v_{ij} \left[(x_i^* - x_j^*)^2 - (x_i - x_j^*)^2 \right]$$
$$B = w_i \left[(x_i^* - x_i^+)^2 - (x_i - x_i^+)^2 \right]$$

Then $\Delta H_i = A + B$ and $\Delta H_0 = 2A + B$. The fact that $\mathbf{y}^*$ is a minimizer for H_0 implies that $\Delta H_0 \leq 0$ otherwise, $\mathbf{y}^*$ could not have been a minimizer. Thus $2A + B \leq 0$. For a rational Player i a change in strategy make sense if (and only if) $A + B > 0$. There are four cases to consider:

Case 1: If $A \leq 0$ and $B \geq 0$, and since $2A + B \leq 0$ and $A + B > 0$, we have $|A| < |B| \leq 2|A|$. That is, Player i has benefitted by moving closer to her comfort value, sacrificing reputation. If this is not the case, then there is no rational reason for Player i to change strategies.

Case 2: If $A, B \leq 0$, then immediately $\Delta H_i \leq 0$ and Player i has not benefitted from changing.

Case 3: If $A \geq 0$ and $B \leq 0$, then $2A + B \leq 0$ implies $|B| \geq |A|$ which implies $A + B \leq 0$ and thus Player i would not change to this alternate strategy.

Case 4: If $A, B \geq 0$, then $2A + B \geq 0$ and $\mathbf{y}^*$ was either not a minimum or (in the case when $A = B = 0$) not a unique minimum.

It follows that only Case 1 prevents a global minimizer for H_0 from being a Nash equilibrium. For $w_i \approx 0$ we have $|B| \approx 0$ and in this case, we see necessarily that $A \leq 0$. Thus the energy minimizing solution is a Nash equilibrium. The following theorem follows naturally from this analysis:

Theorem 1. *For any set of comfort values* $\left\{ x_i^+ \right\}_{i=1}^N$ *and fixed privacy levels* $\mathbf{l} = \langle l_1, \ldots, l_K \rangle$ *there is an* $\epsilon \geq 0$ *so that if* $w_i \leq \epsilon$ *for* $i = 1, \ldots N$, *then there is a pure strategy Nash equilibrium for the following game:*

$$\forall i \begin{cases} H_i(\mathbf{y};\mathbf{l}) = \min_{\mathbf{y}_i} \sum_{j \in N(i)} v_{ij}(x_i - x_j)^2 + w_i(x_i - x_i^+)^2 \\ \\ s.t. \ x_i = \sum_{j=1}^{K} y_{ij} l_j \\ \\ \sum_j y_{ij} = 1 \\ \\ y_{ij} \in \{0,1\} \end{cases} \qquad (15)$$

$\square$

Remark 1. The results in Theorem 1 can be generalized to a game of the form:

$$\forall i \begin{cases} H_i(\mathbf{y};\mathbf{l}) = \min_{\mathbf{y}_i} \sum_{j \in N(i)} v_{ij} f_P(x_i - x_j) + w_i f_C(x_i - x_i^+) \\ \\ s.t. \ x_i = \sum_{j=1}^{K} y_{ij} l_j \\ \\ \sum_j y_{ij} = 1 \\ \\ y_{ij} \in \{0,1\} \end{cases}$$

for appropriately chosen convex functions f_C and f_P with minima at 0. Moreover, for $w_i \approx 0$ the bi-level problem is simply a bi-level combinatorial optimization problem.

Remark 2. If $w_i \gg 0$, then the player will conform more closely to her comfort level and for extremely high values of w_i (for $i = 1, \ldots, N$) there is again a pure strategy Nash equilibrium computed by finding the l_k value as close as possible to Player i's comfort level. Thus, settings with no pure strategy equilibria occur when the Players have values w_i large enough to prevent a pure strategy equilibrium consistent with social conformity, but not large enough to cause all players to follow their own comfort signal.

5   An Approximation Algorithm for Arbitrary Graphs - A Simulation

We have characterized the circumstances under which there exists a pure strategy Nash equilibrium for the bi-level optimization problem which describes the site's task of choosing a discrete set of privacy settings to optimize its payoff. Namely, this equilibrium exists in cases of extremely weak or extremely strong comfort level effects. Even in the case that such an equilibrium exists, we anticipate that finding the solution explicitly is NP-hard. Bi-level optimization problems are NP-hard [13], and even evaluating a solution for optimality is NP-hard [36]. Accordingly, an alternate approach in which we find an approximate solution is needed.

We argue that an approximation algorithm is also a more realistic approach in practice, since real SNs do not typically have the sharing comfort level for each individual user or potentially weighted influences amongst users' peers *a priori*. These parameters of the model are inferred through observation of user behavior under varying constraints, often using similar techniques to those we employ in the sequel; that is a site analyzes users' responses to minor alterations in its policies and recalibrates accordingly.

Here, we present a two-part algorithm for approximately computing the users' and site's utility functions on an arbitrary graphs in order to determine a discrete set of privacy settings beyond the determined lower bound to be made available to users in the SN. The Player Algorithm uses fictitious play simulating the convergence of the players' strategies to a strategy vector dependent on the players' personal comfort levels and the fixed set of privacy options determined by the SN site. Note, from Theorem 1, this may in fact be a pure strategy Nash equilibrium under appropriate assumptions.

To determine the full set l of privacy settings to be offered to users, the Site Algorithm wraps around the Player Algorithm as follows. The site lets $l_1 = 0$. Since players are captive to the site in this model, all players adopt strategy l_1. The level of unhappiness each player experiences for being forced to choose l_1 is calculated. Next, the site makes available a second option $l_2 = l_1 + \delta$. The Player Algorithm uses fictitious play to simulate the convergence of each player's strategy to either l_1 or l_2. A corresponding payoff for the site is calculated. Provided that there is at least one user whose comfort level for sharing is greater than l_1 and δ is small enough, the addition of option l_2 will increase the site's payoff. The site moves l_2 up by increments of δ, monitoring users' responses at each move, recalculating the corresponding site payoff and stopping when this payoff starts to decrease. Intuitively, when l_2 moves too far above individuals' comfort levels, users will become increasingly unhappy and eventually revert back to sharing at l_1 rather than l_2. The local optimum achieved here is taken as $l_2 \in l$. Following this, the site makes available a third option $l_3 = l_2 + \delta$ and allows players to converge on strategies from the set of three options available, incrementing l_3 as before until a local optimum is achieved. At this time, l_3 is added to l. This heuristic is repeated and the set l of privacy options grows by one as each local optimum is discovered until no further gains in site payoff or user happiness can be achieved, which is guaranteed to occur at a value no higher than the comfort level of the site's most privacy-lenient user. Pseudocode for the Player Algorithm and Site Algorithm are given in Figs. 2 and 3, respectively.

Figure 4 visualizes a well-known, real-world social network of members of a karate club [39]. In the absence of any constraints instituted by the site, equivalently in the case that each user may select his optimal privacy setting for a given piece of content, the trajectories of users' selections are guided by inherent personal comfort with sharing and the influence of their peers. Immediate neighbors in the graph are considered peers. We simulate the trajectory of privacy selections for member-users of the karate club network, first given the player algorithm described above in the unconstrained case, namely assuming that users have access to the complete set of options on the interval $[0, 1]$. Figure 5

Site Algorithm

1. Initialize $l_1 = 0, \mathbf{l} = \{l_1\}$
2. Run the Player Algorithm to obtain $\mathbf{x}^*$
3. $i = 1$
4. **while** $l_i < 1$

Player Algorithm

1. Initialize $x_i(0)$ $(i = 1, \ldots, N)$, $t = 0$
2. **while** $t < T_{\max}$
3.   t:=t+1
4.   **for each** $i \in \{1, \ldots, N\}$
5.   Minimize $H_i(\mathbf{y}; \mathbf{l})$ to obtain y_i^*
6. $x_i^*(t+1) := y_i^*(t)$
7.   **end**
8. **end**

5. $l_i = l_{i-1} + \delta l_{i-1}$
6.   Run the Player Algorithm $\{\mathbf{l}, l_i\}$ to obtain $\mathbf{y}^*$
7.   **if** $J_S(\{\mathbf{l}, l_i\}, \mathbf{y}^*) - J_S(\{\mathbf{l}, l_{i-1}\}, \mathbf{x}^*) < 0$
8.   Add l_i to $\mathbf{l}$
9.   **else**
10. $i = i + 1$
11.   **end**
12. **end**

Fig. 2. PlayerAlgorithm

Fig. 3. Site Algorithm

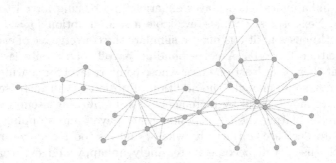

Fig. 4. A visualisation of the karate club network.

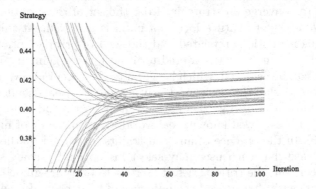

Fig. 5. A visualisation of players' strategies over time, initialised randomly, according to the user model

illustrates players' strategies over time. Strategies are initialized as user's individual sharing comfort levels and comfort levels are selected uniformly randomly from the interval $[0, 1]$. Notice that in this case, the vector of user strategies converges to equilibrium, as guaranteed by Proposition 1.

As described, the site's approximation algorithm influences the user model by iteratively choosing a discrete set of options to be made available to users, simulating user behavior given those constraints, and then adjusting the set of options by small increments until local optima are discovered. A visualization of site payoff during this process simulated over the karate club network is given in Fig. 6. Local optima occur at $x = \{0.4, 0.6, 0.72, 0.88\}$, so the site determines the set of privacy options as $l_1 = 0.4$, $l_2 = 0.6$, $l_3 = 0.72$, $l_4 = 0.88$ and $l_5 = 1$. User comforts are the same as those given in Fig. 5, and we choose $\delta = 0.04$. Note that the choice of δ may indicate a site's willingness to offer a finer granularity of privacy options to its users. A greater value of δ will lead to the discovery of fewer local optima, while smaller delta will yield more. This choice may also depend on the initial set of user comforts and the site business model. To this extent, the general algorithm we present here is the framework for a more personalized algorithm representative of a site's policies, practices and user base.

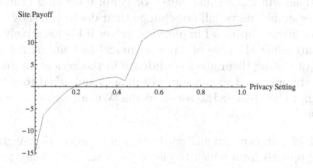

Fig. 6. Site payoff as privacy options are iteratively made available

6   Experimental Results

We designed and executed experiments to evaluate two of our key assumptions with a user study involving 60 participants in a simulated social network. First, our core model assumes that users' sharing decisions are influenced by a weighted sum of peer influence and personal comfort. We aim to determine whether postulated effects peer influence may be observed, even in a simulated context. Second, we seek to determine whether the iterative approach we take in our approximation algorithm may be assumed to in fact approximate the optimal discrete set of privacy options offered by the site. Hypothesizing that it will, we expect to rule out the notion that iterative presentation of an increased number

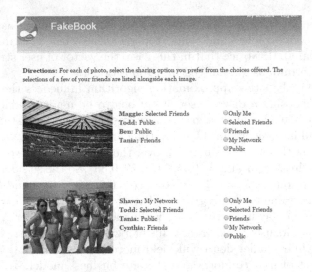

Fig. 7. Sample screenshot from Phase 2

of sharing options will significantly alter or confuse optimal individual preferences. Put more simply, users will not change their decision if they are offered an (optimal) set of privacy options l in one shot verses if l is iteratively built. These two assumptions are at the core of our user model and site model, respectively, and therefore validating them gives confidence in theoretical findings.

Subjects were presented with a series of images and asked to select a privacy setting for each, to be uploaded to social media. We organized the study in three distinct "phases".

1. In Phase 1 of our experiment, subjects were shown 15 images and given five sharing options from which to choose for each, i.e., "only me", "selected friends", "friends", "my network" and "public".
2. In Phase 2, subjects were shown the same images again and asked to choose from amongst the same options, but with the addition of the privacy selections of four of the subject's"friends" listed next to each image (see Fig. 7 for a sample screenshot). In attempt to create a more realistic sense of friendship between the subject and the simulated users, we endowed each simulated user with a profile page including demographic information, photos and other personal details and hyperlinked these profile pages throughout. Subjects were divided into several subgroups and treated to three variations of peer pressure in which friends' selections were skewed towards more private (skew-down), more public (skew-up) or random. In Sect. 6.1, we compare the selections of each user in Phase 1 (which we take as a baseline) with their selections in Phase 2. We expect that users may be influenced to increase their privacy restrictions when seeing that their peers are sharing more conservatively than they are, while on the other hand users may feel comfortable sharing more freely when their friends do the same.

3. Phase 3 was designed to test whether the iterative addition of privacy options (see Sect. 5) would influence users' ultimate privacy selections. Assuming a fixed set of options (i.e., l_1 = "only me", l_2 = "selected friends", l_3 = "friends", l_4 = "my network", l_5 = "public"), we iteratively presented subjects with a subset of photos from Phase 1 and Phase 2. At first, subjects were offered only l_1 and l_2 as privacy settings, next l_1, l_2 and l_3, subsequently l_1 through l_4, and finally l_1 through l_5. Variants of Phase 3 incorporating skew-down, skew-up and random peer pressure, implemented identically as in Phase 2, were also included for subsets of participants. In Sect. 6.2, we compare the selections of each user in Phase 2 with the their selections in the final iteration of Phase 3.

Participants in our study were 68 % female and 32 % male, with mean age 25.6 and standard deviation 2.98. In an initial survey preceding the experiment 100 % of subjects claimed to have an account with at least one social media site, with 92 % asserting that they maintain at least one "comprehensive" social media profile. On average, subjects claimed to participate in 3.4 different social networks, including Facebook, Instagram, Twitter, LinkedIn, Pinterest, Google+ and Vine.

6.1 Experimental Results: Peer Pressure Effects on Privacy Preferences

With respect to peer pressure, subjects were queried during the initial survey on several points related to privacy and peer pressure in content sharing. Over half (54.7 %) of subjects admitted to sometimes, often or always posting content with one privacy setting and later changing their mind and revising this setting, with 70 % of these subjects citing peer pressure as the reason for the revision.

In Table 1, we present the results of a one-factor analysis of variance (ANOVA) on change from baseline privacy selections for users treated with skew-down, skew-up or random peer influence in Phase 2. To quantify privacy options, we let $l_1 = 1$, $l_2 = 2$, $l_3 = 3$, $l_4 = 4$ and $l_5 = 5$. For each subject, for each image, we let change from baseline be defined as *(value of selection in Phase 2)-(value of selection in Phase 1)*. Note that a significant change in user sharing is detected in both subgroups subjected to a consistent peer pressure in either direction of more or less sharing. As might be expected, no significant change in sharing is detected in the random pressure control group. Of note, the most statistically significant change is observed when users are exposed to skew-down peer pressure, that is, when participants observe a change of their friends' privacy settings toward more conservative choices. This finding is consistent with the participants response of change of settings mentioned above, and also in line with existing research in this field [4,37], which has shown how users may change their mind with respect to sharing and may tend to be more conservative once they see the "network" behavior or reactions to their choices.

Follow-on ANOVA analyses blocking on subjects and images also give insight into more subtle user behavior dynamics. In both the skew-up and skew-down groups, subject effects (i.e., the affect a subject's identity had on output privacy settings) were highly significant ($p \approx 0$). This finding is intuitive and serves as strong justification for the inclusion of the parameter v_i in Eq. 2. That is, we must consider individual differences in susceptibility to peer pressure when implementing this type of model. Interestingly, an image effect was present in one of the experimental groups as well. Specifically, a significant effect was observed when image number was treated as an input in the skew-up group ($p = 0.0013$) but not for skew-down ($p = 0.1887$). When considered alongside the strength of skew-down peer pressure effects noted in Table 1, we suggest that these finding may again indicate users' readiness to make more conservative sharing choices for all photos, but hesitance to share more freely for specific images they would prefer to keep private, even when influenced to do so.

Table 1. Change from baseline after exposure to peer influence (Phase 2)

	Subjects	Average Change	p-Value
Skew-Down	17	-0.305	**0.0067**
Skew-Up	19	+0.192	**0.049**
Random	17	-0.086	0.375

6.2   Experimental Results: Iterative Approximation of Privacy Preferences

We have argued that using an approximation algorithm is both necessary and realistic, in the context of our bi-level optimization problem describing the site's task of choosing an optimal set of privacy options to offer its users. We here seek to validate the notion that an iterative approach like the one we take in our proposed algorithm does not disturb players' optimal privacy selections as determined in the theoretical case. Following we present the results of Phase 3 of the experiment, as described above.

For this analysis, we again separate study participants into subgroups by the peer pressure to which they were exposed, if any. Table 2 gives the results of a one-factor analysis of variance (ANOVA) on change from Phase 2 privacy selections for users treated with skew-down, skew-up or random peer influence. As a control group for this Phase, we keep a subset of subjects away from any exposure to peer pressure (that is, these subjects did not participate in Phase 2) and compare their results for Phase 3 with their Phase 1 baseline selections. Findings here indicate no significant change in users' final privacy selections due to the iterative nature of presentation of the options in any of the experimental groups, validating the approximation-algorithm approach as a reasonable alternative for modelling user behavior in cases that closed-form solutions are intractable.

We note here that Phase 3 studies user behavior given that options l_1, l_2 and so forth are presented additively one by one. The approximation algorithm as it

Table 2. Change from Phase 2 selections in the iterated model (Phase 3)

	Subjects	Average Change	p-Value
No Peer Pressure	7	0.086	0.774
Skew-Down	17	-0.28	0.19
Skew-Up	19	-0.2	0.282
Random	17	-0.117	0.527

presented is deployed accordingly, but also includes a routine for the selection of the set of values $\{l_i\}$ making very small, incremental changes to each l_i and monitoring users' responses throughout.

7 Conclusion

In this paper, we have presented a model for privacy decision-making in the context of online social networks. We have modeled the site's role in setting privacy policies that can help to retain users while also optimizing the site's payoff. Our work lays the foundation for further game-theoretic modeling of privacy-related behaviors in online SNs toward the better understanding of the interplay and repercussions of site and user choices.

As future work, we will refine the outlined approximation algorithm, with particular focus on how incremental privacy boundaries could actually be offered to end users. We also plan to investigate how changes to the social network topology and user attitudes towards privacy over time may affect this game. Finally, we plan to carry out more extensive user studies to validate our findings.

Acknowledgements. Portions of Dr. Rajtmajer's, Dr. Griffin's and Dr. Squicciarini's work were supported by the Army Research Office grant W911NF-13-1-0271. Portions of Dr. Squicciarini's work were additionally supported by the National Science Foundation grant 1453080.

References

1. Ackerman, M.S., Cranor, L.F., Reagle, J.: Privacy in e-commerce: Examining user scenarios and privacy preferences. In: Proceedings of the 1st ACM Conference on Electronic Commerce, EC 1999, pp. 1–8. ACM, New York (1999)
2. Acquisti, A., Brandimarte, L., Loewenstein, G.: Privacy and human behavior in the age of information. Science **347**(6221), 509–514 (2015)
3. Bilge, L., Strufe, T., Balzarotti, D., Kirda, E.: All your contacts are belong to us: automated identity theft attacks on social networks. In: Proceedings of the 18th International Conference on World Wide Web, WWW 2009, pp. 551–560. ACM, New York (2009)
4. Caliskan-Islam, A.: How do we decide how much to reveal? SIGCAS Comput. Soc. **45**(1), 14–15 (2015)

5. Chen, J., Brust, M.R., Kiremire, A.R., Phoha, V.V.: Modeling privacy settings of an online social network from a game-theoretical perspective. In: 2013 9th International Conference on Collaborative Computing: Networking, Applications and Worksharing (Collaboratecom), pp. 213–220. IEEE, October 2013

6. Chron. What is the minimum information needed to create a facebook account? (2013). http://smallbusiness.chron.com/minimum-information-needed-create-facebook-account-27690.html

7. Cranor, L.F., Reagle, J., Ackerman, M.S.: Beyond Concern: Understanding Net Users' Attitudes About Online Privacy. MIT Press, Cambridge (2000)

8. Fang, L., LeFevre, K.: Privacy wizards for social networking sites. In: Proceedings of the 19th International Conference on World Wide Web, WWW 2010, pp. 351–360. ACM, New York, April 2010

9. Fu, F., Chen, C., Liu, L., Wag, L.: Social dilemmas in an online social network: The structure and evolution of cooperation. Phys. Lett. A **371**(1–2), 58–64 (2007)

10. Griffin, C., Squicciarini, A.: Toward a game theoretic model of information release in social media with experimental results. In: Proceedings of the 2nd Workshop on Semantic Computing and Security, San Francisco, CA, USA, 28 May 2012

11. Griffin, C., Squicciarini, A., Rajtmajer, S., Tentilucci, M., Li, S.: Site-constrained privacy options for users in social networks through stackelberg games. In: Proceedings of Sixth ASE International Conference on Social Computing, May 2014

12. Gross, R., Acquisti, A.: Information revelation and privacy in online social networks. In: Proceedings of the 2005 ACM Workshop on Privacy in the Electronic Society, WPES 2005, pp. 71–80. ACM, New York (2005)

13. Hansen, P., Jaumard, B., Savard, G.: New branch-and-bound rules for linear bilevel programming. SIAM J. Sci. Stat. Comput. **13**(5), 1194–1217 (1992)

14. Hoadley, C.M., Xu, H., Lee, J.J., Rosson, M.B.: Privacy as information access and illusory control: The case of the facebook news feed privacy outcry. Electron. Commer. Res. Appl. **9**(1), 50–60 (2010)

15. Hui, P., Buchegger, S.: Groupthink and peer pressure: social influence in online social network groups. In: 2009 International Conference on Advances in Social Network Analysis and Mining (ASONAM), pp. 53–59. IEEE, Los Alamitos, July 2009

16. Immorlica, N., Lucier, B., Rogers, B.: Emergence of cooperation in anonymous social networks through social capital. In: Proceedings of the 11th ACM Conference on Electronic Commerce EC (2010)

17. Jagatic, T.N., Johnson, N.A., Jakobsson, M., Menczer, F.: Social phishing. Commun. ACM **50**(10), 94–100 (2007)

18. Kayes, I., Iamnitchi, A.: A survey on privacy and security in online social networks (2015). arXiv preprint arXiv:1504.03342

19. Krishnamurthy, B., Gill, P., Arlitt, M.: A few chirps about twitter. In: Proceedings of the First Workshop on Online Social Networks, WOSN 2008, pp. 19–24. ACM, New York (2008)

20. Krishnamurthy, B., Wills, C.E.: Characterizing privacy in online social networks. In: Proceedings of the First Workshop on Online Social Networks, WOSN 2008, pp. 37–42. ACM, New York (2008)

21. Lindamood, J., Heatherly, R., Kantarcioglu, M., Thuraisingham, B.: Inferring private information using social network data. In: WWW 2009 Proceedings of the 18th International Conference on World Wide Web, pp. 1145–1146. ACM, New York, April 2009

22. Liu, Y., Gummadi, K.P., Krishnamurthy, B., Mislove, A.: Analyzing facebook privacy settings: user expectations vs reality. In: Proceedings of the 2011 ACM SIG-COMM Conference on Internet Measurement Conference, pp. 61–70 (2011)
23. Ohtsuki, H., Hauert, C., Lieberman, E., Nowak, M.A.: A simple rule for the evolution of cooperation on graphs and social networks. Nature 441(7092), 502–505 (2006)
24. Park, N., Kee, K., Valenzuela, S.: Being immersed in social networking environment: Facebook groups, uses and gratifications, and social outcomes. CyberPsychol. Behav. 12(6), 729–733 (2009)
25. Rajtmajer, S.M., Griffin, C., Mikesell, D., Squicciarini, A.: A cooperate-defect model for the spread of deviant behavior in social networks. CoRR, abs/1408.2770 (2014)
26. Rand, D.G., Arbesman, S., Christakis, N.A.: Dynamic social networks promote cooperation in experiments with humans. Proc. Nat. Acad. Sci. 108(48), 19193–19198 (2011)
27. Rosen, J.B.: Existence and uniqueness of equilibrium points for concave n-person games. Econometrica 33(3), 520–534 (1965)
28. Simpson, A.: On the need for user-defined fine-grained access control policies for social networking applications. In: Proceedings of the Workshop on Security in Opportunistic and SOCial Networks, SOSOC 2008. ACM, New York (2008)
29. Squicciarini, A., Griffin, C.: An informed model of personal information release in social networking sites. In: Proceedings of the Fourth IEEE International Conference on Information Privacy, Security, Risk and Trust, September 2012
30. Squicciarini, A., Griffin, C.: An informed model of personal information release in social networking sites. In: 2012 ASE/IEEE Conference on Privacy, Security, Risk and Trust, Amsterdam, Netherlands, September 2012
31. Squicciarini, A., Paci, F., Sundareswaran, S.: PriMa: an effective privacy protection mechanism for social networks. In: Proceedings of the 5th ACM Symposium on Information, Computer and Communications Security, pp. 320–323 (2010)
32. Squicciarini, A., Shehab, M., Wede, J.: Privacy policies for shared content in social network sites. VLDB J. 19(6), 777–796 (2010)
33. Squicciarini, A.C., Griffin, C.: Why and how to deceive: game results with sociological evidence. Soc. Netw. Anal. Min. 4(1), 161 (2014)
34. Squicciarini, A.C., Griffin, C., Sundareswaran, S.: Towards a game theoretical model for identity validation in social network sites. In: PASSAT/SocialCom 2011, Boston, MA, USA, 9–11 October, 2011, pp. 1081–1088 (2011)
35. Squicciarini, A.C., Shehab, M., Paci, F.: Collective privacy management in social networks. In: WWW 2009, pp. 521–530. ACM, New York (2009)
36. Vicente, L., Savard, G., Jdice, J.: Descent approaches for quadratic bilevel programming. J. Optim. Theory Appl. 81(2), 379–399 (1994)
37. Wang, Y., Norcie, G., Komanduri, S., Acquisti, A., Leon, P.G., Cranor, L.F.: "I regretted the minute I pressed share": a qualitative study of regrets on Facebook. In: Proceedings of the Seventh Symposium on Usable Privacy and Security, SOUPS 2011, pp. 10:1–10:16. ACM, New York (2011)
38. Wu, Z.-X., Rong, Z., Yang, H.-X.: Impact of heterogeneous activity and community structure on the evolutionary success of cooperators in social networks. Phys. Rev. E 91, 012802 (2015)
39. Zachary, W.W.: An information flow model for conflict and fission in small groups. J. Anthropol. Res. 33(4), 452–473 (1977)

Approximate Solutions for Attack Graph Games with Imperfect Information

Karel Durkota[1]([✉]), Viliam Lisý[1,2], Branislav Bošanský[3],
and Christopher Kiekintveld[4]

[1] Department of Computer Science, Agent Technology Center,
Czech Technical University in Prague, Prague, Czech Republic
{karel.durkota,viliam.lisy}@agents.fel.cvut.cz

[2] Department of Computing Science, University of Alberta, Edmonton, Canada

[3] Department of Computer Science, Aarhus University, Aarhus, Denmark
bosansky@cs.au.dk

[4] Computer Science Department, University of Texas at El Paso, El Paso, USA
cdkiekintveld@utep.edu

Abstract. We study the problem of network security hardening, in which a network administrator decides what security measures to use to best improve the security of the network. Specifically, we focus on deploying decoy services or hosts called honeypots. We model the problem as a general-sum extensive-form game with imperfect information and seek a solution in the form of Stackelberg Equilibrium. The defender seeks the optimal randomized honeypot deployment in a specific computer network, while the attacker chooses the best response as a contingency attack policy from a library of possible attacks compactly represented by attack graphs. Computing an exact Stackelberg Equilibrium using standard mixed-integer linear programming has a limited scalability in this game. We propose a set of approximate solution methods and analyze the trade-off between the computation time and the quality of the strategies calculated.

1 Introduction

Networked computer systems support a wide range of critical functions in both civilian and military domains. Securing this infrastructure is extremely costly and there is a need for new automated decision support systems that aid human network administrators to detect and prevent the attacks. We focus on network security hardening problems in which a network administrator (defender) reduces the risk of attacks on the network by setting up honeypots (HPs) (fake hosts or services) in their network [30]. Legitimate users do not interact with HPs; hence, the HPs act as decoys and distract attackers from the real hosts. HPs can also send intrusion detection alarms to the administrator, and/or gather detailed information the attacker's activity [13,29]. Believable HPs, however, are costly

© Springer International Publishing Switzerland 2015
MHR Khouzani et al. (Eds.): GameSec 2015, LNCS 9406, pp. 228–249, 2015.
DOI: 10.1007/978-3-319-25594-1_13

to set up and maintain. Moreover, a well-informed attacker anticipates the use of HPs and tries to avoid them. To capture the strategic interactions, we model the problem of deciding which services to deploy as honeypots using a game-theoretic framework.

Our game-theoretic model is motivated in part by the success of Stackelberg models used in the physical security domains [33]. One challenge in network security domains is to efficiently represent the complex space of possible attack strategies, we make use of a compact representation of strategies for attacking computer networks called *attack graphs*. Some recent game-theoretic models have also used attack graphs [12,19], but these models had unrealistic assumptions that the attacker has perfect information about the original network structure. The major new feature we introduce here is the ability to model the imperfect information that the attacker has about the original network (i.e., the network structure before it is modified by adding honeypots). Imperfect information of the attacker about the network have been proposed before [8,28], however, the existing models use very abstract one step attack actions which do not allow the rich analysis of the impact of honeypots on attacker's decision making presented here.

Attack graphs (AGs) compactly represent a rich space of sequential attacks for compromising a specific computer network. AGs can be automatically generated based on known vulnerability databases [15,26] and they are used in the network security to identify the minimal subset of vulnerabilities/sensors to be fixed/placed to prevent all known attacks [24,32], or to calculate security risk measures (e.g., the probability of a successful attack) [14,25]. We use AGs as a compact representation of an attack plan library, from which the rational attacker chooses the optimal contingency plan.

The defender in our model selects which types of fake services or hosts to add to the network as honeypots in order to minimize the trade-off between the costs for deploying HPs and reducing the probability of successful attacks. We assume that the attacker knows the overall number of HPs, but does not know which types of services the defender actually allocated as HPs. This is in contrast to previous work [12], where the authors assumed a simplified version to our game, where the attacker knows the types of services containing HPs. The uncertainty in the existing model is only about which specific service/computer is real among the services/computers of the same type. Our model captures more general (and realistic) assumptions about the knowledge attackers have when planning attacks, and we show that the previous perfect information assumptions can lead to significantly lower solution quality.

Generalizing the network hardening models to include imperfect information greatly increases the computational challenge in solving the models, since the models must now consider the space of all networks the attacker believes are possible, which can grow exponentially. Computing Stackelberg equilibria with stochastic events and imperfect information is generally NP-hard [18] and algorithms that compute the optimal solution in this class of games typically do not scale to real-world settings [7]. Therefore we (1) present a novel collection of polynomial time algorithms that compute approximate solutions by relaxing certain aspects of the game, (2) experimentally show that the strategies computed in the approximated models are often very close to the optimal strategies in the original model, and (3) propose novel algorithms to compute upper bounds on the expected utility of the defender in the original game to allow the evaluation of the strategies computed by the approximate models even in large games.

2   Background and Definitions

We define a computer network over a set of host types T, such as firewalls, workstations, etc. Two hosts are of the same type if they run the same services, have the same vulnerabilities and connectivity in the network and have the same value for the players (i.e., a collection of identical workstations is modeled as a single type). All hosts of the same type present the same attack surface, so they can be represented only once in an attack graph. Formally, a computer network $x \in \mathbb{N}_0^T$ contains x_t hosts of type $t \in T$. An example network instance is depicted in Fig. 1a, where, e.g., host type WS_1 represents 20 workstations of the same type. We first define attack graphs for the case where attackers have perfect information about the network.

Attack Graph. There are multiple representations of AGs common in the literature. *Dependency AGs* are more compact and allow more efficient analysis than the alternatives [21]. Formally, they are a directed AND/OR graph consisting of fact nodes and action nodes. Fact nodes represent logical statements about the network and action nodes correspond to the attacker's atomic actions. Every action a has *preconditions*, set of facts that must be true in order to preform the action, and *effects*, set of facts that become true if action succeeds, in which case the attacker obtains the rewards of corresponding facts. Moreover, action a has probability of being performed successfully $p_a \in [0, 1]$, cost $c_a \in \mathbb{R}^+$ that the attacker pays regardless whether the action succeeded or not, and a set of host types $\tau_a \subseteq T$ that action a interacts with. The first time the attacker interacts with a type $t \in T$, a specific host of that type is selected with a uniform probability. Since we assume a rational attacker, future actions on the same host type interact with the same host. Interacting with different host of the same type (1) has no additional benefit for the attacker as rewards are defined based on the types and (2) can only increases the probability of interacting with a honeypot and ending the game. The attacker can terminate the attack any time. We use the common monotonicity assumption [1,23,26] that once a fact becomes true during an attack, it can never become false again as an effect of any action.

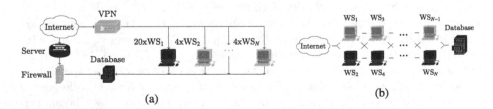

(a)

(b)

Fig. 1. Simple (a) business-like and (b) chain network topology.

AGs can be automatically generated using various tools. We use the Mul-VAL [27] to construct dependency AGs from information automatically collected using network scanning tools, such as Nessus[1] or OpenVAS[2]. These AGs consist of an attacker's atomic actions, e.g., exploit actions for each vulnerability of each host, pivoting "hop" actions between the hosts that are reachable from each other, etc. Previous works (e.g., [31]) show that probabilistic metrics can be extracted from the Common Vulnerability Scoring System [22], National Vulnerability Database [2], historical data, red team exercises, or be directly specified by the network administrator.

Attack Policy. In order to fully characterize the attacker's attack, for a given AG we compute a *contingent attack policy* (AP), which defines an action from the set of applicable actions according to the AG for each situation that may arise during an attack. This plan specifies not only the actions likely to be executed by a rational attacker, but also the *order* of their execution. Linear plans that may be provided by classical planners (e.g., [5,21]) are not sufficient as they cannot represent attacker's behavior after action failures. The *optimal AP* is the AP with maximal expected reward for the attacker. See [12] for more details on the attack graphs and attack policies and explanatory examples.

3 Imperfect Information HP Allocation Game

A real attacker does not know the network topology deployed in the company, but may have prior beliefs about the set of networks that the organization would realistically deploy. We assume that the attacker's prior belief about the set of networks that the organization is likely to deploy is common knowledge to both players. However, the attacker may know a subset of host types used by the organization, we refer to as a *basis* of a network, e.g., server, workstation, etc. To capture the set of networks we model the game as an extensive-form game with a specific structure. *Nature* selects a network from the set of possible networks (extensions of the basis network) with the probabilities corresponding to the prior attacker's beliefs about the likelihood of the different networks. The defender observes the actual network and hardens it by adding honeypots to it. Different networks selected by nature and hardened by the defender may lead to networks that look identical to the attacker. The attacker observes the network resulting from the choices of both, nature and the defender, and attacks it optimally based on the attack graph for the observed network. We explain each stage of this three stage game in more detail for the simple example in Fig. 2.

3.1 Nature Actions

For the set of host types T, total number of hosts $n \in \mathbb{N}$ and basis network $b \in \mathbb{N}_0^T$, we generate set of possible networks X including all possible

---

[1] http://www.nessus.org
[2] http://www.openvas.org

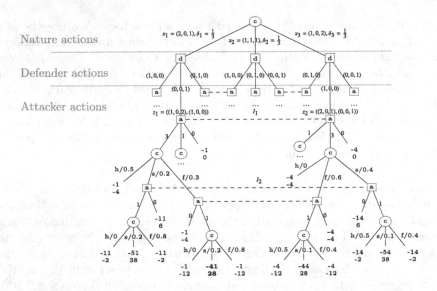

Fig. 2. Simple game tree with $|T| = 3$ host types, basis $b = (1, 0, 1)$, number of hosts $n = 3$ and $k = 1$ HP. The defender's costs for HPs are $c_1^h = 4$ and $c_3^h = 1$. The attacker's attack action 1 (resp. 3) exploits vulnerability of host type 1 (resp. 3), costs $c_1 = 8$ (resp. $c_3 = 4$); reward is $r_1 = 40$ (resp. $r_3 = 10$); and success probability $p_1 = 0.2$ (resp. $p_3 = 0.4$). The action's probabilities of interacting with honeypot (h) depend on defender's honeypot allocations and probabilities of succeeding (s) and failing (f) are accordingly normalized. Attacker's action 0 denotes the attacker ends his attack, which leads to the terminal state. In the chance nodes (except the one in the root) nature chooses weather the previous action: interacts with the HP (h), did not interact with HP and succeeded (s) or failed (f) with the given probabilities.

combinations of assigning n hosts into T host types that contain basis in it ($\forall x \in X : \forall t \in T : x_t \geq b_t$). E.g., in Fig. 2 the set of types is $T = \{D, W, S\}$ (e.g., database, workstation, server), and the network basis is $b = (1, 0, 1)$, a database and a server. Nature selects a network $x \in X = \{(2, 0, 1), (1, 1, 1), (1, 0, 2)\}$ with uniform probability $\delta_x = \frac{1}{3}$.

3.2   Defender's Actions

Each network $x \in X$ the defender further extends by adding k honeypots of types from T. Formally, set of all defender's actions is $Y = \{y \in \mathbb{N}_0^T | \sum_{t \in T} y_t = k\}$. Performing action $y \in Y$ on network $x \in X$ results in network $z = (x, y)$, where each host type t consist of x_t real hosts and y_t HPs. The attacker's action on host type t interacts with a honeypot with probability $h_t = \frac{y_t}{x_t + y_t}$. Let $Z = X \times Y$ be the set of all networks created as fusion of $x \in X$ with $y \in Y$. We also define $c_t^h \in \mathbb{R}_+$ to be the cost that the defender pays for adding and maintaining a HP of type t. In the example in Fig. 2 the defender adds $k = 1$ HP and set of the defender's actions is $Y = \{(1, 0, 0), (0, 1, 0), (0, 0, 1)\}$. Extending each network $x \in X$ by every choice from Y results in $|Z| = 9$ different networks.

3.3   Attacker's Actions

The attacker observes the number of hosts of each type, but not whether they are real or honeypots. The attacker's imperfect observation is modeled using *information sets* $\mathcal{I}$ that form a partition over the networks in Z. Networks in an information set are indistinguishable for the attacker. Two networks $z = (x, y)$ and $z' = (x', y')$ belong to the same information set $I \in \mathcal{I}$ if and only if $\forall t \in T$: $x_t + y_t = x'_t + y'_t$ holds. Networks $z, z' \in I$ have the same attack graph structure and differ only in the success probabilities and probabilities of interacting with a honeypot, therefore, they produce the same set of attack policies. Let S_I denote the set of valid attack policies in information set I. We also define $I(z)$ (resp. $I(x, y)$) to be a function that for a given network z (resp. (x, y)) returns the information set $I \in \mathcal{I}$ such that $z \in I$ (resp. $(x, y) \in I$). Executing the AP $s \in S_I$ leads to the terminal state of the game. In the example in Fig. 2, the attacker observes 6 different information sets, three singletons (contain only one network), e.g., $\{((2, 0, 1), (1, 0, 0))\}$, and three information sets that contain two networks (denoted with dashed lines), e.g., $I_1 = \{z_1 = ((2, 0, 1), (0, 0, 1)), z_2 = ((1, 0, 2), (1, 0, 0))\}$. An example of AP is: perform action 3 in I_1; if it succeeds, continue with action 1 and if fails then 0.

3.4   Players' Utilities

The players *utilities* in terminal state $l \in L$ with path P from the root of the game tree to l is computed based on three components: R_l - the sum of the rewards $\sum_{t \in T^s} r_t$ for successfully compromising host types $T^s \subseteq T$ along P; C_l - the sum of the performed action costs by the attacker along P, and H_l - the defender's cost for allocating the HPs along P. The defender's utility is then $u_d(l) = -R_l - H_l$ and attacker's utility is $u_a(l) = R_l - C_l$. Utility for an attack policy is expected utility of the terminal states. Although we assume that R_l is a zero-sum component in the utility, due to player private costs H_l and C_l the game is general-sum.

In our example in Fig. 2, utilities are at the leaf of the game tree labeled with two values. The value at the top is the defender's utility and at the bottom is the attacker's utility in that terminal state. We demonstrate the player's utility computations for the terminal state, the bold one in Fig. 2, we refer as to l_1. The three components are as follows: $R_{l_1} = r_1 = 40$ (only action 1 succeeded), $C_{l_1} = c_1 + c_3 = 12$ (attempted actions were 1 and 3) and $H_{l_1} = c_3^h = 1$ (for allocating HP in as type $t = 3$); thus the attacker's utility is $u_a(l_1) = R_{l_1} - C_{l_1} = 28$ and the defender's $u_d(l_1) = -R_{l_1} - H_{l_1} = -41$.

3.5   Solution Concepts

Formally, we define the Stackelberg solution concept, where the leader (the *defender* in our case) commits to a publicly known strategy and the follower (the *attacker* in our case) plays a best response to the strategy of the leader. The motivated attacker may be aware of the defender's use of game-theoretic approach, in which case the attacker can compute or learn from past experiences the defender's strategy and optimize against it. We follow the standard assumption of breaking the ties in favor of the leader (often termed as *Strong Stackelberg Equilibrium*, (SSE); e.g. [11,33]).

We follow the standard definition of strategies in extensive-form games. A *pure strategy* $\pi_i \in \Pi_i$ for player $i \in \{d, a\}$ is an action selection for every information set in the game (Π_i denotes the set of all pure strategies). *Mixed strategy* $\sigma_i \in \Sigma_i$ for player i is a probability distribution over the pure strategies and Σ_i is the set of all mixed strategies. We overload the notation for the utility function and use $u_i(\sigma_i, \sigma_{-i})$ to denote the expected utility for player i if the players are following the strategies in $\sigma = (\sigma_i, \sigma_{-i})$. *Best response* pure strategy for player i against the strategy of the opponent σ_{-i}, denoted $BR_i(\sigma_{-i}) \in \Pi_i$, is such that $\forall \sigma_i \in \Sigma_i : u_i(\sigma_i, \sigma_{-i}) \le u_i(BR_i(\sigma_{-i}), \sigma_{-i})$. Let d denote the defender and a the attacker, then Stackelberg equilibrium is a strategy profile

$$(\sigma_d, \pi_a) = \underset{\sigma'_d \in \Sigma_d; \pi'_a \in BR_a(\sigma'_d)}{\arg \max} u_d(\sigma'_d, \pi'_a).$$

In our game, the defender chooses honeypot types to deploy in each network $x \in X$ and the attacker chooses pure strategy $\pi_a \in \Pi_a = \times_{I \in \mathcal{I}} S_I$, an attack policy to follow in each information set $I \in \mathcal{I}$.

4   Game Approximations

The general cases of computing Stackelberg equilibria of imperfect information games with stochastic events is NP-hard [18]. The state-of-the-art algorithm for solving this general class of games uses mixed-integer linear programming and the sequence-form representation of strategies [7]. Our case of attack graph games is also hard because the size of the game tree representation is exponential in natural parameters that characterize the size of a network (number of host types T, number of hosts n, or number of honeypots k), which further limits the scalability of algorithms. We focus here on a collection of *approximations* that find strategies close to SSE in polynomial time w.r.t. the size of the game tree. We present the general idea of several approximation methods first, and discuss the specific details of new algorithms in the next section.

4.1   Perfect Information Game Approximation

A straightforward game approximation is to remove the attacker's uncertainty about the actions of nature and the defender, which results in a perfect information (PI) game. Although the authors in [18] showed that in general the PI game with chance nodes is still NP-hard to solve, the structure of our game allows us to find a solution in polynomial time. The nature acts only once and only at the beginning of game. After nature's move the game is a PI game without chance nodes, which can be solved in polynomial time w.r.t. the size of the game [18]. To solve the separate subgames, we use the algorithm proposed in [12]. It computes the defender's utility for each of the defender's actions followed by attacker's best response. Next, the algorithm selects the best action to be played in each sub-game by selecting the action with maximal utility for the defender. In Sect. 5.2 we discuss the algorithm that computes the optimal attack policy.

4.2   Zero-Sum Game Approximation

In [17] the authors showed that under certain conditions approximating the general sum (GS) game as a zero-sum (ZS) game can provide an optimal strategy for the GS game. In this section we use a similar idea for constructing ZS game approximations, for which we compute a NE that coincides with SSE in ZS games. A NE can be found in polynomial time in the size of the game tree using the LP from [16].

Recall that in our game the defender's utility is $u_d(l) = -R_l - H_l$ and the attacker's utility is $u_a(l) = R_l - C_l$ for terminal state $l \in L$. In the payoff structure R_l is a ZS component and the smaller $|H_l - C_l|$, the closer our game is to a ZS game. We propose four ZS game approximations: (ZS1) players consider only the expected rewards of the attack policy $u_d(l) = -R_l$; (ZS2) consider only the attacker's utility $u_d(l) = -R_l + C_l$; (ZS3) consider only the defender's utility $u_d(l) = -R_l - H_l$; and (ZS4) keep the player's original utilities with motivation to harm the opponent $u_d(l) = -R_l - H_l + C_l$.

We also avoid generating the exponential number of attack policies by using a single oracle algorithm (Sect. 5.1). This algorithm has two subroutines: (i) computing a SSE of a ZS game and (ii) finding the attacker's best response strategy to the defender's strategy. The attacker's best response strategy we find by translating the problem into the *Partially Observable Markov Decision Process* (POMDP), explained in Sect. 5.2.

4.3   Commitment to Correlated Equilibrium

The main motivation for this approximation is the concept of correlated equilibria and an extension of the Stackelberg equilibrium, in which the leader commits to a correlated strategy. It means that the leader not only commits to a mixed strategy but also to signal the follower an action the follower should take such that the follower has no incentive to deviate. This concept has been used in normal-form games [10] and stochastic games [18]. By allowing such a richer set of strategies, the leader can gain at least the same utility as in the standard Stackelberg solution concept.

Unfortunately, computing commitments to correlated strategies is again an NP-hard problem in general extensive-form games with imperfect information and chance nodes (follows from Theorem 1.3 in [34]). Moreover, the improvement of the expected utility value for the leader can be arbitrarily large if commitments to correlated strategies are allowed [18]. On the other hand, we can exploit these ideas and the linear program for computing the Stackelberg equilibrium [10], and modify it for the specific structure of our games. This results in a novel linear program for computing an upper bound on the expected value of the leader in a Stackelberg equilibrium in our game in Sect. 5.3.

5 Algorithms

5.1 Single Oracle

The single oracle (SO) algorithm is an adaptation of the double oracle algorithm introduced in [6]. It is often used when one player's action space is very large (in our case the attacker's). The SO algorithm uses the concept of a *restricted game* $\hat{G}$, which contains only a subset of the attacker's actions from the full game G.

In iteration m the SO algorithm consists of the following steps: (i) compute SSE strategy profile $(\hat{\sigma}_d^m, \hat{\pi}_a^m)$ (if $m = 1$ then $\hat{\sigma}_d^1$ is a strategy where the defender plays every action with uniform probability) of the restricted game $\hat{G}$ and compute the attacker's best response $\pi_a^m = BR_a(\hat{\sigma}_d^m)$ in the full game G. If all actions from π_a^m are included in the restricted game $\hat{G}$, the algorithm returns strategy profile $(\hat{\sigma}_d^m, \hat{\pi}_a^m)$ as a SSE of the full game G. Otherwise, (ii) it extends the restricted game $\hat{G}$ by including the attacker's policies played in $\hat{\pi}_a^m$ and goto (i) with incremented iteration counter m. Initially $\hat{G}$ contains all nature's and the defender's actions and none of the attacker's actions. We use this algorithm to solve all four variants of the ZS approximations proposed in Sect. 4.2. We refer to this approach as SOZS.

The SO algorithm is also well defined for GS games and can be directly applied to the original game. However, it does not guarantee that the computed SSE of the $\hat{G}$ is also the SSE of G. The reason is that the algorithm can converge prematurely and $\hat{G}$ may not contain all the attacker's policies played in SSE in G. Nevertheless, the algorithm may find a good strategy in a short time. We apply this algorithm to our GS game and use *mixed integer linear program* (MILP) formulation ([7]), to compute the SSE of $\hat{G}$ in each iteration. Finding a solution for a MILP is an NP-complete problem, so this algorithm is not polynomial. We refer to this approach as SOGS.

5.2 Attacker's Optimal Attack Policy

The attacker's best response $\pi_a = BR_a(\sigma_d)$ to the defender's strategy σ_d is computed by decomposing the problem into the subproblems of computing the optimal AP for each of the attacker's information set separately. We can do that because subgames of any two informations sets do not interleave (do not have any common state). We calculate the probability distribution of the networks in an information set based on σ_d, which is the attacker's prior belief about the probabilities about the states in the information set. The networks in an information set produce the same attack graph structure. However, the actions may have different probabilities of interacting with the honeypots depending on the defender's honeypot deployment on the path to that network.

Single Network. First, we describe an existing algorithm that finds the optimal AP for a single AG for a network, introduced in [12]. The algorithm translates the problem of finding the optimal AP of an AG into a (restricted) finite horizon

Markov Decision Process (MDP) and uses backward induction to solve it. A state in the MDP is represented by: (i) the set of executable attack actions α in that state (according to the AG), (ii) the set of compromised host types and (iii) the set of host types that the attacker interacted with so far. In each state the attacker can execute an action from α. Each action has a probabilistic outcome of either succeeding (s), failing (f), or interacting with a honeypot (h), described in detail in [12]. After each action, the sets that represent the MDP state are updated based on the AG, the performed action and its outcome (e.g., the actions that became executable are added to α, the performed action and actions no longer needed are removed, etc.), which represents a new MDP state. If the action successfully compromises a host type t, reward r_t is assigned to that transition. The MDP can be directly incorporated into the game tree, where the attacker chooses an action/transition in each of his states and stochastic events are modeled as chance nodes. The rewards are summed and presented in the terminal states of the game. The authors propose several pruning to generate only promising and needed part of the MDP such as branch and bound and sibling-class theorem and speed-up techniques, such as dynamic programming, which we also adopt.

Multiple Networks. The previous algorithm assumes that the MDP states can be perfectly observed. One of our contributions in this paper is an extension of the existing algorithm that finds the optimal AP for a set of networks with a prior probability distribution over them. The attacker has imperfect observation about the networks. We translate the problem into a POMDP. Instead of computing the backward induction algorithm on single MDP, we compute it concurrently in all MDPs, one per network in the information set. In Fig. 2 we show a part of the POMDP for information set I_1, which consists of two MDPs, one for network z_1 and another for z_2.

The same action in different MDPs may have different transition probabilities, so we use Bayes rule to update the probability distribution among the MDPs based on the action probabilities. Let J be the number of MDPs and let $\beta_j(o)$ be the probability that the attacker is in state o in MDP $j \in \{1, \dots, J\}$. Performing action a leads to state o' with probability $P_j(o, o', a)$. The updated probability of being in j-th MDP given state o' is $\beta_j(o') = \frac{P_j(o, o', a)\beta_j(o)}{\sum_{j'=1}^{J} P_{j'}(o, o', a)\beta_{j'}(o)}$. This algorithm returns the policy with the highest expected reward given the probability distribution over the networks. During the optimal AP computation, we use similar pruning techniques to those described in [12].

5.3   Linear Program for Upper Bounds

In [10] the authors present a LP that computes SSE of a matrix (or normal-form) game in polynomial time. The LP finds the probability distribution over the outcomes in the matrix with maximal utility for the defender under the condition that the attacker plays a best response. We represent our game as a collection of matrix games, one for each of the attacker's IS, and formulate it as a one LP problem.

Formally, for each attacker's information set $I \in \mathcal{I}$ we construct a matrix game M_I where the defender chooses network $z \in I$ (more precisely an action $y \in Y$ that leads to network $z \in I$) and the attacker chooses an AP $s \in S_I$ for information set I. The outcomes in the matrix game coincide with the outcomes in the original extensive-form game. The LP formulation follows:

$$\max \sum_{x \in X} \sum_{y \in Y} \sum_{s \in S_{I(x,y)}} p_{xys} u_{\mathrm{d}}(x, y, s) \tag{1a}$$

$$\text{s.t.} :(\forall I \in \mathcal{I}, s, s' \in S_I) : \sum_{(x,y) \in I} p_{xys} u_{\mathrm{a}}(x, y, s) \geq \sum_{(x,y) \in I} p_{xys} u_{\mathrm{a}}(x, y, s') \tag{1b}$$

$$(\forall x \in X, y \in Y) : \sum_{x \in X} \sum_{y \in Y} \sum_{s \in S_{I(x,y)}} p_{xys} = 1 \tag{1c}$$

$$(\forall x \in X, y \in Y, s \in S_{I(x,y)}) : p_{xys} \geq 0 \tag{1d}$$

$$(\forall x \in X) : \sum_{y \in Y} \sum_{s \in S_{I(x,y)}} p_{xys} = \delta_x, \tag{1e}$$

where the only variables are p_{xys}, which can be interpreted as probability that natures play x, the defender plays y and the attacker is recommended to play s. The objective is to maximize the defender's expected utility. Constraint 1b ensures that the attacker is recommended (and therefore plays) best response. It states that deviation from the recommended action s by playing any other action s' does not increase the attacker's expected utility. Equations 1c and 1d are standard probability constraints and 1e restricts the probabilities of the outcomes to be coherent with the probabilities of the chance node.

We demonstrate our approach on game in Fig. 3a. The game consists of two ISs: $I_1 = \{z_{11}, z_{23}\}$ and $I_2 = \{z_{12}, z_{24}\}$ each corresponds to a matrix game in Fig. 3b. The defender's actions y_1 and y_3 lead to I_1 and y_2 and y_4 lead to I_2. The attacker's attack policies are $S_{I_1} = \{s_1, s_2\}$ and $S_{I_2} = \{s_3, s_4\}$. The probabilities of the terminal states of the game tree correspond to the outcome probabilities in the matrix games ($p_{x,y,s}$). Moreover, the probabilities $p_{111}, p_{112}, p_{123}$ and p_{124} sum to δ_1, as they root from nature's action x_1 played with probability δ_1. The same holds for the other IS.

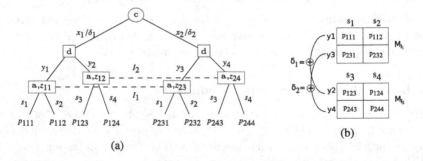

(a)

(b)

Fig. 3. The extensive-form game in (a) translated into two normal-form games in (b).

This LP has weaker restrictions on the solution compared to the MILP formulation for SSE [7] since it does not restrict the attacker to play a pure best response strategy. The objective is to maximize the defender's utility, as in the MILP. Therefore, it does not exclude any SSE of the game. The value of this LP, referred to as SSEUB, is an upper bound on the defender's expected utility when playing an SSE.

The drawback of formulating our game as a LP is that it requires finding all (exponential many) AP for each network in advance. We reduce this number by considering only *rationalizable* (in [4]) APs for each information set. An AP is rationalizable if and only if it is the attacker's best response to some belief about the networks in an IS. The set of all rationalizable APs is called *Closed Under Rational Behaviour* (CURB) set [3]. By considering only the CURB set for the attacker, we do not exclude any SSE with the following rationale. Any AP that is in SSE is the set of attacker best responses, so it must be rationalizable and therefore it must be in the CURB set.

From the LP result we extract the defender's strategy as a marginal probability for each defender's action: the probability that defender plays action $y \in Y$ in state $x \in X$ is $\sum_{s \in S_{I(x,y)}} P_{xys}$. We will refer to this mixed strategy as σ_d^{CCE} and to the defender's utility in the strategy profile $u_d(\sigma_d^{CSE}, BR_a(\sigma_d^{CSE}))$ as CSE.

CURB for Multiple Networks. We further extend the best response algorithms to compute the CURB set. We use the *incremental pruning* algorithm [9], a variant of the backward induction algorithm that in every attacker decision state propagates the CURB set of attack policies for the part of the POMDP that begins in that decision state. Let A be a set of actions in a decision state o. The algorithm is defined recursively as follows. (i) Explore each action $a \in A$ in state o and obtain the CURB set of policies S^a for the part of the POMDP after the action a; (ii) for every action $a \in A$ extend each policy $s_b \in S^a$ to begin with action a in the current state o and then continue with policy s_b; (iii) return the CURB set from the union of all policies $\cup_{a \in A} S_a$ for state o. In step (iii) we use the standard *feasibility linear program* to check whether policy s_b is in the CURB set by finding if there exists a probability distribution between MDPs where s_b yields the highest utility among $\cup_{a \in A} S^a \setminus s_j$, as described in [3,9].

6 Experiments

We experimentally evaluate and compare our proposed approximation models and the corresponding algorithms. Namely we examine: (i) Perfect information (PI) approximation solved with backward induction (Sect. 4.1), (ii) the ZS approximation games solved with SO algorithm, which we refer to as to SOZS1 through SOZS4 (number corresponds to the variant of the ZS approximation), (iii) SO algorithm applied on GS game (SOGS), and (iv) Correlated Stackelberg Equilibrium (CSE) (Sect. 5.3). We also compute the defender's upper bound utility SSEUB and use it as reference point to evaluate the strategies found by the other approximations.

The structure of the experimental section is as follows: in Sect. 6.1 we describe networks we use to generate the action graph game, in Sect. 6.2 we discuss an issue of combinatorially large CURB sets for one of the networks, in Sect. 6.3 we analyze the scalability of the approximated models, in Sect. 6.4 we analyze the quality of the strategies found by the approximated models and in Sect. 6.5 we analyze how the strategies calculated by the approximated models of ZS games depend on how close the games are to being zero-sum. In Sect. 6.6 we investigate the defender's regret for imprecisely modeling the attack graph, and conclude with a case-study analysis in Sect. 6.7.

6.1   Networks and Attack Graphs

We use three different computer network topologies. Two of them are depicted in Fig. 1, *small business* (Fig. 1a) and *chain* network (Fig. 1b). Connections between the host types in the network topology correspond to pivoting actions for the attacker in the attack graph (from the compromised host the attacker can further attack the connected host types). We vary the number of vulnerabilities of each host type, which is reflected in the attack graph as an attack action per vulnerability. We generate the actions' success probabilities p_a using the MulVAL that uses Common Vulnerability Scoring System. Action costs c_a are drawn randomly in the range from 1 to 100, and host type values r_t and the cost for honeypot c_t^h of host type t are listed in Table 1. We assume that the more valuable a host type is the more expensive it is to add a HP of that type. We derive honeypot costs linearly from the host values with a factor of 0.02. The basis network b for the business and chain network consists of the black host types in Fig. 1. We scale each network by adding the remaining depicted host types and then by additional workstations. We also scale the total number of hosts n in the network and the number of honeypots k. Each parameter increases combinatorially the size of the game.

The third network topology is the *unstructured* network, where each host type is directly connection only to the internet (not among each other). The attack graph consists of one attack action t per host type T, which attacker can perform at any time. For the unstructured network we create diverse attack graphs by drawing: host types values uniformly from $r_t \in [500, 1500]$, action success probabilities uniformly from $p_t \in [0.05, 0.95]$ and action costs uniformly from $c_t \in [1, r_t p_t]$. We restrict the action costs from above with $r_t p_t$ to avoid the situations where an action is not worth performing for the attacker, in which case the attack graph can be reduced to a problem with $|T| - 1$ types. The basis network b consists of two randomly chosen host types.

Table 1. Host type values and costs for deploying them as honeypots.

Host type t	Database	Firewall	WS$_n$	Server
Value of host type r_t	5000	500	1000	2500
Cost for deploying HP of host type c_t^h	100	10	20	50

All experiments were run on a 2-core 2.6 GHz processor with 32 GB memory limitation and 2 h of runtime.

6.2   Analytical Approach for CURB for Unstructured Network

The incremental pruning algorithm described in Sect. 5.3 generates a very large number of attack policies in the CURB set for the unstructured network. In order to be able to compute the upper bound for solution quality for larger game and in order to understand the complexities hidden in CURB computation, we analyze this structure of the curb for this simplest network structure formally. In Fig. 4a we show an example of the attacker's utilities for the policies in a CURB set generated for an information set with two networks. Recall $h_t = \frac{y_t}{x_t + y_t}$ is the probability that action that interacts with host type t (in this case action t) will interact with a honeypot. On the x-axis is probability distribution space between two networks, one with $h_t = 0$ ($y_t = 0$ and $x_t > 0$) and other with $h_t = 1$ ($y_t > 0$ and $x_t = 0$). The y-axis is the attacker's utility for each attack policy in the CURB. The algorithm generates the attack policies known as *zero optimal area policies* (ZAOPs) [20], denoted with dashed lines in the figure. A policy is ZAOP if and only if it is an optimal policy at a single point in the probability space (dashed policies in Fig. 4a). The property of ZAOP is that there is always another policy in the CURB set with strictly larger undominated area. It raises two questions: (i) can we remove ZAOPs from the CURB set and (ii) how to detect them. Recall that in SSE the attacker breaks ties in favour of the defender. Therefore, we can discard ZAOP as long as we keep the best option for the defender.

Further analysis showed that ZAOPs occur when $h_t = 1 - \frac{c_t}{p_t r_t}$ (at probability 0.69 and 0.99 in Fig. 4a). It is because the expected reward of action t at that point is $p_t(1 - h_t)r_t - c_t = p_t r_t \frac{c_t}{p_t r_t} - c_t = 0$, which means that the attacker is indifferent whether to perform action t or not. The algorithm at probability $h_t = 1 - \frac{c_t}{p_t r_t}$ generates the set of attack policies with all possible combinations where the attacker can perform action t in the attack policy. Let $P(t)$ be the probability that the attacker performs action t in an attack policy. The defender's utility for action t is $-r_t p_t(1 - h_t)P(t) = -c_t P(t)$. Because the attacker breaks ties in favour of the defender, at $h_t = 1 - \frac{c_t}{p_t r_t}$ the attacker will choose not to perform action t and we can keep only the policy that does not contain action t.

Furthermore, we categorize each action t based on h_t to one of three classes: to class A if $h_t = 0$, to class B if $0 < h_t < 1 - \frac{c_t}{p_t r_t}$ and to class C if $1 - \frac{c_t}{p_t r_t} < h_t$. In an optimal attack policy: (i) all actions from A are performed first and any of their orderings yield the same expected utility for the attacker, (ii) all actions from B are performed and their order is in increasing order of $\frac{p_t(1 - h_t)r_t - c_t}{h_t}$ ratios, and (iii) none of the actions from C are performed, as they yield negative expected reward for the attacker. We partition all probability distributions into regions $h_t = 1 - \frac{c_t}{p_t r_t}$ and in each region we assign actions to the classes. We find the set of optimal attack policies for each region. The attack policies in one region differ from each other in ordering of the actions in B.

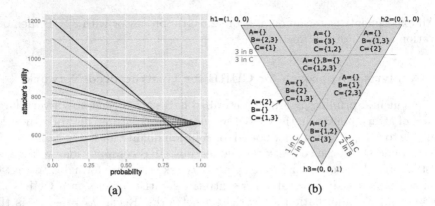

Fig. 4. (a) Attack policies from a CURB set for an information set for the unstructured network. (b) Probability space partitioning by action belonging into the categories.

In Fig. 4b we show an example of the probability distribution space of three networks in an information set. The probabilities that actions $1, 2$ and 3 interact with a honeypot represent a point (h_1, h_2, h_3). We partition the probability space and assign each action to a class. In all experiments we use this approach to generate the CURB set without ZAOPs in games for unstructured networks.

6.3 Scalability

In this section we compare the runtimes of the algorithms. We present the mean runtimes (x-axis in logarithmic scale) for each algorithm on business (Fig. 5a), chain (Fig. 5b top) and unstructured (Fig. 5b bottom) with of 5 runs (the runtimes were almost the same for each run). We increased the number of host types T, number of hosts n and number of honeypots k. The missing data indicate that the algorithm did not finish within a 2 h lime limit. From ZS approximations we show only SOZS4 since the others (SOZS1, SOZS2 and SOZS3) had almost identical runtimes.

From the results we see that least scalable are SOGS and CSE approach. SOGS is very slow due to the computation time of the MILP. Surprisingly, in some cases the algorithm solved more complex game (in Fig. 5b $T = 7, n = 7, k = 3$) and not the simpler game (in Fig. 5b $T = 7, n = 7, k = 1$). The reason is that the more complex game requires 3 iterations to converge, while the simpler games required over 5 iterations, after which the restricted game became too large to solve the MILP. The CSE algorithm was the second worst. The bottle-neck is in the incremental pruning algorithm subroutine, which took on average 91 % of the total runtime for the business network and 80 % for the chain network. In the unstructured network the problem specific CURB computation took only about 4 % of total runtime. The algorithms for ZS1–ZS4 and PI approximation showed the best scalability. Further scaling was prohibited due to memory restrictions.

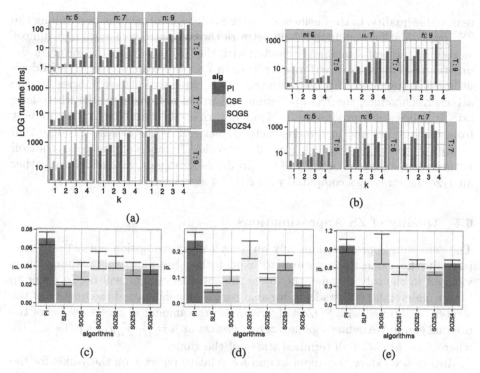

Fig. 5. Comparison of approximations scalability for (a) business, and (b top) chain and (b bottom) unstructured network. In (c), (d) and (e) we compare the defender's upper bound of relative regret of strategies computed with approximation algorithms business, chain and unstructured network, respectively.

6.4 Solution Quality

In this section we analyze the quality of the strategy that each approximation algorithm found. We use the concept of *relative regret* to capture the relative difference in the defender's utilities for using one strategy instead of another. Formally, the relative regret of strategy σ_d w.r.t. the optimal strategy σ_d^* is $\rho(\sigma_d, \sigma_d^*) = \frac{u_d(\sigma_d^*, BR_a(\sigma_d^*)) - u_d(\sigma_d, BR_a(\sigma_d))}{u_d(\sigma_d^*, BR_a(\sigma_d^*))}$. The higher the regret $\rho(\sigma_d, \sigma_d^*)$ the worse strategy σ_d is compared to strategy σ_d^* for the defender. We calculate the defender's *upper bound for relative regret* $\bar{\rho}$ by comparing the utilities of the computed strategies to SSEUB. Notice that the results are overestimation of the worst-case relative regrets for the defender. In Fig. 5 we show the means and standard errors $\bar{\rho}$ of 200 runs for the business network (Fig. 5c), chain network (Fig. 5d) and unstructured network (Fig. 5e), with $T = 5, n = 5$ and $k = 2$. In each instance we altered the number of vulnerabilities of the host types and host type values. The action costs c_i we draw randomly from $[1, 100]$ and action success probabilities p_i from $[0, 1]$.

The CSE algorithm computed the best strategies with lowest $\bar{\rho}$. The SOGS is second best in all networks except unstructured. Unfortunately, these algorithms are least scalable. The strategies computed with SOZS algorithm are within reasonable quality. In the business network SOZS4 performed the best among the ZS approximations and in the chain network the computed strategies were almost as good as the best strategies computed with the CSE algorithm. However, in the unstructured network it performed worse. In ZS4 approximations the defender's utility is augmented to prefer outcomes with expensive attack policies for the attacker. Therefore, the ZS4 approximation works well for networks where long attack policies are produced. In chain networks the database is the furthest from the internet and in the unstructured network is the closest. PI algorithm computed the worst strategies in all networks. Because of the good tradeoff between scalability and quality of the produces strategies, we decided to further analyze the strategies computed with SOZS4 algorithm.

6.5   Quality of ZS Approximations

The zero sum approximations rely on a zero-sum assumption not actually satisfied in the game. It is natural to expect that the more this assumption is violated in the solved game, the lower the solution quality will be. In order to better understand this tradeoff, we analyze the dependence of the quality of the strategy computed with SOZS4 algorithm on the amount of *zero-sumness* of the original game. We define a game's *zero-sumness* as $\bar{u} = \frac{1}{|L|} \sum_{l \in L} (|u_{\mathrm{d}}(l) + u_{\mathrm{a}}(l)|)$, where L is the set of all terminal states of the game.

In Fig. 6 we show the upper bound for relative regret $\bar{\rho}$ on the y-axis for the strategies computed by SOZS4 and amount of game zero-sumness $\bar{u}$ on the x-axis for 300 randomly generated game instances for the business network (Fig. 6a), chain network (Fig. 6b) and unstructured network (Fig. 6c) with $T = 5, n = 5$ and $k = 2$. In each instance we randomly chose the attacker's action costs $c_a \in [1, 500]$ and honeypot costs $c_t^h \in [0, 0.1r_t]$, while host type values r_t were fixed. We also show the means and the standard errors of the instances partitioned by step sizes of 50 for $\bar{u}$.

We show that the games with low zero-sumness can be approximated as zero-sum games and the computed strategies have low relative regret for the defender. For example, in a general sum game with $\bar{u} = 250$ the defender computes a strategy at most 6 % worse than the optimal strategy.

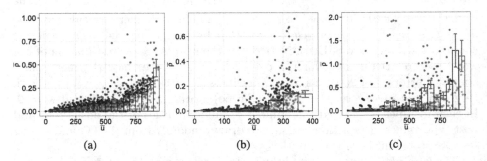

(a)   (b)   (c)

Fig. 6. The defender's relative regret dependence on game *zero-sumness* (computed as average $u_{\mathrm{a}}(l) + u_{\mathrm{d}}(l)$) for (a) business, (b) chain and (c) unstructured networks.

6.6   Sensitivity Analysis

The defender's optimal strategy depends on the attack graph structure, the action costs, success probabilities and rewards. In real-world scenarios the defender can only estimate these values. We analyze the defender's strategy sensitivity computed with SOZS4 to perturbations in action costs, probabilities and rewards in attack graphs.

We generate the defender's estimate of the attack graph by perturbing the original attack graph actions as follows: (i) action success probability are chosen uniformly from the interval $[p_a - \delta_p, p_a + \delta_p]$ restricted to $[0.05, 0.95]$ to prevent it becoming impossible or infallible, (ii) action costs are chosen uniformly from interval $[c_a(1 - \delta_c), c_a(1 + \delta_c)]$, and (iii) rewards for host t from uniformly from the interval $[r_t(1 - \delta_r), r_t(1 + \delta_r)]$, where p_a, c_a and r_t are the original values and δ_p, δ_c and δ_r is the amount of perturbation. The action probabilities are perturbed absolutely (by $\pm \delta_p$), but the costs and rewards are perturbed relative to their original value (by $\pm \delta_c c_a$ and $\pm \delta_r r_f$). The intuition behind this is that the higher the cost or reward values the larger the errors the defender could have made while estimating them, which cannot be assumed for the probabilities.

We compute (i) the defender's strategy σ_d on the game with the original attack graphs and (ii) the defender's strategy σ_d^p on the game with the perturbed attack graph. Figure 7 presents the dependence of the relative regret $\rho(\sigma_d, \sigma_d^p)$ on the perturbations of each parameter individually ($\delta_p, \delta_c, \delta_r$) and altogether ($\delta_a$). The results suggest that the regret depends significantly on the action success probabilities and the least on the action costs. E.g., the error of 20 % ($\delta_a = 0.2$) in the action probabilities results in a strategy with 25 % lower expected utility for the defender than the strategy computed based on the

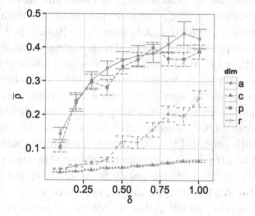

Fig. 7. The defender's utility regret for perturbed action success probabilities, action costs, and host type values.

true values. The same imprecision in action costs or host type rewards result in only 5 % lower utility.

6.7   Case Study

In order to understand what types of errors individual approximations make, we analyze the differences in strategies computed by the algorithms on a specific game for business network with $T = 5$, $n = 5$ and $k = 2$. The network basis is $b = (1, 0, 1, 0, 1)$, where the elements correspond to the number of databases, firewalls, WS1, WS2 and servers, respectively. There are $|X| = 15$ possible networks,

Table 2. Algorithm comparison for the case-study.

Algorithm	SSEUB	CSE	SOGS	SOZS1	SOZS2	SOZS3	SOZS4	PI
Defender's utility	−643	−645	−654	−689	−665	−676	−656	−699
Runtime [s]	2.9	3.2	6027	1.3	1.5	1.3	1.5	1.4

each selected with probability $\delta_x = \frac{1}{15}$. The defender can deploy honeypots in $|Y| = 15$ ways and with honeypot costs as showed in Table 1. There are 225 network settings partitioned into 70 information sets for the attacker. Table 2 presents the comparison of the strategy qualities computed with the algorithms and their runtime in seconds. The upper bound for the defender's optimal utility is SSEUB = −643. The best strategy was computed with CSE algorithm with utility $u_d = -645$. Although the difference between the utilities is very small, it suggests that the CSE strategy in not necessary optimal. We compare the strategies of the other algorithms to the CSE strategy.

SOGS computed the second best strategy ($u_d = -654$) and failed to compute the optimal strategy because the restricted game lacks strategies played by attacker in SSE. For example, both strategies from SOGS and CSE in the network $x_1 = (3, 0, 1, 0, 1)$ play $y_1 = (0, 0, 1, 0, 1)$ (adds a WS1 and a server as HPs) with probability 1. The attacker aims to attack the most valuable host (database) either via WS1 (policy s_1) or server (policy s_2). Both have the same probability of interacting with a honeypot 0.5 and a rational attacker will choose s_2 to compromise the server as well. Attack policy s_2 leads to a terminal state with the defender's expected utility −600. The strategy from CSE, in contrast to strategy from SOGS, additionally plays $y_2 = (1, 0, 0, 0, 1)$ in network $x_2 = (2, 0, 2, 0, 1)$ with probability 0.037, which leads to the same information set as action y_1 in x_1. The attacker's uncertainty between the two states in the information set changes his optimal attack policy from s_2 to s_1 for that information set. Attacking via the WS1 host type has a lower probability of interacting with the HP than via a server, which yields the defender expected utility −538, since the server will not be compromised. The restricted game in SOGS algorithm did not contain strategy s_1, so the computed strategy did not play y_2 in x_2 at all.

The PI strategy has the lowest defender's utility as it does not exploit the attacker's imperfect information at all. In this game the defender always adds a server host type as a honeypot to try to stop the attacker at the beginning. The second honeypot is added by a simple rule: (i) if the database can be compromised only via server and WS1, add honeypot of WS1 host type, otherwise (ii) as a database host type.

SOZS4 computed the best strategy among the ZS approximations. However, each of them have certain drawbacks. In SOZS1 and SOZS2 the defender ignores his costs for deploying the honeypots; these strategies often add database hosts as honeypots, which is in fact the most expensive honeypot to deploy. In SOZS2 and SOZS4 the defender prefers outcomes where the attacker has expensive attack policies. They often deploy honeypots with motivation for the attacker to have an expensive costs for attack policies (e.g., a strategy computed with SOZS2 adds database as a honeypot in 74 % while the strategy from CSE only in 43 %). Strategies computed with SOZS3 and SOZS4 are difficult to analyze. The strategies often miss the precise probability distribution between the networks where the attacker is indifferent between the attack policies and therefore chooses the one in favour for the defender. There is no general error they make in placing the honeypots as with the previous strategies.

7   Conclusion

We study a class of attack graph games which models the problem of optimally hardening a computer network against a strategic attacker. Previous work on attack graph games has made simplifying assumptions that the attacker has perfect information about the original structure of the network, before any actions are taken to harden the network. We consider the much more realistic case where the attacker only observes the current network, but is uncertain about how the network has been modified by the defender. We show that modeling imperfect information in this domain has a substantial impact on the optimal strategies for the game.

Unfortunately, modeling the imperfect information in attack graph games leads to even larger and more computationally challenging games. We introduce and evaluate several different approaches for solving these games approximately. One of the most interesting approaches uses a relaxation of the optimal MILP solution method into an LP by removing the constraint that attackers play pure strategies. This results in a polynomial method for finding upper bounds on the defender's utility that are shown to be quite tight in our experiments. We are able to use this upper bound to evaluate the other approximation techniques on relatively large games. For games that are close to zero-sum games, the zero-sum approximations provide a good tradeoff between scalability and solution quality, while the best overall solution quality is given the by the LP relaxation method. Several of these methods should generalize well to other classes of imperfect information games, including other types of security games.

Acknowledgments. This research was supported by the Office of Naval Research Global (grant no. N62909-13-1-N256), the Danish National Research Foundation and the National Science Foundation of China (under the grant 61361136003) for the Sino-Danish Center for the Theory of Interactive Computation. Viliam Lisý is a member of the Czech Chapter of The Honeynet Project.

References

1. Ammann, P., Wijesekera, D., Kaushik, S.: Scalable, graph-based network vulnerability analysis. In: Proceedings of CCS, pp. 217–224 (2002)
2. Bacic, E., Froh, M., Henderson, G.: Mulval extensions for dynamic asset protection. Technical report, DTIC Document (2006)

3. Benisch, M., Davis, G.B., Sandholm, T.: Algorithms for closed under rational behavior (curb) sets. J. Artif. Int. Res. **38**(1), 513–534 (2010)
4. Bernheim, B.D.: Rationalizable strategic behavior. Econometrica **52**, 1007–1028 (1984)
5. Boddy, M.S., Gohde, J., Haigh, T., Harp, S.A.: Course of action generation for cyber security using classical planning. In: Proceedings of ICAPS, pp. 12–21 (2005)
6. Bošanský, B., Kiekintveld, C., Lisý, V., Pěchouček, M.: An exact double-oracle algorithm for zero-sum extensive-form games with imperfect information. J. Artif. Int. Res. **51**, 829–866 (2014)
7. Bošanský, B., Čermak, J.: Sequence-form algorithm for computing stackelberg equilibria in extensive-form games. In: Proceedings of AAAI Conference on AI, pp. 805–811 (2015)
8. Carroll, T.E., Grosu, D.: A game theoretic investigation of deception in network security. Secur. Commun. Netw. **4**(10), 1162–1172 (2011)
9. Cassandra, A., Littman, M.L., Zhang, N.L.: Incremental pruning: a simple, fast, exact method for partially observable markov decision processes. In: Proceedings of UAI, pp. 54–61. Morgan Kaufmann Publishers Inc. (1997)
10. Conitzer, V., Korzhyk, D.: Commitment to correlated strategies. In: Proceedings of AAAI, pp. 632–637 (2011)
11. Conitzer, V., Sandholm, T.: Computing the optimal strategy to commit to. In: Proceedings of ACM EC, pp. 82–90. ACM (2006)
12. Durkota, K., Lisý, V., Bošanský, B., Kiekintveld, C.: Optimal network security hardening using attack graph games. In: Proceedings of IJCAI, pp. 7–14 (2015)
13. Grimes, R.A., Nepomnjashiy, A., Tunnissen, J.: Honeypots for windows (2005)
14. Homer, J., Zhang, S., Ou, X., Schmidt, D., Du, Y., Rajagopalan, S.R., Singhal, A.: Aggregating vulnerability metrics in enterprise networks using attack graphs. J. Comput. Secur. **21**(4), 561–597 (2013)
15. Ingols, K., Lippmann, R., Piwowarski, K.: Practical attack graph generation for network defense. In: Proceedings of ACSAC, pp. 121–130 (2006)
16. Koller, D., Megiddo, N., Von Stengel, B.: Efficient computation of equilibria for extensive two-person games. Games Econ. Behav. **14**(2), 247–259 (1996)
17. Korzhyk, D., Yin, Z., Kiekintveld, C., Conitzer, V., Tambe, M.: Stackelberg vs. nash in security games: An extended investigation of interchangeability, equivalence, and uniqueness. J. Artif. Int. Res. **41**(2), 297–327 (2011)
18. Letchford, J., Conitzer, V.: Computing optimal strategies to commit to in extensive-form games. In: Proceedings of ACM EC, pp. 83–92 (2010)
19. Letchford, J., Vorobeychik, Y.: Optimal interdiction of attack plans. In: Proceedings of AAMAS, pp. 199–206 (2013)
20. Littman, M.L.: The witness algorithm: Solving partially observable markov decision processes. Technical report, Providence, RI, USA (1994)
21. Lucangeli Obes, J., Sarraute, C., Richarte, G.: Attack planning in the real world. In: Working notes of SecArt 2010 at AAAI, pp. 10–17 (2010)
22. Mell, P., Scarfone, K., Romanosky, S.: Common vulnerability scoring system. Secur. Priv. **4**, 85–89 (2006)
23. Noel, S., Jajodia, S.: Managing attack graph complexity through visual hierarchical aggregation. In: Proceedings of ACM VizSEC/DMSEC, pp. 109–118. ACM (2004)
24. Noel, S., Jajodia, S.: Optimal ids sensor placement and alert prioritization using attack graphs. J. Netw. Syst. Manage. **16**, 259–275 (2008)
25. Noel, S., Jajodia, S., Wang, L., Singhal, A.: Measuring security risk of networks using attack graphs. Int. J. Next-Gener. Comput. **1**(1), 135–147 (2010)

26. Ou, X., Boyer, W.F., McQueen, M.A.: A scalable approach to attack graph generation. In: Proceedings of ACM CCS, pp. 336–345. ACM (2006)
27. Ou, X., Govindavajhala, S., Appel, A.W.: Mulval: a logic-based network security analyzer. In: Proceedings of USENIX SSYM. pp. 113–128. USENIX Association, Berkeley (2005)
28. Píbil, R., Lisý, V., Kiekintveld, C., Bošanský, B., Pěchouček, M.: Game theoretic model of strategic honeypot selection in computer networks. In: Grossklags, J., Walrand, J. (eds.) GameSec 2012. LNCS, vol. 7638, pp. 201–220. Springer, Heidelberg (2012)
29. Provos, N.: A virtual honeypot framework. In: Proceedings of USENIX SSYM, pp. 1–14. Berkeley, CA, USA (2004)
30. Qassrawi, M.T., Hongli, Z.: Deception methodology in virtual honeypots. In: Proceedings of NSWCTC, vol. 2, pp. 462–467. IEEE (2010)
31. Sawilla, R.E., Ou, X.: Identifying critical attack assets in dependency attack graphs. In: Jajodia, S., Lopez, J. (eds.) ESORICS 2008. LNCS, vol. 5283, pp. 18–34. Springer, Heidelberg (2008)
32. Sheyner, O., Haines, J., Jha, S., Lippmann, R., Wing, J.M.: Automated generation and analysis of attack graphs. In: IEEE Symposium Security and Privacy, pp. 273–284. IEEE (2002)
33. Tambe, M.: Security and Game Theory: Algorithms, Deployed Systems, Lessons Learned, 1st edn. Cambridge University Press, New York (2011)
34. Von Stengel, B., Forges, F.: Extensive form correlated equilibrium: definition and computational complexity. Math. Oper. Res. 33(4), 1002–1022 (2008)

When the Winning Move is Not to Play: Games of Deterrence in Cyber Security

Chad Heitzenrater[1,2], Greg Taylor[3], and Andrew Simpson[2]([✉])

[1] U.S. Air Force Research Laboratory Information Directorate,
525 Brooks Road, Rome, NY 13441, USA
[2] Department of Computer Science, University of Oxford,
Wolfson Building, Parks Road, Oxford OX1 3QD, UK
Andrew.Simpson@cs.ox.ac.uk
[3] Oxford Internet Institute, University of Oxford,
1 St. Giles, Oxford OX1 3JS, UK

Abstract. We often hear of measures that promote traditional security concepts such as 'defence in depth' or 'compartmentalisation'. One aspect that has been largely ignored in computer security is that of 'deterrence'. This may be due to difficulties in applying common notions of strategic deterrence, such as attribution — resulting in previous work focusing on the role that deterrence plays in large-scale cyberwar or other esoteric possibilities. In this paper, we focus on the operational and tactical roles of deterrence in providing everyday security for individuals. As such, the challenge changes: from one of attribution to one of understanding the role of attacker beliefs and the constraints on attackers and defenders. To this end, we demonstrate the role deterrence can play as part of the security of individuals against the low-focus, low-skill attacks that pervade the Internet. Using commonly encountered problems of spam email and the security of wireless networks as examples, we demonstrate how different notions of deterrence can complement well-developed models of defence, as well as provide insights into how individuals can overcome conflicting security advice. We use dynamic games of incomplete information, in the form of screening and signalling games, as models of users employing deterrence. We find multiple equilibria that demonstrate aspects of deterrence within specific bounds of utility, and show that there are scenarios where the employment of deterrence changes the game such that the attacker is led to conclude that the best move is not to play.

1 Introduction

When seeking advice on computer security, any combination of the terms 'computer' and 'security' will produce myriad results from academics, businesses looking to sell products, governments at local and national levels, 'hackers' (of the black- and white-hatted varieties), and bloggers; these results are often then moderated by input from friends, family and colleagues. From this conflicting

Approved for Public Release; Distribution Unlimited: 88ABW-2015-1336 20150323.

© Springer International Publishing Switzerland 2015
MHR Khouzani et al. (Eds.): GameSec 2015, LNCS 9406, pp. 250–269, 2015.
DOI: 10.1007/978-3-319-25594-1_14

guidance emerge the choices and decisions made by individuals. This can be further complicated by a lack of knowledge or evidence of utility, as some topics are still a matter of active discussion among even the most knowledgeable of practitioners.

Recent years have seen the emergence of security economics, which seeks to augment such discussions with the insight that these failures are often not the result of engineering challenges, but of economic challenges: misaligned incentives, information asymmetries, and externalities [15]. Given this landscape, what can be asked (and expected) of those who lack a technical background, technical staff, and a security budget? This is the question posed by many small businesses and home users, who often must make security decisions based upon their limited resources (with respect to time and money) and their ability to search related terms, digesting the information that appears in (at best) the first few hits. The answer is important, as it is precisely these decisions that affect us all: we all deal with the results of these failures [15].

In examining the source of much of our modern concept of cyber security — the doctrine of the military, an entity whose primary role is security — we see that the concepts that lead to security are well-defined, but multi-faceted. With respect to current research and practice, many concepts have been widely adopted as paradigms for cyber security [20]: "defence in depth", "compartmentalisation", etc. However, one aspect of security that has been largely ignored (outside of military doctrine) is the "first line of defence": deterrence [2,14]. In examining the role deterrence might play for individuals, we move towards a principled discussion of deterrence through the lens of information security economics. We conclude that, for a set of adversaries that can be defined in the economic context of utility, deterrence as an aspect of a comprehensive security stance is rational, contributory, and quantifiable against specific actor groups.

Section 2 introduces the various concepts at play: the notion of deterrence, the problems of signalling and screening in economics, and conceptual scenarios that are employed to provide context. Section 3 presents two concepts of deterrence as information asymmetries, formed as games of imperfect information. Section 4 provides a discussion of related work, placing this contribution within the broader context of deterrence and security economics. Finally, Sect. 5 summarises the contribution of this paper.

2   Background

2.1   Concepts of Deterrence

The concept of deterrence has a long history in our collective consciousness, primarily confined to our notions of national security. Throughout the Cold War, our collective security relied on a deterrence strategy of Mutually Assured Destruction (MAD) [14], with much being written on the topic of strategic deterrence: even our definition of the word is linked to this notion, including qualifiers such as "especially deterring a nuclear attack by the capacity or threat of retaliating" [7]. This emphasis on the threat of retaliation would seem to be an unnecessary deviation in concept from the simple act of *deterring*, where *to deter* is "to discourage or restrain from acting or proceeding" or "to prevent; check; arrest" [6]. We will refer to 'deterrence' in the context of deterring attacks, using the more general notion without emphasis on retaliation (requiring attribution) that is embodied in the former definition. One may argue that this is where the concept of deterrence in cyberspace has been stymied, as attribution is a known hard problem [14]. In decoupling attribution from deterrence, we examine the latter in a sense not possible when the concepts are intertwined.

In conjunction with this line of thought is the movement from concepts of strategic deterrence towards deterrence that results from more commonplace interactions: the deterrence that leads to the everyday security of individuals. In this spirit, Morral and Jackson [16] consider deterrence at the strategic, operational and tactical levels. In [16], strategic deterrence is defined by reducing the net expected utility of a particular means (e.g. attack) for a group to achieve their objective against another group. This is differentiated from operational deterrence by an emphasis on specific operations (or classes of operations), ideally leading to abandonment for that particular operation. Tactical deterrence then refers to the alteration of net utility after the attack is initiated. These definitions map nicely to current concepts of cyber security: strategically deterring attacks against internet users via the installation of various technical protections and procedures; operationally deterring against malware via the use of antivirus; and tactically thwarting the exfiltration of information from a machine via the use of a firewall. This also highlights the obvious links between deterrence and broader security, in that being secure can be a deterrent itself. Morral and Jackson [16] offer interesting insights regarding the nature and role of deterrence in these contexts, with relevance to information technologies and cyber deterrence. One point involves the role of complexity; all else being equal (regarding the utility of the target to the attacker, or other factors such as accessibility), a more complex attack is less appealing to an attacker. The resulting increase in complexity gives rise to an increase in observable signatures, resources expended, etc. — all of which lead to a less attractive target. This is tempered with the caveat that the deterrence cannot be trivial to overcome, no matter the likelihood of engagement by the attacker.

Bunn [4] considers the distinction between deterrence and dissuasion: dissuasion is related to the aim to convince a potential adversary from engaging at all. Using the above example of deterrence measures, dissuasion would be akin to laws against malware-writing and campaigns to warn potential attackers of computer misuse. Relevant to this discussion is the distinction between something that is more closely related to the psychological with respect to dissuasion, against measures that may have a more distinct technical aspect of deterrence. Bunn additionally contributes the notion that one deters *someone* from doing *something*, implying that actors and actions are of importance when considering deterrence. This leads one to conclude that deterrence will manifest itself differently given different scenarios; this is a central tenet of our contribution.

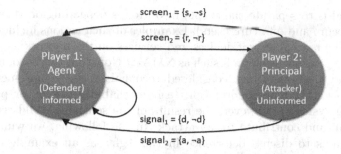

Fig. 1. The Agent–Principal model of deterrence enacted through screening and signalling. In screening, the principal moves first via a screening action $\{s, \neg s\}$ in an attempt to classify the agent's type (e.g. a viable or non-viable target). The agent may choose to respond $\{r, \neg r\}$, potentially betraying their type. In signalling, the agent moves first, broadly signalling $\{d, \neg d\}$, which may or may not be indicative of their type. The principal, observing this signal, chooses to react $\{a, \neg a\}$. The arrows differentiate a directed action by the defender to the attacker (in screening), and the broader action visible to all parties, including other defenders and non-players (in signalling).

2.2   Information Asymmetries in Security

One increasingly popular view of security is that of an information asymmetry between two entities: a user, who has the ability to take some action to secure themselves against attack, and therefore has information regarding the state of security; and an attacker, who seeks to identify targets for exploitation, but lacks information regarding the security of any given user. Information asymmetries arise when two entities engage in a transaction with each having access to different levels of information; they are a common source of *market failures* [21], which arise when inefficiencies in markets lead to suboptimal operation, such as one side gaining a distinct advantage in the marketplace. In our construct, the market for security is represented by this interaction between the attacker and the user; this differs from other characterisations that focus on the information asymmetry between users and security products, e.g. [3].

As with other forms of security, we can formally describe deterrence in terms of information asymmetry. We define this market as having an agent and a principal where the user (as the agent) has more information regarding their security level than the attacker (the principal). In this case, the information that is asymmetric is the type of the user, who might (through various actions undertaken prior to this point) be of type 'secure' (t_s), or type 'unsecure' (t_u).

Short of resolution through regulation (a factor for computer security, but something that thus far failed to resolve this market), there are two primary means of dealing with information asymmetries [21]: screening and signalling. Figure 1 depicts these concepts as sets of moves between agents and principals. We consider each in turn.

Screening involves the principal moving first to resolve the asymmetry via an action that (potentially) prompts a response from the agent. The goal of the principal is to separate the agents into groups depending on their type (in this case, secure and unsecure users). Examples of such actions include pings to determine the reachability of devices and services on the network, or operating system fingerprinting using tools such as NMAP.[1] Note that this is not an 'attack' as such, and is perhaps best considered reconnaissance — movement by the principal to gather more information (e.g. reachability of IPs, or patch level of operating systems). However, the results of the screening could certainly be employed in, and contribute to, an attack. In the following, we will use email phishing scams to discuss deterrence in this light, as an example of tactical deterrence of an 'attack' in progress.

Signalling involves the agent moving first via an observable action, prompting the principal to make a decision as to their type and react accordingly. The agent may be honest or dishonest regarding their type, forcing the principal to react based on belief. A good example from cyber security is the bug bounty offered by software providers to indicate the security of their systems. Here, poor software would not be able to offer such a bounty lest the software provider go bankrupt. Therefore the existence of such a scheme both signals to consumers that the software is of high quality, and increases that quality through the awards that are made — which, in turn, prompts further bugs to be found. We will look at the role of 'weak' security mechanisms, such as SSID hiding and MAC filtering, as signals of security that have an operational deterrence effect.

2.3   Adversary Scenarios

Having established the notion of deterrence as an information asymmetry, we now construct two adversarial scenarios corresponding to our concepts of operational and tactical deterrence. As a starting point, we consider attackers as falling on a spectrum, as postulated by Schneier [18]. Schneier characterises attackers along two axes: focus (interest in a specific victim) and skill (technical ability, such as use of existing scripts/tools vice development). Schneier maintains that the majority of attacks are "low-skill and low-focus — people using common hacking tools against thousands of networks world-wide" [18]. It is precisely these kinds of attacks on which we will focus our attention.

In the first scenario we consider a phishing scam, where the attacker (as the principal) moves first. The user, as the agent, drives the beliefs of the attacker through their response (or lack thereof). This is depicted in the upper part of Fig. 1. Using the construct of [12], we frame the scenario as an attacker sending spam emails to unwitting users in order to examine tactical deterrence. In [12], Herley conjectures that attackers who profit from attacks that depend on economies of scale (such as the infamous Nigerian scams) face the same economic and technological challenges as other disciplines. In constructing the scam, attackers must overcome statistical problems such as thresholding and binary classification when selecting victims, and therefore must weigh various

---

[1] The "Network Mapper". See http://nmap.org/ for a discussion on using NMAP for operating system fingerprinting.

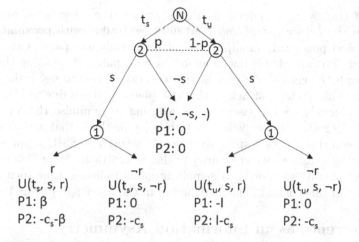

Fig. 2. Extensive form of the deterrence screening game.

aspects in order to make the attack profitable. Herley shows how success for an attacker depends on the density of viable users, d, as a fraction of viable victims M within a population N, $d = M/N$. With each attack costing the attacker C and yielding a net profit of G, it is obvious that, as the density d is small, it is important for the attacker that C is kept low and that G is maximised. To this extent, the attacker must use some criterion to select those to attack, which Herley terms 'viability'. Therefore, in order to identify d, the attacker utilises a 'viability score' x to separate users into a class $\{viable, non\text{-}viable\}$. Herley provides two insights regarding the role of beliefs in such attacks that has implications to deterrence. First, binary classification of users is as much a problem for attack as for defence. Thus, as the attacker's ability to separate viable from non-viable targets decreases, the effect on the true positive rate t_p versus the effect on false positive rate f_p can lead to dramatic shifts in the action of the attacker. Second, optimism on the part of the attacker does not pay, as over-estimation can quickly lead to unprofitability due to the non-zero cost of carrying out the attack. Thus, it is to the attacker's benefit to reduce costs and to be conservative in the choice of thresholding x, which drives both t_p and f_p.

The second scenario covers operational deterrence, and uses the example of an attacker attempting to undermine wireless connections. This network could be the responsibility of a small business proprietor utilising wireless connectivity for their business network, or a home user in a densely occupied space such as an apartment building in a large city. The key to this scenario is that the proprietor or user, acting as the agent, has a wireless network which they seek to secure from eavesdropping and unauthorised use by an attacker, acting as the principal. The security level of the user ('secure' or 'insecure') will serve to distinguish types of users, corresponding to the user having taken steps to protect against attacks against information disclosure or unauthorised use (e.g. having enabled WPA2 security). In this case, the attacker is assumed to be capable of employing 'standard' measures against the network — attempt to connect to the network, sniff and read message traffic (with proximity to the network), and potentially manipulate and retransmit any packets transmitted in the clear. The attacker is not assumed to be capable of breaking the WPA2 key, although the goal of the user in this context will be to deter the attacker from attempting such an attack in the first place (perhaps due to the user not using a sufficiently secure password, or wanting to minimise the log of failed attempts). As such, the user will seek to employ methods that are widely cited as recommended practices despite being 'weak' security — SSID suppression and MAC filtering — as 'signals' of security to dissuade attacks. We will demonstrate how modelling this scenario as a signalling game indicates that such methods have utility in this context. This is depicted in the lower half of Fig. 1.

3   Deterrence as an Information Asymmetry

3.1   Deterrence as Screening: An Example of Tactical Deterrence

We first look at the concept of tactical deterrence (deterrence of an 'attack' that is underway) though the lens of a screening game, using Herley's construct of the Nigerian scammer [12]. The game as conceived is depicted in Fig. 2, and unfolds as follows.

1. Player 'Nature' moves, allocating the distribution of the types of users t_s and t_u. We assume a distribution of Player 1 types $(p, 1 - p)$ but that neither player observes the realisation of this random variable, as this is reliant on the nature of the scam.
2. Player 2 (the attacker/spammer) makes the first move, not knowing the type of Player 1 (a given victim/user). Player 2 chooses to initiate a screening action s at a cost c_s (the spamming email), or chooses not to engage $(\neg s)$ and thus incurs no cost. This is done according to a belief p that Player 2 holds regarding the type of Player 1. It is assumed in this case that c_s is relatively small, but this is not necessarily the case in other scenarios.
3. Player 1's recognition of the scam then dictates their type, as either type secure (t_s) or of type unsecure (t_u). As a result, Player 1 may choose to respond or not to respond to the screening action and this choice may or may not be indicative of their type. Choosing not to respond has no loss or gain — a payoff of 0. Note that, following this exchange, Player 1's type is inferred by both players and the game unfolds similar to that of a game of complete information.
4. For simplicity in this game, the payoff for Player 2 is modelled as capturing l (Player 1's 'loss') upon successfully generating a response from an unsecure user, while Player 1 incurs the same loss $-l$ if unsecure and responding to the scam. Alternatively, a secure Player 1 exacts a benefit β from the scammer (the consumption of attacker resources that are not employed elsewhere; a 'social benefit'), while Player 2 loses that benefit along with the plus cost of screen $-c_s - \beta$ if prompting a response that does not result in a payoff (due to missing out on the potential profit from another user).

Table 1. Ex ante expected payoffs for the deterrence screening game of Fig. 2.

	s	$\neg s$
r	$(p(\beta + l) - l, l - p(\beta + l) - c_s)$	$(0,0)$
$\neg r$	$(0, -c_s)$	$(0,0)$

We make the simplifying assumption that, in getting a user type t_u to respond, the ruse is played out and the attacker captures l. As such, we are not considering instances in which an unsecure user engages but the transaction is thwarted (they instead appear as t_s users, with $\beta < l$).

The payoffs for a distribution of players p are provided in Table 1. We see that the strategy for Player 2 hinges on the value of p: if Player 2 believes Player 1 to be of type t_u ($p = 0$), it is beneficial for Player 2 to attempt the game as long as $l \gg c_s$ (as presumably would be the case for a spam email). In fact, as $p \to 0$ if the spammer is able to push the marginal cost of the attack $c_s = 0$, the strategy to attack is weakly dominant for the attacker. At the other end, as $p \to 1$ the payoffs for secure players are either positive or 0, while Player 2 has only losses (assuming $\beta \geq 0$). In this case, a strictly dominant strategy emerges in which the attacker avoids loss by not incurring any cost; Player 2 chooses not to engage in the game by choosing not to screen in the first place, forming a pure Nash equilibrium at $(\neg s, -), p = 1$.

Between these extremes ($0 < p < 1$), we find the attacker decision driven by both p and potential lost benefit β. In order for the scam to be viable, the attacker must believe that both the attack cost and potential for failed followthrough are sufficiently low to justify the effort of identifying unsecure users ($c_s \leq l - p(\beta + l)$ for $p < 1$). As it is to the attacker's benefit for this distribution to be in their favour ($p < \frac{1}{2}$), as education with respect to such scams grows (e.g. p increases) attackers must also carefully consider c_s. However, even with more unsecure than secure players, as $\beta \to l$ the ability for the scam to be profitable is quickly constrained by the potential payout and the attacker's cost ($c_s \leq l - 2pl$ for $p < \frac{1}{2}$). The attacker relies on Player 1 to find $\beta < \frac{l}{p} - l$ so as not to invoke a response from a secure user (e.g. one who does not complete the transaction), resulting in the consumption of resources for no gain. As well as introducing the potential deterrent of secure users purposefully engaging in the scam in order to consume resources, this threat of engaging with a non-viable target speaks to the heart of Herley's finding: it is to the attacker's best interest to utilise devices in order to identify the most gullible. As per [12], optimism on the part of the attacker is not a viable long-term strategy.

At this point, our findings are mere restatements of the results of [12]. We see evidence to support the conclusion that "at very low densities certain attacks pose no economic threat to anyone, even though there may be many viable targets" [12]. As shown, the belief of the attacker is critical; as viable target density decreases the attacker's belief that a potential target is secure rationally rises, leading to an attacker trade-space that must consider attacker costs and user benefit — with an increase in either quickly pushing the equilibrium towards

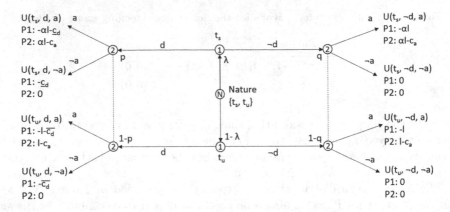

Fig. 3. Extensive form of the deterrence signalling game.

deterrence $(\neg s, -)$. We next look at a more complex game in which Player 1 moves first to signal their type and thus deter the attack at the onset. This will serve to account for the actions Player 1 might take in a more interactive defensive posture.

3.2 Deterrence as Signalling: An Example of Operational Deterrence

We now examine operational deterrence (in which a class of operations is deterred, but not the attackers themselves) within the context of a signalling game, as depicted in Fig. 3. This construct is based upon the concept of actions signalling a particular security level (secure or unsecure) for the purpose of deterring an attack. Using our conceptual scenario of a wireless network, we examine the employment of 'weak' security constructs (such as SSID hiding) as a means to signal that a user is secure. This game proceeds as follows:

1. Player 'Nature' moves, allocating the distribution of the types of users. As the real distribution of secure versus unsecure users is scenario-specific, we represent this as a probability λ of being secure (t_s), and a corresponding probability of $(1 - \lambda)$ of being unsecure (t_u).
2. Player 1 (the agent) then chooses to send (d) or not to send $(\neg d)$ a 'message' — that is, chooses to deter (e.g. hiding the SSID) or not — with the former action implying a cost that differs between types of user. Thus, the action costs secure users a low amount $\underline{c_d}$, while unsecure users will incur a higher cost of $\overline{c_d}$. In this model, messages have no meaningful effect on security; the question to be addressed is whether they can nevertheless deter attacks.
3. Player 2 (the principal) observes the message (deterrent) and subsequently chooses to attack or not attack, a or $\neg a$. Attacking incurs a cost of attack, c_a. Attacking a user of type t_u will be assumed to succeed, resulting in a gain of l (Player 1's loss); whereas attacking a user of type t_s will be assumed to succeed only with some small probability α, resulting in a gain of αl. At any point the attacker chooses not to attack $(\neg a)$, the resulting cost is 0.

We assume a difference in the cost to the secure user t_s and to the unsecure user t_u to send this signal, in which the latter is significantly higher $(t_u \gg t_s)$. The cost for an unsecure user to act secure (without actually being secure) warrants such a distinction, with experience costs being the primary differentiator. For instance, setting up wireless security on a modern home router can often be accomplished through a single action, as manufacturers seek to improve user experience. We can identify 'secure' users as those with experience enough to either use such mechanisms or by virtue of their own knowledge of how to do this themselves, and 'unsecure' users who may simply plug in the device and take no action — and who, presumably, are so due to a lack of understanding that would impose significant awareness costs if they were to only hide their SSID, but not implement any further security.

One important aspect of this type of game is Player 2's inability to discern the type of the user. As such, the best that Player 2 can do is to form a set of beliefs as to which type of agent $(t_s$ or $t_u)$ they are playing. This is represented by the value p, corresponding to the belief probability that a message d (that is, deterrent action) corresponds to a user of type t_s, and the corresponding belief probability $(1 - p)$ it indicates a player of type t_u. The belief probability q (and $(1 - q)$) serves the same function for $\neg d$.

We now analyse this game for equilibria, which for such games is defined by Perfect Bayesian Nash Equilibria (PBNE). Here, multiple conditions have to be met for a strategy profile to be in equilibrium: players must have a belief (probability distribution) as to the node reached any time they make a decision; players must act optimally given their beliefs and the continuation of the game; and beliefs are determined by Bayes' rule on the equilibrium path (as well as those off the path where possible). There are three types of equilibria that can come into play in such games:

- *Separating equilibria*, where a message (the deterrence action) perfectly separates the types of users.
- *Hybrid equilibria*, where a user type may receive the same expected utility from multiple actions and therefore randomise their response.
- *Pooling equilibria*, whereby one or both types find it profitable to take the same action (deter or not deter).

We start by examining for a separating equilibrium, noting that two types of such equilibria are possible: secure users deter, while unsecure users do not; and unsecure users deter, while secure users do not. Looking first at the latter, we note that this corresponds to beliefs of $p = 0$ and $q = 1$. We examine the utilities to Player 2 and see that, given these beliefs, we examine the strategy for Player 2 and find a likely course of action to be a in the case of seeing d, in that

$$E[U_{Player2}(d, a)] \geq E[U_{Player2}(d, \neg a)] \Rightarrow l - c_a \geq 0$$

where $l > c_a$.

Likewise, Player 2 may attack upon seeing $\neg d$ according to the value of $l > \frac{c_a}{\alpha}$:

$$E[U_{Player2}(\neg d, a)] \geq E[U_{Player2}(\neg d, \neg a)] \Rightarrow \alpha l - c_a \geq 0$$

However, in this instance we see that there exists a profitable deviation by Player 1. Given a by Player 2, while a type t_s player has no motivation to deviate (since $-\alpha l > -\alpha l - c_d$), a player of type t_u finds it beneficial to switch and play $\neg d$ as $-l > -l - \overline{c_d}$. As such, a separating equilibrium cannot exist in this case since a profitable deviation exists. In general, we can see from the game that, due to the symmetry of the payoff to Player 2 in the case of t_u, Player 1 of this type will always find it profitable to deviate and play $\neg d$ when $p = 0$ and $q = 1$ due to the cost of deterrence.

Looking now at the case where secure users deter and unsecure users do not, we employ beliefs $p = 1$ and $q = 0$. In this instance it is beneficial for Player 2 to refrain from attack upon seeing the signal d, as when $p = 1$,

$$E[U_{Player2}(d, \neg a)] \geq E[U_{Player2}(d, a)] \Rightarrow 0 \geq \alpha l - c_a$$

where $c_a \geq \alpha l$.

Likewise, consistent with $q = 0$, Player 2 finds the best move to be a upon failing to see a deterrent, as long as the gain from attack (e.g. Player 1's loss) is more than the cost of attack, $l > c_a$:

$$E[U_{Player2}(d, a)] \geq E[U_{Player2}(\neg d, \neg a)] \Rightarrow l - c_a \geq 0$$

Examining for deviation, we consider types t_s and see that a deviation to $\neg d$ may be desirable, since $-\alpha l > -\alpha l - \overline{c_d}$. While in this case Player 1 would no longer incur the additional cost of deterring $\underline{c_d}$, consistent with belief $q = 0$, Player 2 should now respond with a since $l - c_a > 0$. As such, deviation is only profitable for Player 1 if $-\alpha l > -\underline{c_d}$; that is, the potential loss (with small probability α) is greater than the cost to deter.

Looking now at type t_u players, we see that in any event a switch from $\neg d$ to d is going to incur an additional cost $\overline{c_d}$. As such, we can conclude that such an equilibrium exists under the condition $\frac{c_d}{\alpha} < l < \overline{c_d}$. Put another way, this equilibrium exists as long as it is inexpensive for secure users to implement a deterrence mechanism (specifically, less than αl), and the cost to unsecure users is greater than their loss l (given attacker beliefs $p = 1$ and $q = 0$). The meaning of this result is somewhat nuanced and requires further exposition; as such, the implication will be further discussed in Sect. 5. For now, we note that an equilibrium exists under these beliefs and conditions.

Considering hybrid equilibria, we note that the existence of such equilibria would require that actions exist between d and $\neg d$ such that the payoff is the same for one of the user types t_s or t_u. We can see from the game's construct that no such equilibrium exists. This is due to the cost of deterring which, despite presumably being small (at least for the case of secure users), changes the payoff function for Player 1. It is important to note that if the cost of deterrence to Player 1 or Player 2 reduces towards 0, this game becomes somewhat symmetric in its payoffs and multiple hybrid equilibria become possible. In such an instance the best course of action for the attacker is to randomise their attacks. Such a game would more closely follow the notion of a 'Cheap Talk' game [8], and arguably may have correspondence to current reality. However, we point out that the asymmetry induced serves to strengthen the case for deterrence measures having utility in a comprehensive defensive posture — but only when they impose an attacker cost that is non-negligible. This is consistent with the conceptualisation of deterrence presented by Morral and Jackson in [16].

We now examine the possibility of pooling equilibria, and first consider the case of an equilibrium at d under the assumption that both player types benefit from deterring. Consistent with the belief upon seeing d that $p = \lambda$:

$$E[(U(d, a)] = \lambda(\alpha l - c_a) + (1 - \lambda)(l - c_a) = l + \lambda \alpha l - \lambda l - c_a$$

while

$$E[U(d, \neg a)] = 0$$

Therefore we can see this will hold in instances where $\lambda > \frac{c_a - l}{\alpha l - l}$, rendering this possibility plausible with Player 2 playing $\neg a$. However, as we now look at potential deviation, we see that Player 1 has a potential profit in both instances: type t_s players can find a profitable deviation with $0 > -c_d$, as can type t_u players with $0 > -\overline{c_d}$. Put another way, Player 1 can get the same amount of payoff (security) without incurring the cost of deterring (consistent with the idea that deterrents have no security value themselves).

Considering now Player 2's move given these potential deviations, we compare the cost of $\neg a$ and a under the belief $q = \lambda$ and find that Player 2 attacks only as:

$$E[(U(d, a)] \geq E[U(d, \neg a)] \Rightarrow l - \lambda \alpha l - \lambda l - c_a \geq 0$$

Therefore, Player 2 would only change from $\neg a$ to a in the event that $\lambda > \frac{c_a - l}{\alpha l - l}$, which is inconsistent with the belief stated previously. Given this, Player 1 has found a profitable deviation and so we can conclude that a pooling equilibrium does not exist at this point.

Next, we examine the possibility of pooling equilibria existing at $\neg d$ (both players finding it beneficial not to deter), noting that the attacker's a posteriori belief in this case must now be $q = 1 - \lambda$. Upon seeing a play of $\neg d$, it is always to the benefit of Player 1 to play a (as $\alpha l - c_a > 0$ and $l - c_a > 0$), with the consideration that

$$E[U(\neg d, a) \geq E[U(\neg d, \neg a) \Rightarrow 2\lambda c_a - c_a + l(\alpha - \lambda \alpha - \lambda) \geq 0$$

We see that this indeed holds in the event that $\lambda < \frac{1}{2}$ (or $\frac{\alpha}{\alpha + 1} > \lambda$), with Player 1 payoffs of $-\alpha l$ for t_s and $-l$ for type t_u. Put another way, this is true only when the distribution of unsecure users is dominant, or the probability of success against a secure user is much greater than the instances of secure users. Examining now for deviation, we see in both cases that the payoff for Player 1 is reduced in each case (secure and unsecure users), as each faces the same potential loss and additionally incurs the cost of deterring. Therefore, a pooling equilibrium potentially exists whereby Player 1 chooses not to deter and Player 2 chooses to attack, with the beliefs $p = \lambda$, $q = 1 - \lambda$, and $\lambda < \frac{1}{2}$.

Finally, using the same approach, it is straightforward to show that another potential pooling equilibrium exists, with both types of Player 1 choosing not to deter and Player 2 choosing not to attack, as

$$E[U(\neg d, \neg a) \geq E[U(\neg d, a) \Rightarrow 0 > 2\lambda c_a - c_a + l(\alpha - \lambda \alpha - \lambda)$$

when $\lambda > \frac{1}{2}$, all else being the same.

Discussion on the realism of these beliefs is saved for Sect. 3.3; for now, we summarise that we have identified the following potential equilibria:

- A separating equilibrium when $\frac{c_d}{\alpha} < l < \overline{c_d}$, with

$$(P1_s(d), P1_u(\neg d), P2_d(\neg a), P2_{\neg d}(a), p = 1, q = 0).$$

- A pooling equilibrium when $\lambda < \frac{1}{2}$, with

$$(P1_s(\neg d), P1_u(\neg d), P2_d(a), P2_{\neg d}(a), p = \lambda, q = (1 - \lambda)).$$

- A pooling equilibrium when $\lambda > \frac{1}{2}$, with

$$(P1_s(\neg d), P1_u(\neg d), P2_d(\neg a), P2_{\neg d}(\neg a), p = \lambda, q = (1 - \lambda)).$$

3.3   Discussion

Starting with the screening game, the salient question that emerges is: how do we represent shifting attacker beliefs? In [12], Herley touches on this through the notion that the attacker would employ a series of one or more observables for which they can base a value for x in an attempt to classify the victim. We can think of x as now encompassing the necessary information for the choice of belief of the attacker. In this particular scenario since there is only one move by each player this is fully based upon the response of Player 1 to Player 2's screening message s, such that the choice of Player 1 to respond (r) or not to respond $(\neg r)$ corresponds to a belief $p = 1$ or $p = 0$, respectively. However, in other scenarios we can conceive of how this might be a combination of positive observables o_+ and negative observables o_-, such that these observations raise or lower the overall value of x and affect the attacker's assessment of viability. In this construct, we can now think of o_- observables as taking on the role of deterrents. Since the value Player 2 assigns to x is directly tied to the true and false positive rates of their classifier, this affects the risk to the attacker, who as noted cannot afford optimism. Minimising the value Player 2 assigns to x will result in two inter-related effects that will contribute to unprofitability: as a given assessment x is decreased (via such negative observables), the associated user is more likely to be placed into the category of 'not viable' and thus not subject to attack; and as the perceived set of viable users becomes smaller, attackers are faced with having to find ways to increase true positive and reduce false positive rates, or be faced with decreased attacker profits in the ways described in [12]. This rests not on the user type actually being secure or unsecure (i.e. the 'truth' of Player 1's response), but rather on the belief of the attacker. The response (or lack thereof) represents a single measurement upon which the attacker must infer viability.

We could conceive of a more general game, in which multiple measures beyond a single exchange result in complex screening scenarios (e.g. multiple emails) using the notion of positive and negative observables. Such a construct could be useful in characterising activities such as 'reconnaissance' leading to an attack, port probing (reporting open ports or services running on those ports), information contained within a DNS response that may lead the attacker to believe the system is up to date or of a specific type, or system fingerprinting (reporting specific patch levels, installed applications, etc.).

The separating equilibrium in the signalling version of our deterrence game tells us exactly what we might expect: there is a benefit for players to deter, as it conveys belief that the user's type is t_s. Note that for a user of type t_u playing d is off the equilibrium path, and so no information can be ascertained. In fact, due to this equilibrium, such a move is likely to swing the belief of the attacker towards inferring that the user is of type t_s and refrain from attack, thereby providing a type t_u player the best outcome. This equilibrium required beliefs that seeing a deterrent indicated security, and likewise not seeing such deterrents indicated a lack of security; we claim that this is a reasonable assumption, given the abundance of websites advocating such measures. Users who have taken the time to acquire such devices and follow recommendations on their set-up have likely completed true security tasks as well, such as setting up WPA2 encryption. Additionally, this result requires the constraint that $c_a \geq \alpha l$, such that the expected result of attacking a secure player is less than the cost to attack. This is in line with accepted notions of security.

This result shows that the deterrent must also meet the requirement that $\frac{c_d}{\alpha} < l < \overline{c_d}$, so that the cost of deterring for an unsecure user is higher than the expected loss. This may or may not hold, depending on the conceptualisation employed in the game analysis: in our scenario of a wireless user, someone with a lack of equipment, or improper or unusable equipment, might have a hardware investment to overcome. A lack of technical expertise might result in a user finding that developing an understanding of what an SSID is, or how to find a MAC address and set up filtering, to simply be too burdensome — more so than having to cancel a credit card and deal with the removal of a few charges. This strays into aspects such as time valuation and technical expertise, which is clearly going to vary based on the specifics of the scenario. However, for two users with similar routers — one of whom has set up security, and the other who has simply plugged in out-of-the-box — this becomes more reliant on the user's perceptions and how they value their time. We note that, as the deterrence costs converge $\overline{c_d} \rightarrow \underline{c_d}$, the asymmetry in payoffs between deterring and not deterring disappears, and Player 1 becomes agnostic (as discussed in Sect. 3.2). This leads to various hybrid equilibria in which secure players are attacked. As $\overline{c_d} \rightarrow \underline{c_d} \rightarrow 0$, this will only hold if the value of the loss decreases as well, and thus nothing of value is being protected. Thus, one result that can be interpreted from this inquiry is that as such 'security' mechanisms become more user-friendly, they may also lose value in their utility to signal security if they don't result in a sufficient cost to the attacker; this is again consistent with accepted concepts of deterrence.

Turning to the pooling equilibrium, we see that the nature of the equilibrium depends on the distribution of secure users λ. Hard metrics of this type are often scarce and difficult to estimate reliably. Fortunately, some empirical research for the wireless network security scenario exists, placing the occurrence of secure routers at 61 % in 2006 [13]. While such analyses are fraught with difficulty and only temporally relevant, this result allows us to assert that instances of secure router set-up are (at least somewhat) more common than not. We can now place a value on our a posteriori beliefs (e.g. $\lambda = 0.61$), and find that our first pooling equilibrium is unlikely to hold as it was dependent on $\lambda < \frac{1}{2}$. However, this distribution is consistent with our second equilibrium, in which neither Player 1 type is deterred but Player 2 chose not to attack. This reflects a belief held by Player 2 that secure players are more prevalent (backed by empirical evidence), and that the likelihood of successful attack is small.

All of these outcomes naturally rely on the attacker incurring a sufficient cost $c_a \geq \alpha l$, as with a small c_a the attacker becomes indifferent to various plays (since they incur little or no cost). As $c_a \rightarrow 0$, we again expect a number of hybrid equilibria situations, leading to probabilistic attack strategies. This results in interesting ramifications, especially as network-sniffing software reduces this to a point-and-click exercise.

Combining these results, we can see that changing the outcome of the game involves changing one or more of the salient parameters. Focusing first on costs, we see that in the screening game the key inequality is between the attacker cost (c_s) and the potential payout (l) or benefit (β). In the signalling case, while a sufficient attacker cost $(c_a \geq \alpha l)$ must still exist, the key cost relationship shifts to the defender cost $(\underline{c_d}$ or $\overline{c_d})$ and payout (l), driving a similar inequality that is also conditioned on the attacker's success probability (α). In both cases, this finding reinforces our current notions of security — and forms the basis for much of the effort to combat such crimes. In the case of spamming, efforts in the form of botnet take-down, capture, fines and jail time dominate; probabilistic costs which the attacker must consider within c_s, and when considered explicitly are a confirmation of the role law enforcement in a specific country/region has in deterrence. In the case of signalling, the focus within wireless security has been towards improving usability, and thereby lowering user costs. These respective costs represent government and industry actions in response to these issues.

Ultimately, in both of these games it is the perpetuation of the information asymmetry that is of benefit to the user. This of course stands to reason: the less the attacker can determine of the user's security, the greater the benefit to the user's security. What additionally becomes clear through this analysis is that the effect of such mechanisms can be either direct, by signalling the type or viability of a victim, or indirect, leaving the attacker without actionable information. It is here that the user (defender) appears to have the most direct impact on the resulting security, regardless of prior investment or external constructs. Most directly, in the case of screening the action (or inaction) of the defender provides the conditions to drive a binary attacker belief ($p = 1$ or $p = 0$), and, coupled with the threat of a failed engagement, forces equilibrium. It is this adherence to recommended 'good practice' that sets attacker beliefs, and one could conceive different scenarios in which continued iterations require the defender to continually follow such advice (as characterised by the 'the user is the weakest link' ethos). This reinforces the findings of [12] that it is the small, gullible minority who respond to spam that enables the perpetuation of such scams by allowing attackers to believe it is profitable, given its low cost of entry.

In the case of signalling, while by the construct of the game the signal itself (i.e. SSID hiding) fails to have any security impact, the equilibrium found indicates that the value it provides is in affecting attacker belief. This may help explain the continued endorsement of the practice despite widespread understanding that it does little to affect wireless security, and would appear to provide the justification of heeding such advice. Again, this appears to perpetuate the continued adherence to security guidance even if it has dubious contribution to the actual security stance — as long as the good advice is also followed, and the rest 'looks like' security and comes at a sufficiently low cost.

Naturally, these results only hold in specific circumstances. In these games, Player 1 has knowledge of their type, which may not be the case in many circumstances (or is arguably more likely only in that a 'secure' type would identify as such, with all others falling into the 'unsecure' category). Additionally, these results are in the presence of attacks at scale, as wholly different constructs (with different utilities) are required for examination of directed, focused attacks. Given these conditions, from these results we come to the conclusion foreshadowed by the title of the paper: in both cases of games constructed here, there exists a deterrence outcome in which the winning move is not to play.

4   Related Work

The work described in this paper is intertwined with the wider literature on deterrence, cyber security, and adversarial behaviour, although to the authors' knowledge it is the first to tackle the concept of deterrence from the operational and tactical level in cyberspace.

The role of game theory as the construct for examining deterrence is well studied. Relevant to this work is that of Robbins et al. [17], in which they present an extension of the 1960's US–USSR game-theoretic model for strategic nuclear deterrence. Their concept of decision criteria being in the "mind's eye" of the adversary and leading to probability assessments has synergy with the signalling game as defined in this paper. Other attempts at defining deterrence mathematically have also employed game-theoretic constructs to measure reductions in intent to attack [19], although it is not clear how this is to be employed when the potential target set is not specifically known. Generally, the interplay between adversary belief manipulation and cost–benefit realisation are the common themes in definitions of deterrence [4,16].

Attempts to define cyber deterrence typically stem from these traditional military concepts of strategic deterrence. Much of this literature is focused around cyber attack and notions of 'cyberwar' likened to approaches deterrence in the nuclear era; there is no lack of examples of such treatments [9]. Regarding the role of deterrence as a part of the larger concept of cyber defence, Jabbour and Ratazzi [14] discuss deterrence as a combination of denial, increased costs and decreased benefits, noting that the first of these aspects (denial) relies on a defence-dominated environment — which cyberspace is not. This links the second and third aspects (increased costs and decreased benefits) to the notions of assurance and avoidance, but the authors do not specify how this might be exacted or quantified. While this characterisation soundly dismisses the notion that deterrence can be thought of exclusively in traditional terms of 'Mutually Assured Destruction (MAD)' or retaliatory action, it doesn't reach the level of describing how this could be measurably performed — noting only that it will vary with the domain and application.

In the field of security economics, research involving deterrence has thus far focused primarily on the role it plays to dissuade large-scale malicious behaviour. Previous treatments have included deterrence of employee behaviour with regards to information security (to include employee knowledge of the security mechanisms in play) [11], as well as the application of various theories of deterrence with respect to combatting specific cyber crimes, such as online fraud [1]. These contributions represent a growing trend towards examining deterrence in various perspectives outside of war, but retain the emphasis on larger-scale engagements (e.g. many potential bad actors) and are generally abstracted beyond specific interactions between actors.

Grossklags et al. [10] investigate the application of static game-theoretic models to cyber security problems. This scope permits the authors to investigate security concerns ranging from redundant network defence, software vulnerability introduction, and insider threats. The primary focus is in the analysis of the trade-off between 'security' and 'insurance' measures in these instances, and on decisions regarding approach rather than allocation. As such, their results lead to conclusions regarding the role of centralised policy planning and organisational structure in defensive posturing. Differences in approach and emphasis aside, our work follows the same vein of utilising such models to provide insights to enhance development, planning and decision-making processes.

Finally, the contribution of Cremonini and Nizovtsev [5] examines the role of security investment on attacker decisions. While never using the term 'deterrence' to describe this concept, the authors examine the duality of the security contribution ("ability to withstand attacks of a given intensity") and the behavioural contribution ("change in attacker's perception of the target in question") present in any given security investment; as with our work, they rely on the presence of alternative targets. Cremonini and Nizovstev argue that this second component is often ignored, and develop a model in four scenarios to capture this effect — the fourth of which is an incomplete, asymmetric information game of similar construct to our operational deterrence model. With their focus on the investment aspects, Cremonini and Nizovstev come to the conclusion that the magnitude of the behavioural contribution can greatly exceed that of the direct investment. Additionally, they find that in the incomplete, asymmetric information case attacker treatment of each target is the same, and thus this behavioural component is removed. They argue that this lack of "transparency" in favour of "opacity" is a benefit for the less secure, at the detriment of the more secure users who as a result are disincentivized to invest in security. It is this phenomenon to which they attribute the failure of econometrics such as Annual Loss Expectancy (ALE) to properly capture the security investment, as it fails to account for such indirect contribution and may lead to underinvestment or misallocation of resources.

The model of [5] shares many common themes and concepts with our construct, with both drawing conclusions along complementary lines. In addition to considering the role of screening within potential behavioural contributions, our model most identifies a concrete example of such a mechanism. This addresses a concern of Cremonini and Nizovstev [5] as to what can "serve as a credible signal of strong inner security and not undermine that security at the same time". In addition, our construct further extends the discussion of 'transparency' and 'opacity' to more fully characterise instances (in the form of game equilibria) where the role of belief can be observed. In the instance that the signal is not seen as an indicator of security, the two resulting pooling equilibria are then driven by the attacker beliefs (and therefore can be considered to be related to the prior probabilities). The first equilibrium is analogous to the findings of Cremonini and Nizovstev, where 'opacity' leads to each defender being attacked equally when $\lambda < \frac{1}{2}$. We also find that in the case that the distribution shifts toward secure users $(\lambda > \frac{1}{2})$ another equilibrium is possible, whereby the situation flips such that everyone benefits as the attacker chooses not to move. This result is not considered by Cremonini and Nizovstev, although it serves to support their conclusions regarding the role of the behavioural component within security. In addition, we also find that when the cost of signalling by a less secure player is sufficiently expensive (coupled with attacker beliefs regarding the role of such signals) a 'transparent' environment with a separating equilibrium emerges, which clearly benefits investment in security. Through a more descriptive treatment of signalling by user types within this environment, we complement [5] with a description that relates cost to loss l and loss probability α. This allows the actions of both low and high security users to be more granular with respect to the desired outcome. As such, our construct suggests that the behavioural component to security and its benefit is indeed still present in these cases, and is reliant on attacker beliefs. These findings further bolster the arguments made in the conclusion of [5] regarding the rejection of 'security through obscurity' and the role of layered defence.

5 Conclusions

We have demonstrated the explanatory power gained by treating the concept of deterrence as an information asymmetry, which is then modelled as a set of games: a screening game, where the attacker moves first and attempts to identify targets for attack, and a signalling game in which the user undertakes measures to attempt to deter potential attackers. In both cases, we showed how the propagation of the asymmetry through the action (or inaction) of the user provided security benefits that can be measured in terms of utility.

We do not attempt to make an argument for deterrence to replace security (to, for example, forgo WPA2 encryption and merely hide one's SSID). In fact, the results show that such constructs have no value in the absence of secure users. Notably, this construct has relevance only to low focus, low skill attacks. As such, they operate as part of a filter for the 'background noise' of internet attacks, but as noted don't hold for directed attacks. The model as presented is highly simplified in its consideration of the cost to the attacker and the user. In addition to the various parameters of the model that may vary from case to case, there are assumptions (such as the equality in the loss of Player 1 and the gain of Player 2) that would be far more complex in reality.

We plan to further investigate the effects of more detailed modelling; however, it is the authors' belief that the value of such concepts lie not in more complex models, but in their explanatory power to describe alternative and complementary concepts of security. As such, such models are expected to have an impact on the formation of requirements and the approaches to security engineering that result from such insights. Movement from the existing paradigms require that we think differently about security throughout the engineering life cycle, and expand our ability to conceptualise and quantify our approaches.

Acknowledgements. We would like to thank Paul Ratazzi and Kamal Jabbour for sharing their previous work, Kasper Rasmussen for the discussion that led to this line of investigation, and Luke Garratt for his insights. We are also grateful to the anonymous reviewers for their helpful and constructive comments.

References

1. Alanezi, F., Brooks, L.: Combatting online fraud in Saudi Arabia using General Deterrence Theory (GDT). In: Proceedings of the 20th Americas Conference on Information Systems (AMCIS 2014) (2014)
2. Alberts, D.S.: Defensive Information Warfare. National Defense University Press, Washington, D.C. (1996)

3. Anderson, R.: The economics of information security. Science **314**(5799), 610–613 (2006)
4. Bunn, M.E.: Can deterrence be tailored? Technical report 225, Institute for National Strategic Studies National Defense University, January 2007. http://www.ndu.edu/inss/nduhp
5. Cremonini, M., Nizovtsev, D.: Understanding and influencing attackers' decisions: Implications for security investment strategies. In: Proceedings of the 5th Annual Workshop on the Economics of Information Security (WEIS 2006) (2006)
6. Dictionary.com: Deter — define deter at dictionary.com, January 2014. http://dictionary.reference.com/browse/deter
7. Dictionary.com: Deterrence — define deterrence at dictionary.com, January 2014. http://dictionary.reference.com/browse/deterrence?s=t
8. Gibbons, R.: Game Theory for Applied Economists. Princeton University Press, Princeton (1992)
9. Gray, C.S.: Deterrence and the nature of strategy. Small Wars Insurgencies **11**(2), 17–26 (2000)
10. Grossklags, J., Christin, N., Chuang, J.: Secure or insure?: A game-theoretic analysis of information security games. In: Proceedings of the 17th International Conference on World Wide Web (WWW 2008), pp. 209–218. ACM (2008). http://doi.acm.org/10.1145/1367497.1367526
11. Herath, T., Rao, H.R.: Protection motivation and deterrence: A framework for security policy compliance in organisations. Eur. J. Inf. Syst. **18**(2), 106–125 (2009)
12. Herley, C.: Why do Nigerian scammers say they are from Nigeria? In: Proceedings of the 11th Annual Workshop on the Economics of Information Security (WEIS 2012) (2012). http://research.microsoft.com/apps/pubs/default.aspx?id=167719
13. Hottell, M., Carter, D., Deniszczuk, M.: Predictors of home-based wireless security. In: Proceedings of the 5th Annual Workshop on the Economics of Information Security (WEIS 2006) (2006)
14. Jabbour, K.T., Ratazzi, E.P.: Deterrence in cyberspace. In: Lowther, A. (ed.) Thinking About Deterrence: Enduring Questions in a Time of Rising Powers, Rogue Regimes, and Terrorism, pp. 37–47. Air University Press (2013)
15. Moore, T., Anderson, R.: Economics and internet security: A survey of recent analytical, empirical and behavioral research. Technical report TR-03-11, Computer Science Group, Harvard University (2011)
16. Morral, A.R., Jackson, B.A.: Understanding the role of deterrence in counterterrorism security. Technical report OP-281-RC, RAND Corporation, Santa Monica, CA (2009)
17. Robbins, E.H., Hustus, H., Blackwell, J.A.: Mathematical foundaitons of strategic deterrence. In: Lowther, A. (ed.) Thinking About Deterrence: Enduring Questions in a Time of Rising Powers, Rogue Regimes, and Terrorism, pp. 137–165. Air University Press (2013)
18. Schneier, B.: Schneier on security: Lessons from the Sony hack, December 2014. www.schneier.com/blog/archives/2014/12/lessons_from_th_4.html
19. Taquechel, E.F., Lewis, T.G.: How to quantify deterrence and reduce critical infrastructure risk. Homeland Security Affairs 8, Article 12 (2012)
20. Tirenin, W., Faatz, D.: A concept for strategic cyber defense. In: IEEE Military Communications Conference 1999 (MILCOM 1999), vol. 1, pp. 458–463 (1999)
21. Varian, H.R.: Intermediate Microeconomics: A Modern Approach, 7th edn. W.W. Norton and Company, New York (2005)

Sequentially Composable Rational Proofs

Matteo Campanelli and Rosario Gennaro[✉]

The City University of New York, New York, USA
mcampanelli@gradcenter.cuny.edu, rosario@ccny.cuny.edu

Abstract. We show that Rational Proofs do not satisfy basic compositional properties in the case where a large number of "computation problems" are outsourced. We show that a "fast" incorrect answer is more remunerable for the prover, by allowing him to solve more problems and collect more rewards. We present an enhanced definition of Rational Proofs that removes the economic incentive for this strategy and we present a protocol that achieves it for some uniform bounded-depth circuits.

1 Introduction

The problem of securely outsourcing data and computation has received widespread attention due to the rise of *cloud computing:* a paradigm where businesses lease computing resources from a service (the *cloud provider*) rather than maintain their own computing infrastructure. Small mobile devices, such as smart phones and netbooks, also rely on remote servers to store and perform computation on data that is too large to fit in the device.

It is by now well recognized that these new scenarios have introduced new security problems that need to be addressed. When data is stored remotely, outside our control, how can we be sure of its integrity? Even more interestingly, how do we check that the results of outsourced computation on this remotely stored data are correct. And how do perform these tests while preserving the efficiency of the client (i.e. avoid retrieving the whole data, and having the client perform the computation) which was the initial reason data and computations were outsourced.

Verifiable Outsourced Computation is a very active research area in Cryptography and Network Security (see [8] for a survey), with the goal of designing protocols where it is impossible (under suitable cryptographic assumptions) for a provider to "cheat" in the above scenarios. While much progress has been done in this area, we are still far from solutions that can be deployed in practice.

A different approach is to consider a model where "cheating" might actually be possible, but the provider would have no motivation to do so. In other words while cryptographic protocols prevent any adversary from cheating, one considers protocols that work against rational adversaries whose motivation is to maximize a well defined utility function.

This work was supported by NSF grant CNS-1545759

MHR Khouzani et al. (Eds.): GameSec 2015, LNCS 9406, pp. 270–288, 2015.
DOI: 10.1007/978-3-319-25594-1_15

Previous Work. An earlier work in this line is [3] where the authors describe a system based on a scheme of rewards [resp. penalties] that the client assesses to the server for computing the function correctly [resp. incorrectly]. However in this system checking the computation may require re-executing it, something that the client does only on a randomized subset of cases, hoping that the penalty is sufficient to incentivize the server to perform honestly. Morever the scheme might require an "infinite" budget for the rewards, and has no way to "enforce" payment of penalties from cheating servers. For these reasons the best application scenario of this approach is the incentivization of volunteer computing schemes (such as SETI@Home or Folding@Home), where the rewards are non-fungible "points" used for "social-status".

Because verification is performed by re-executing the computation, in this approach the client is "efficient" (i.e. does "less" work than the server) only in an amortized sense, where the cost of the subset of executions verified by the client is offset by the total number of computations performed by the server. This implies that the server must perform many executions for the client.

Another approach, instead, is the concept of Rational Proofs introduced by Azar and Micali in [1] and refined in subsequent papers [2,6]. This model captures, more accurately, real-world financial "pay-for-service" transactions, typical of cloud computing contractual arrangements, and security holds for a single "stand-alone" execution.

In a Rational Proof, given a function f and an input x, the server returns the value $y = f(x)$, and (possibly) some auxiliary information, to the client. The client will in turn pay the server for its work with a reward which is a function of the messages sent by the server and some randomness chosen by the client. The crucial property is that this reward is maximized in expectation when the server returns the correct value y. Clearly a rational prover who is only interested in maximizing his reward, will always answer correctly.

The most striking feature of Rational Proofs is their simplicity. For example in [1], Azar and Micali show single-message Rational Proofs for any problem in #P, where an (exponential-time) prover convinces a (poly-time) verifier of the number of satisfying assignment of a Boolean formula.

For the case of "real-life" computations, where the Prover is polynomial and the Verifier is as efficient as possible, Azar and Micali in [2] show d-round Rational Proofs for functions computed by (uniform) Boolean circuits of depth d, for $d = O(\log n)$ (which can be collapsed to a single round under some well-defined computational assumption as shown in [6]). The problem of rational proofs for any polynomial-time computable function remains tantalizingly open.

Our Results. Motivated by the problem of volunteer computation, our first result is to show that the definition of Rational Proofs in [1,2] does not satisfy a basic compositional property which would make them applicable in that scenario. Consider the case where a large number of "computation problems" are outsourced. Assume that solving each problem takes time T. Then in a time interval of length T, the honest prover can only solve and receive the reward for a single problem. On the other hand a dishonest prover, can answer up to

T problems, for example by answering at random, a strategy that takes $O(1)$ time. To assure that answering correctly is a rational strategy, we need that at the end of the T-time interval the reward of the honest prover be larger than the reward of the dishonest one. But this is not necessarily the case: for some of the protocols in [1,2,6] we can show that a "fast" incorrect answer is more remunerable for the prover, by allowing him to solve more problems and collect more rewards.

The next questions, therefore, was to come up with a definition and a protocol that achieves rationality both in the stand-alone case, and in the composition described above. We first present an enhanced definition of Rational Proofs that removes the economic incentive for the strategy of fast incorrect answers, and then we present a protocol that achieves it for the case of some (uniform) bounded-depth circuits.

2   Rational Proofs

In the following we will adopt a "concrete-security" version of the "asymptotic" definitions and theorems in [2,6]. We assume the reader is familiar with the notion of interactive proofs [7].

Definition 1 (Rational Proof). *A function $f : \{0,1\}^n \rightarrow \{0,1\}^n$ admits a rational proof if there exists an interactive proof (P,V) and a randomized reward function* rew $: \{0,1\}^* \rightarrow \mathbb{R}_{\geq 0}$ *such that*

1. *(Rational completeness) For any input $x \in \{0,1\}^n$, $\Pr[\text{out}((P,V)(x)) = f(x)] = 1$.*
2. *For every prover $\widetilde{P}$, and for any input $x \in \{0,1\}^n$ there exists a $\delta_{\widetilde{P}}(x) \geq 0$ such that $\mathbb{E}[\text{rew}((\widetilde{P},V)(x))] + \delta_{\widetilde{P}}(x) \leq \mathbb{E}[\text{rew}((P,V)(x))]$.*

The expectations and the probabilities are taken over the random coins of the prover and verifier.

Let $\epsilon_{\widetilde{P}} = \Pr[\text{out}((P,V)(x)) \neq f(x)]$. Following [6] we define the reward gap as

$$\Delta(x) = min_{P^*:\epsilon_{P^*}=1}[\delta_{P^*}(x)]$$

i.e. the minimum reward gap over the provers that always report the incorrect value. It is easy to see that for arbitrary prover $\widetilde{P}$ we have $\delta_{\widetilde{P}}(x) \geq \epsilon_{\widetilde{P}} \cdot \Delta(x)$. Therefore it suffices to prove that a protocol has a strictly positive reward gap $\Delta(x)$ for all x.

Examples of Rational Proofs. For concreteness here we show the protocol for a single threshold gate (readers are referred to [1,2,6] for more examples).

Let $G_{n,k}(x_1, \ldots, x_n)$ be a threshold gate with n Boolean inputs, that evaluates to 1 if at least k of the input bits are 1. The protocol in [2] to evaluate this gate goes as follows. The Prover announces the number $\tilde{m}$ of input bits equal to 1, which allows the Verifier to compute $G_{n,k}(x_1, \ldots, x_n)$. The Verifier select a random index $i \in [1..n]$ and looks at input bit $b = x_i$ and rewards the Prover using Brier's Rule $BSR(\tilde{p}, b)$ where $\tilde{p} = \tilde{m}/n$ i.e. the probability claimed by the Prover that a randomly selected input bit be 1. Then

$$BSR(\tilde{p}, 1) = 2\tilde{p} - \tilde{p}^2 - (1 - \tilde{p})^2 + 1 = 2\tilde{p}(2 - \tilde{p})$$

$$BSR(\tilde{p}, 0) = 2(1 - \tilde{p}) - \tilde{p}^2 - (1 - \tilde{p})^2 + 1 = 2(1 - \tilde{p}^2)$$

Let m be the true number of input bits equal to 1, and $p = m/n$ the corresponding probability, then the expected reward of the Prover is

$$pBSR(\tilde{p}, 1) + (1 - p)BSR(\tilde{p}, 0) \tag{1}$$

which is easily seen to be maximized for $p = \tilde{p}$ i.e. when the Prover announces the correct result. Moreover one can see that when the Prover announces a wrong $\tilde{m}$ his reward goes down by $2(p - \tilde{p})^2 \geq 2/n^2$. In other words for all n-bit input x, we have $\Delta(x) = 2/n^2$ and if a dishonest Prover $\tilde{P}$ cheats with probability $\epsilon_{\tilde{P}}$ then $\delta_{\tilde{P}} > 2\epsilon_{\tilde{P}}/n^2$.

3 Profit vs. Reward

Let us now define the profit of the Prover as the difference between the reward paid by the verifier and the cost incurred by the Prover to compute f and engage in the protocol. As already pointed out in [2,6] the definition of Rational Proof is sufficiently robust to also maximize the profit of the honest prover and not the reward. Indeed consider the case of a "lazy" prover $\tilde{P}$ that does not evaluate the function: even if $\tilde{P}$ collects a "small" reward, his total profit might still be higher than the profit of the honest prover P.

Set $R(x) = \mathbb{E}[\text{rew}((P, V)(x))]$, $\tilde{R}(x) = \mathbb{E}[\text{rew}((\tilde{P}, V)(x))]$ and $C(x)$ [resp. $\tilde{C}(x)$] the cost for P [resp. $\tilde{P}$] to engage in the protocol. Then we want

$$R(x) - C(x) \geq \tilde{R}(x) - \tilde{C}(x) \implies \delta_{\tilde{P}}(x) \geq C(x) - \tilde{C}(x)$$

In general this is not true (see for example the previous protocol), but it is always possible to change the reward by a multiplier M. Note that if $M \geq C(x)/\delta_{\tilde{P}}(x)$ then we have that

$$M(R(x) - \tilde{R}(x)) \geq C(x) \geq C(x) - \tilde{C}(x)$$

as desired. Therefore by using the multiplier M in the reward, the honest prover P maximizes its profit against all provers $\tilde{P}$ except those for which $\delta_{\tilde{P}}(x) \leq C(x)/M$, i.e. those who report the incorrect result with a "small" probability $\epsilon_{\tilde{P}}(x) \leq \frac{C(x)}{M\Delta(x)}$.

We note that M might be bounded from above, by budget considerations (i.e. the need to keep the total reward $MR(x) \leq B$ for some budget B). This point out to the importance of a large reward gap $\Delta(x)$ since the larger $\Delta(x)$ is, the smaller the probability of a cheating prover $\tilde{P}$ to report an incorrect result must be, in order for $\tilde{P}$ to achieve an higher profit than P.

Example. In the above protocol we can assume that the cost of the honest prover is $C(x) = n$, and we know that $\Delta(x) = n^2$. Therefore the profit of the honest prover is maximized against all the provers that report an incorrect result with probability larger than n^3/M, which can be made sufficiently small by choosing the appropriate multiplier.

Remark 1. If we are interested in an asymptotic treatment, it is important to notice that as long as $\Delta(x) \geq 1/\mathsf{poly}(|x|)$ then it is possible to keep a polynomial reward budget, and maximize the honest prover profit against all provers who cheat with a substantial probability $\epsilon_{\tilde{P}} \geq 1/\mathsf{poly}'(|x|)$.

4   Sequential Composition

We now present the main results of our work. First we informally describe our notion of sequential composition of rational proof, via a motivating example and show that the protocols in [1,2,6] do not satisfy it. Then we present our definition of sequential rational proofs, and a protocol that achieves it for circuits of bounded depth.

4.1   Motivating Example

Consider the protocol in the previous section for the computation of the function $G_{n,k}(\cdot)$. Assume that the honest execution of the protocol (including the computation of $G_{n,k}(\cdot)$) has cost $C = n$.

Assume now that we are given a sequence of n inputs $x^{(1)}, \ldots, x^{(i)}, \ldots$ where each $x^{(i)}$ is an n-bit string. In the following let m_i be the Hamming weight of $x^{(i)}$ and $p_i = m_i/n$.

Therefore the honest prover investing $C = n$ cost, will be able to execute the protocol only once, say on input $x^{(i)}$. By setting $p = \tilde{p} = p_i$ in Eq. 1, we see that P obtains reward

$$R(x^{(i)}) = 2(p_i^2 - p_i + 1) \leq 2$$

Consider instead a prover $\tilde{P}$ which in the execution of the protocol outputs a random value $\tilde{m} \in [0..n]$. The expected reward of $\tilde{P}$ on any input $x^{(i)}$ is (by setting $p = p_i$ and $\tilde{p} = m/n$ in Eq. 1 and taking expectations):

$$\tilde{R}(x^{(i)}) = \mathop{\mathbb{E}}_{m,b}\left[BSR(\frac{m}{n}, b)\right]$$

$$= \frac{1}{n+1}\sum_{m=0}^{n}\mathbb{E}_{b}[BSR(\frac{m}{n}, b]$$

$$= \frac{1}{n+1}\sum_{m=0}^{n}(2(2p_i \cdot \frac{m}{n} - \frac{m^2}{n^2} - p_i + 1))$$

$$= 2 - \frac{2n+1}{3n} > 1 \text{ for } n > 1.$$

Therefore by "solving" just two computations $\widetilde{P}$ earns more than P. Moreover t the strategy of $\widetilde{P}$ has cost 1 and therefore it earns more than P by investing a lot less cost[1].

Note that "scaling" the reward by a multiplier M does not help in this case, since both the honest and dishonest prover's rewards would be multiplied by the same multipliers, without any effect on the above scenario.

We have therefore shown a rational strategy, where cheating many times and collecting many rewards is more profitable than collecting a single reward for an honest computation.

4.2   Sequentially Composable Rational Proofs

The above counterexample motivates the following Definition which formalizes that the reward of the honest prover P must always be larger than the total reward of any prover $\widetilde{P}$ that invests less computation cost than P.

Technically this is not trivial to do, since it is not possible to claim the above for *any* prover $\widetilde{P}$ and *any* sequence of inputs, because it is possible that for a given input $\tilde{x}$, the prover $\widetilde{P}$ has "hardwired" the correct value $\tilde{y} = f(\tilde{x})$ and can compute it without investing any work. We therefore propose a definition that holds for inputs randomly chosen according to a given probability distribution $\mathcal{D}$, and we allow for the possibility that the reward of a dishonest prover can be "negligibly" larger than the reward of the honest prover (for example if $\widetilde{P}$ is lucky and such "hardwired" inputs are selected by $\mathcal{D}$).

Definition 2 (Sequential Rational Proof). *A rational proof (P, V) for a function $f : \{0,1\}^n \rightarrow \{0,1\}^n$ is ϵ-sequentially composable for an input distribution $\mathcal{D}$, if for every prover $\widetilde{P}$, and every sequence of inputs $x, x_1, \ldots, x_k \in \mathcal{D}$ such that $C(x) \geq \sum_{i=1}^{k} \tilde{C}(x_i)$ we have that $\sum_i \tilde{R}(x_i) - R \leq \epsilon$.*

A few sufficient conditions for sequential composability follow.

Lemma 1. *Let (P, V) be a rational proof. If for every input x it holds that $R(x) = R$ and $C(x) = C$ for constants R and C, and the following inequality holds for every $\widetilde{P} \neq P$ and input $x \in \mathcal{D}$:*

$$\frac{\tilde{R}(x)}{R} \leq \frac{\tilde{C}(x)}{C} + \epsilon$$

then (P, V) is $kR\epsilon$-sequentially composable for $\mathcal{D}$.

Proof. It suffices to observe that, for any k inputs $x_1, ..., x_k$, the inequality above implies

$$\sum_{i=1}^{k} \tilde{R}(x_i) \leq R[\sum_{i=1}^{k}(\frac{\tilde{C}(x_i)}{C} + \epsilon)] \leq R + kR\epsilon$$

where the last inequality holds whenever $\sum_{i=1}^{k} \tilde{C}(x_i) \leq C$ as in Definition 2.

---

[1] If we think of cost as time, then in the same time interval in which P solves one problem, $\widetilde{P}$ can solve up to n problems, earning a lot more money, by answering fast and incorrectly.

Corollary 1. *Let (P,V) and* rew *be respectively an interactive proof and a reward function as in Definition 1; if* rew *can only assume the values 0 and R for some constant R, let $\tilde{p}_x = \Pr[\text{rew}((\tilde{P},V)(x)) = R]$. If for $x \in \mathcal{D}$*

$$\tilde{p}_x \le \frac{\tilde{C}(x)}{C} + \epsilon$$

then (P,V) is $kR\epsilon$-sequentially composable for $\mathcal{D}$.

Proof. Observe that $\tilde{R}(x) = \tilde{p}_x \cdot R$ and then apply Lemma 1.

4.3   Sequential Rational Proofs in the PCP Model

We now describe a rational proof appeared in [2] and prove that is sequentially composable. The protocol assumes the existence of a trusted memory storage to which both Prover and Verifier have access, to realize the so-called "PCP" (Probabilistically Checkable Proof) model. In this model, the Prover writes a very long proof of correctness, that the verifier checks only in a few randomly selected positions. The trusted memory is needed to make sure that the prover is "committed" to the proof before the verifier starts querying it.

The following protocol for proofs on a binary logical circuit $\mathcal{C}$ appeared in [2]. The Prover writes all the (alleged) values α_w for every wire $w \in \mathcal{C}$, on the trusted memory location. The Verifier samples a single random gate value to check its correctness and determines the reward accordingly:

1. The Prover writes the vector $\{\alpha_w\}_{w \in \mathcal{C}}$
2. The Verifier samples a random gate $g \in \mathcal{C}$.
   - The Verifier reads $\alpha_{g_{out}}, \alpha_{g_L}, \alpha_{g_R}$, with g_{out}, g_L, g_R being respectively the output, left and right input wires of g; the verifier checks that $\alpha_{g_{out}} = g(\alpha_{g_L}, \alpha_{g_R})$;
   - If g in an input gate the Verifier also checks that $\alpha_{g_L}, \alpha_{g_R}$ correspond to the correct input values;

   The Verifier pays R if both checks are satisfied, otherwise it pays 0.

Theorem 1 ([2]). *The protocol above is a rational proof for any boolean function in $P^{\|NP}$, the class of all languages decidable by a polynomial time machine that can make non-adaptive queries to NP.*

We will now show a cost model where the rational proof above is sequentially composable. We will assume that the cost for any prover is given by the number of gates he writes. Thus, for any input x, the costs for honest and dishonest provers are respectively $C(x) = S$, where $S = |\mathcal{C}|$, and $\tilde{C}(x) = \tilde{s}$ where $\tilde{s}$ is the number of gates written by the dishonest prover. Observe that in this model a dishonest prover may not write all the S gates, and that not all of the $\tilde{s}$ gates have to be correct. Let $\sigma \le \tilde{s}$ the number of correct gates written by $\tilde{P}$.

Theorem 2. *In the cost model above the PCP protocol in [2] is sequentially composable.*

Proof. Observe that the probability $\tilde{p}_x$ that $\tilde{P} \neq P$ earns R is such that

$$\tilde{p}_x = \frac{\sigma}{S} \leq \frac{\tilde{s}}{S} = \frac{\tilde{C}}{C}$$

Applying Corollary 1 completes the proof.

The above cost model, basically says that the cost of writing down a gate dominates everything else, in particular the cost of *computing* that gate. In other cost models a proof of sequential composition may not be as straightforward. Assume, for example, that the honest prover pays \$1 to compute the value of a single gate while writing down that gate is "free". Now $\tilde{p}_x$ is still equal to $\frac{\sigma}{S}$ but to prove that this is smaller than $\frac{\tilde{C}}{C}$ we need some additional assumption that limits the ability for $\tilde{P}$ to "guess" the right value of a gate without computing it (which we will discuss in the next Section).

4.4 Sequential Composition and the Unique Inner State Assumption

Definition 2 for sequential rational proofs requires a relationship between the reward earned by the prover and the amount of "work" the prover invested to produce that result. The intuition is that to produce the correct result, the prover must run the computation and incur its full cost. Unfortunately this intuition is difficult, if not downright impossible, to formalize. Indeed for a specific input x a "dishonest" prover $\tilde{P}$ could have the correct $y = f(x)$ value "hardwired" and could answer correctly without having to perform any computation at all. Similarly, for certain inputs x, x' and a certain function f, a prover $\tilde{P}$ after computing $y = f(x)$ might be able to "recycle" some of the computation effort (by saving some state) and compute $y' = f(x')$ incurring a much smaller cost than computing it from scratch.

A way to circumvent this problem was suggested in [3] under the name of *Unique Inner State Assumption*: the idea is to assume a distribution $\mathcal{D}$ over the input space. When inputs x are chosen according to $\mathcal{D}$, then we assume that computing f requires cost C from any party: this can be formalized by saying that if a party invests $\tilde{C} = \gamma C$ effort (for $\gamma \leq 1$), then it computes the correct value only with probability negligibly close to γ (since a party can always have a "mixed" strategy in which with probability γ it runs the correct computation and with probability $1 - \gamma$ does something else, like guessing at random).

Assumption 1. *We say that the (C, ϵ)-Unique Inner State Assumption holds for a function f and a distribution $\mathcal{D}$ if for any algorithm $\tilde{P}$ with cost $\tilde{C} = \gamma C$, the probability that on input $x \in \mathcal{D}$, $\tilde{P}$ outputs $f(x)$ is at most $\gamma + (1 - \gamma)\epsilon$.*

Note that the assumption implicitly assumes a "large" output space for f (since a random guess of the output of f will be correct with probability 2^{-n} where n is the binary length of $f(x)$).

More importantly, note that Assumption 1 immediately yields our notion of sequential composability, if the Verifier can detect if the Prover is lying or not. Assume, as a mental experiment for now, that given input x, the Prover claims that $\tilde{y} = f(x)$ and the Verifier checks by recomputing $y = f(x)$ and paying a reward of R to the Prover if $y = \tilde{y}$ and 0 otherwise. Clearly this is not a very useful protocol, since the Verifier is not saving any computation effort by talking to the Prover. But it is sequentially composable according to our definition, since $\tilde{p}_x$, the probability that $\tilde{P}$ collects R, is equal to the probability that $\tilde{P}$ computes $f(x)$ correctly, and by using Assumption 1 we have that

$$\tilde{p}_x = \gamma + (1 - \gamma)\epsilon \le \frac{\tilde{C}}{C} + \epsilon$$

satisfying Corollary 1.

To make this a useful protocol we adopt a strategy from [3], which also uses this idea of verification by recomputing. Instead of checking every execution, we check only a random subset of them, and therefore we can amortize the Verifier's effort over a large number of computations. Fix a parameter m. The prover sends to the verifier the values $\tilde{y}_j$ which are claimed to be the result of computing f over m inputs $x_1, \ldots, x_m$. The verifier chooses one index i randomly between 1 and m, and computes $y_i = f(x_i)$. If $y_i = \tilde{y}_i$ the verifier pays R, otherwise it pays 0.

Let T be the total cost by the honest prover to compute m instances: cleary $T = mC$. Let $\tilde{T} = \Sigma_i \tilde{C}_i$ be the total effort invested by $\tilde{P}$, by investing $\tilde{C}_i$ on the computation of x_i. In order to satisfy Corollary 1 we need that $\tilde{p}_x$, the probability that $\tilde{P}$ collects R, be less than $\tilde{T}/T + \epsilon$.

Let $\gamma_i = \tilde{C}_i/C$, then under Assumption 1 we have that $\tilde{y}_i$ is correct with probability at most $\gamma_i + (1 - \gamma_i)\epsilon$. Therefore if we set $\gamma = \sum_i \gamma_i/m$ we have

$$\tilde{p}_x = \frac{1}{m} \sum_i [\gamma_i + (1 - \gamma_i)\epsilon] = \gamma + (1 - \gamma)\epsilon \le \gamma + \epsilon$$

But note that $\gamma = \tilde{T}/T$ as desired since

$$\tilde{T} = \sum_i \tilde{C}_i = \sum_i \gamma_i C = T \sum_i \gamma_i/m$$

Efficiency of the Verifier. If our notion of "efficient Verifier" is a verifier who runs in time $o(C)$ where C is the time to compute f, then in the above protocol m must be sufficiently large to amortize the cost of computing one execution over many (in particular a constant – in the input size n – value of m would not work). In our "concrete analysis" treatment, if we requests that the Verifier runs in time δC for an "efficiency" parameter $\delta \le 1$, then we need $m \ge \delta^{-1}$.

Therefore we are still in need of a protocol which has an efficient Verifier, and would still works for the "stand-alone" case ($m = 1$) but also for the case of sequential composability over any number m of executions.

5 Our Protocol

We now present a protocol that works for functions $f : \{0,1\}^n \rightarrow \{0,1\}^n$ expressed by an arithmetic circuit $\mathcal{C}$ of size C and depth d and fan-in 2, given as a common input to both Prover and Verifier together with the input x.

Intuitively the idea is for the Prover to provide the Verifier with the output value y and its two "children" y_L, y_R in the gate, i.e. the two input values of the last output gate G. The Verifier checks that $G(y_L, y_R) = y$, and then asks the Prover to verify that y_L or y_R (chosen a random) is correct, by recursing on the above test. The protocol description follows.

1. The Prover evaluates the circuit on x and sends the output value y_1 to the Verifier.
2. **Repeat r times:** The Verifier identifies the root gate g_1 and then invokes $Round(1, g_1, y_1)$,

where the procedure $Round(i, g_i, y_i)$ is defined for $1 \le i \le d$ as follows:

1. The Prover sends the value of the input wires z_i^0 and z_i^1 of g_i to the Verifier.
2. The Verifiers performs the following
   - Check that y_i is the result of the operation of gate g_i on inputs z_i^0 and z_i^1. If not **STOP** and pay a reward of 0.
   - If $i = d$ (i.e. if the inputs to g_i are input wires), check that the values of z_i^0 and z_i^1 are equal to the corresponding bits of x. Pay reward R to Merlin if this is the case, nothing otherwise.
   - If $i < d$, choose a random bit b, send it to Merlin and invoke $Round(i + 1, g_{i+1}^b, z_i^b)$ where g_{i+1}^b is the child gate of g_i whose output is z_i^b.

5.1 Efficiency

The protocol runs at most in d rounds. In each round, the Prover sends a constant number of bits representing the values of specific input and output wires; The Verifier sends at most one bit per round, the choice of the child gate. Thus the communication complexity is $O(d)$ bits.

The computation of the Verifier in each round is: (i) computing the result of a gate and checking for bit equality; (ii) sampling a child. Gate operations and equality are $O(1)$ per round. We assume our circuits are T-uniform, which allows the Verifier to select the correct gate in time $T(n)^2$. Thus the Verifier runs in $O(rd \cdot T(n))$ with $r = O(\log C)$.

---

[2] We point out that the Prover can provide the Verifier with the requested gate and then the Verifier can use the uniformity of the circuit to check that the Prover has given him the correct gate at each level in time $O(T(n))$.

5.2   Proofs of (Stand-Alone) Rationality

Theorem 3. *The protocol in Sect. 5 for $r = 1$ is a Rational Proof according to Definition 1.*

We prove the above theorem by showing that for every input x the reward gap $\Delta(x)$ is positive.

Proof. Let $\widetilde{P}$ a prover that always reports $\tilde{y} \neq y_1 = f(x)$ at Round 1.

Let us proceed by induction on the depth d of the circuit. If $d = 1$ then there is no possibility for $\widetilde{P}$ to cheat successfully, and its reward is 0.

Assume $d > 1$. We can think of the binary circuit $\mathcal{C}$ as composed by two subcircuits $\mathcal{C}_L$ and $\mathcal{C}_R$ and the output gate g_1 such that $f(x) = g_1(\mathcal{C}_L(x), \mathcal{C}_R(x))$. The respective depths d_L, d_R of these subcircuits are such that $0 \leq d_L, d_R \leq d - 1$ and $max(d_L, d_R) = d - 1$. After sending $\tilde{y}$, the protocol requires that $\widetilde{P}$ sends output values for $\mathcal{C}_L(x)$ and $\mathcal{C}_R(x)$; let us denote these claimed values respectively with $\tilde{y}_L$ and $\tilde{y}_R$. Notice that at least one of these alleged values will be different from the respective correct subcircuit output: if it were otherwise, V would reject immediately as $g(\tilde{y}_L, \tilde{y}_R) = f(x) \neq \tilde{y}$. Thus at most one of the two values $\tilde{y}_L$, $\tilde{y}_R$ is equal to the output of the corresponding subcircuit. The probability that the $\widetilde{P}$ cheats successfully is:

$$\Pr[V \text{ accepts}] \leq \frac{1}{2} \cdot (\Pr[V \text{ accepts on } \mathcal{C}_L] + \Pr[V \text{ acceptson } \mathcal{C}_R]) \tag{2}$$

$$\leq \frac{1}{2} \cdot (1 - 2^{-\max(d_L, d_R)}) + \frac{1}{2} \tag{3}$$

$$\leq \frac{1}{2} \cdot (1 - 2^{-d+1}) + \frac{1}{2} \tag{4}$$

$$= 1 - 2^{-d} \tag{5}$$

At Eq. 3 we used the inductive hypothesis and the fact that all probabilities are at most 1.

Therefore the expected reward of $\widetilde{P}$ is $\tilde{R} \leq R(1 - 2^{-d})$ and the reward gap is $\Delta(x) = 2^{-d}R$ (see Remark 2 or an explanation of the equality sign).

The following useful corollary follows from the proof above.

Corollary 2. *If the protocol described in Sect. 5 is repeated $r \geq 1$ times a prover can cheat with probability at most $(1 - 2^{-d})^r$.*

Remark 2. We point out that one can always build a prover strategy P^* which always answers incorrectly and achieves exactly the reward $R^* = R(1 - 2^{-d})$. This prover outputs an incorrect $\tilde{y}$ and then computes one of the subcircuits that results in one of the input values (so that at least one of the inputs is correct). This will allow him to recursively answer with values z_i^0 and z_i^1 where one of the two is correct, and therefore be caught only with probability 2^{-d}.

Remark 3. In order to have a non-negligible reward gap (see Remark 1) we need to limit ourselves to circuits of depth $d = O(\log n)$.

5.3 Proof of Sequential Composability

General Sufficient Conditions for Sequential Composability

Lemma 2. *Let C be a circuit of depth d. If the (C, ϵ) Unique Inner State Assumption (see Assumption 1) holds for the function f computed by C, and input distribution D, then the protocol presented above with r repetitions is a $k\epsilon R$-sequentially composable Rational Proof for C for D if the following inequality holds*

$$(1 - 2^{-d})^r \leq \frac{1}{C}$$

Proof. Let $\gamma = \frac{\tilde{C}}{C}$. Consider $x \in D$ and prover $\tilde{P}$ which invests effort $\tilde{C} \leq C$. Under Assumption 1, $\tilde{P}$ gives the correct outputs with probability $\gamma + \epsilon$ – assume that in this case $\tilde{P}$ collects the reward R. If $\tilde{P}$ gives an incorrect output we know (following Corollary 2) that he collects the reward R with probability at most $(1 - 2^{-d})^r$ which by hypothesis is less than γ. So either way we have that $\tilde{R} \leq (\gamma + \epsilon)R$ and therefore applying Lemma 1 concludes the proof.

The problem with the above Lemma is that it requires a large value of r for the result to be true resulting in an inefficient Verifier. In the following sections we discuss two approaches that will allows us to prove sequential composability even for an efficient Verifier:

– Limiting the class of provers we can handle in our security proof;
– Limiting the class of functions/circuits.

Limiting the Strategy of the Prover: Non-adaptive Provers. In proving sequential composability it is useful to find a connection between the amount of work done by a dishonest prover and its probability of cheating. The more a dishonest prover works, the higher its probability of cheating. This is true for our protocol, since the more "subcircuits" the prover computes correctly, the higher is the probability of convincing the verifier of an incorrect output becomes. The question then is: how can a prover with an "effort budget" to spend maximize its probability of success in our protocol?

As we discussed in Remark 2, there is an *adaptive* strategy for the $\tilde{P}$ to maximize its probability of success: compute one subcircuit correctly at every round of the protocol. We call this strategy "adaptive", because the prover allocates its "effort budget" $\tilde{C}$ on the fly during the execution of the rational proof. Conversely a *non-adaptive* prover $\tilde{P}$ uses $\tilde{C}$ to compute some subcircuits in C before starting the protocol. Clearly an adaptive prover strategy is more powerful, than a non-adaptive one (since the adaptive prover can direct its computation effort where it matters most, i.e. where the Verifier "checks" the computation).

Is it possible to limit the Prover to a non-adaptive strategy? This could be achieved by imposing some "timing" constraints to the execution of the protocol: to prevent the prover from performing large computations while interacting with the Verifier, the latter could request that prover's responses be delivered "immediately", and if a delay happens then the Verifier will not pay the reward. Similar timing constraints have been used before in the cryptographic literature, e.g. see the notion of *timing assumptions* in the concurrent zero-knowledge protocols in [5].

Therefore in the rest of this subsection we assume that non-adaptive strategies are the only rational ones and proceed in analyzing our protocol under the assumption that the prover is adopting a non-adaptive strategy.

Consider a prover $\tilde{P}$ with effort budget $\tilde{C} < C$. A *DFS* (for "depth first search") prover uses its effort budget $\tilde{C}$ to compute a whole subcircuit of size $\tilde{C}$ and maximal depth d_{DFS}. Call this subcircuit $\mathcal{C}_{DFS}$. $\tilde{P}$ can answer correctly any verifier's query about a gate in $\mathcal{C}_{DFS}$. During the interaction with V, the behavior of a DFS prover is as follows:

– At the beginning of the protocol send an arbitrary output value y_1.
– During procedure $Round(i, g_i, y_i)$:
  • If $g_i \in \mathcal{C}_{DFS}$ then $\tilde{P}$ sends the two correct inputs z_i^0 and z_i^1.
  • If $g_i \notin \mathcal{C}_{DFS}$ and neither of g_i's input gate belongs to $\mathcal{C}_{DFS}$ then $\tilde{P}$ sends two arbitrary z_i^0 and z_i^1 that are consistent with y_i, i.e. $g_i(z_i^0, z_i^1) = y_i$.
  • $g_i \notin \mathcal{C}_{DFS}$ and one of g_i's input gates belongs to $\mathcal{C}_{DFS}$, then $\tilde{P}$ will send the correct wire known to him and another arbitrary value consistent with y_i as above.

Lemma 3 (Advantage of a DFS Prover). *In one repetition of the protocol above, a DFS prover with effort budget $\tilde{C}$ investment has probability of cheating $\tilde{p}_{DFS}$ bounded by*

$$\tilde{p}_{DFS} \leq 1 - 2^{-d_{DFS}}$$

The proof of Lemma 3 follows easily from the proof of the stand-alone rationality of our protocol (see Theorem 3).

If a DFS prover focuses on maximizing the depth of a computed subcircuit given a certain investment, *BFS* provers allot their resources to compute all subcircuits rooted at a certain height. A BFS prover with effort budget $\tilde{C}$ computes the value of all gates up to the maximal height possible d_{BFS}. Note that d_{BFS} is a function of the circuit $\mathcal{C}$ and of the effort $\tilde{C}$. Let $\overline{\mathcal{C}}_{BFS}$ be the collection of gates computed by the BFS prover. The interaction of a BFS prover with V throughout the protocol resembles that of the DFS prover outlined above:

– At the beginning of the protocol send an arbitrary output value y_1.
– During procedure $Round(i, g_i, y_i)$:
  • If $g_i \in \overline{\mathcal{C}}_{BFS}$ then $\tilde{P}$ sends the two correct inputs z_i^0 and z_i^1.
  • If $g_i \notin \overline{\mathcal{C}}_{BFS}$ and neither of g_i's input gate belongs to $\overline{\mathcal{C}}_{BFS}$ then $\tilde{P}$ sends two arbitrary z_i^0 and z_i^1 that are consistent with y_i, i.e. $g_i(z_i^0, z_i^1) = y_i$.
  • $g_i \notin \overline{\mathcal{C}}_{BFS}$ and both g_i's input gates belong to $\mathcal{C}_{DFS}$, then $\tilde{P}$ will send one of the correct wires known to him and another arbitrary value consistent with y_i as above.

As before, it is not hard to see that the probability of successful cheating by a BFS prover can be bounded as follows:

Lemma 4 (Advantage of a BFS Prover). *In one repetition of the proto-col above, a BFS prover with effort budget $\tilde{C}$ has probability of cheating $\tilde{p}_{BFS}$ bounded by*

$$\tilde{p} \le 1 - 2^{-d_{BFS}}$$

BFS and DFS provers are both special cases of the general non-adaptive strategy which allots its investment $\tilde{C}$ among a general collection of subcircuits $\overline{C}$. The interaction with V of such a prover is analogous to that of a BFS/DFS prover but with a collection of computed subcircuits not constrained by any specific height. We now try to formally define what the success probability of such a prover is.

Definition 3 (Path Experiment). *Consider a circuit C and a collection $\overline{C}$ of subcircuits of C. Perform the following experiment: starting from the output gate, flip a unbiased coin and choose the "left" subcircuit or the "right" subcircuit at random with probability $1/2$. Continue until the random path followed by the experiment reaches a computed gate in $\overline{C}$. Let i be the height of this gate, which is the output of the experiment. Define with Π_i the probability that this experiment outputs i.*

The proof of the following Lemma is a generalization of the proof of security of our scheme. Once the "verification path" chosen by the Verifier enters a fully computed subcircuit at height i (which happens with probability $\Pi_i^{\overline{C}}$), the prob-ability of success of the Prover is bounded by $(1 - 2^{-i})$

Lemma 5 (Advantage of a Non Adaptive Prover). *In one repetition of the protocol above, a generic prover with effort budget $\tilde{C}$ used to compute a collection $\overline{C}$ of subcircuits, has probability of cheating $\tilde{p}_{\overline{C}}$ bounded by*

$$\tilde{p}_{\overline{C}} \le \sum_{i=0}^{d} \Pi_i (1 - 2^{-i})$$

where Π_i-s are defined as in Definition 3.

Limiting the Class of Functions: Regular Circuits. Lemma 5 still does not produce a clear bound on the probability of success of a cheating prover. The reason is that it is not obvious how to bound the probabilities $\Pi_i^{\overline{C}}$ that arise from the computed subcircuits $\overline{C}$ since those depends in non-trivial ways from the topology of the circuit C.

We now present a type of circuits for which it can be shown that the BFS strategy is optimal. The restriction on the circuit is surprisingly simple: we call them regular circuits. In the next section we show examples of interesting func-tions that admit regular circuits.

Definition 4 (Regular Circuit). *A circuit C is said to be regular if the following conditions hold:*

- *C is layered;*
- *every gate has fan-in 2;*
- *the inputs of every gate are the outputs of two distinct gates.*

The following lemma states that, in regular circuits, we can bound the advantage of any prover investing $\tilde{C}$ by looking at the advantage of a BFS prover with the same investment.

Lemma 6 (A Bound for Provers' Advantage in Regular Circuits). *Let $\widetilde{P}$ be a prover investing $\tilde{C}$. Let C be the circuit being computed and $\delta = d_{BFS}(C, \tilde{C})$. In one repetition of the protocol above, the advantage of $\widetilde{P}$ is bounded by*

$$\tilde{p} \leq \tilde{p}_{BFS} = 1 - 2^{-\delta}$$

Proof. Let $\overline{C}$ be the family of subcircuits computed by $\widetilde{P}$ with effort $\tilde{C}$. As pointed out above the probability of success for $\widetilde{P}$ is

$$\tilde{p} \leq \sum_{i=0}^{d} \Pi_i^{\overline{C}}(1 - 2^{-i})$$

Consider now a prover $\widetilde{P}'$ which uses $\tilde{C}$ effort to compute a different collection of subcircuits $\overline{C}'$ defined as follows:

- Remove a gate from a subcircuit of height j in $\overline{C}$: this produces two subcircuits of height $j - 1$. This is true because of the regularity of the circuit: since the inputs of every gate are the outputs of two distinct gates, when removing a gate of height j this will produce two subcircuits of height $j - 1$;
- Use that computation to "join" two subcircuits of height k into a single subcircuit of height $k + 1$. Again we are using the regularity of the circuit here: since the circuit is layered, the only way to join two subcircuits into a single computed subcircuit is to take two subcircuits of the same height.

What happens to the probability $\tilde{p}'$ of success of $\widetilde{P}'$? Let ℓ be the number of possible paths generated by the experiment above with $\overline{C}$. Then the probability of entering a computed subcircuit at height j decreases by $1/\ell$ and that probability weight goes to entering at height $j - 1$. Similarly the probability of entering at height k goes down by $2/\ell$ and that probability weight is shifted to entering at height $k + 1$. Therefore

$$\tilde{p}' \leq \sum_{i \neq j, j-1, k, k+1} \Pi_i(1 - 2^{-i})$$
$$+ (\Pi_j - \frac{1}{\ell})(1 - 2^{-j}) + (\Pi_{j-1} + \frac{1}{\ell})(1 - 2^{-j+1})$$
$$+ (\Pi_k - \frac{2}{\ell})(1 - 2^{-k}) + (\Pi_{k+1} + \frac{2}{\ell})(1 - 2^{-k-1})$$

$$= \tilde{p} + \frac{1}{\ell 2^j} - \frac{1}{\ell 2^j} \mathbf{1} + \frac{1}{\ell 2^{k-1}} - \frac{1}{\ell 2^k}$$

$$= \tilde{p} + \frac{2^k - 2^{k+1} + 2^{j+1} - 2^j}{\ell 2^{j+k}} = \tilde{p} + \frac{2^j - 2^k}{\ell 2^{j+k}}$$

Note that $\tilde{p}'$ increases if $j > k$ which means that it's better to take "computation" away from tall computed subcircuits to make them shorter, and use the saved computation to increase the height of shorter computed subtrees, and therefore that the probability is maximized when all the subtrees are of the same height, i.e. by the BFS strategy which has probability of success $\tilde{p}_{BFS} = 1 - 2^{-\delta}$.

The above Lemma, therefore, yields the following.

Theorem 4. *Let $\mathcal{C}$ be a regular circuit of size C. If the (C, ϵ) Unique Inner State Assumption (see Assumption 1) holds for the function f computed by $\mathcal{C}$, and input distribution $\mathcal{D}$, then the protocol presented above with r repetitions is a $k\epsilon R$-sequentially composable Rational Proof for $\mathcal{C}$ for $\mathcal{D}$ if the prover follows a non-adaptive strategy and the following inequality holds for all $\tilde{C}$*

$$(1 - 2^{-\delta})^r \le \frac{\tilde{C}}{C}$$

where $\delta = d_{BFS}(\mathcal{C}, \tilde{C})$.

Proof. Let $\gamma = \frac{\tilde{C}}{C}$. Consider $x \in \mathcal{D}$ and prover $\tilde{P}$ which invests effort $\tilde{C} \le C$. Under Assumption 1, $\tilde{P}$ gives the correct outputs with probability $\gamma + \epsilon$ – assume that in this case $\tilde{P}$ collects the reward R.

If $\tilde{P}$ gives an incorrect output we can invoke Lemma 6 and conclude that he collects reward R with probability at most $(1 - 2^{-\delta})^r$ which by hypothesis is less than γ. So either way we have that $\tilde{R} \le (\gamma + \epsilon)R$ and therefore applying Lemma 1 concludes the proof.

6   Results for FFT Circuits

In this section we apply the previous results to the problem of computing FFT circuits, and by extension to polynomial evaluations.

6.1   FFT Circuit for Computing a Single Coefficient

The Fast Fourier Transform is an almost ubiquitous computational problem that appears in many applications, including many of the volunteer computations that motivated our work. As described in [4] a circuit to compute the FFT of a vector of n input elements, consists of $\log n$ levels, where each level comprises $n/2$ *butterflies* gates. The output of the circuit is also a vector of n input elements.

Let us focus on the circuit that computes a single element of the output vector: it has $\log n$ levels, and at level i it has $n/2^i$ butterflies gates. Moreover the circuit is regular, according to Definition 4.

Theorem 5. *Under the (C, ϵ)-unique inner state assumption for input distribution $\mathcal{D}$, the protocol in Sect. 5, when repeated $r = O(1)$ times, yields sequentially composable rational proofs for the FFT, under input distribution $\mathcal{D}$ and assuming non-adaptive prover strategies.*

Proof. Since the circuit is regular we can prove sequential composability by invoking Theorem 4 and proving that for $r = O(1)$, the following inequality holds

$$\tilde{p} = (1 - 2^{-\delta})^r \leq \frac{\tilde{C}}{C}$$

where $\delta = d_{BFS}(\mathcal{C}, \tilde{\mathcal{C}})$.

But for any $\tilde{\delta} < d$, the structure of the FFT circuit implies that the number of gates below height $\tilde{\delta}$ is $\tilde{C}_{\tilde{\delta}} = \Theta(C(1 - 2^{-\tilde{\delta}}))$. Thus the inequality above can be satisfied with $r = \Theta(1)$.

6.2   Mixed Strategies for Verification

One of the typical uses of the FFT is to change representation for polynomials. Given a polynomial $P(x)$ of degree $n - 1$ we can represent it as a vector of n coefficients $[a_0, \ldots, a_{n-1}]$ or as a vector of n points $[P(\omega_0), \ldots, P(\omega_{n-1})]$. If ω_i are the complext n-root of unity, the FFT is the algorithm that goes from one representation to the other in $O(n \log n)$ time, rather than the obvious $O(n^2)$.

In this section we consider the following problem: given two polynomial P, Q of degree $n - 1$ in point representation, compute the inner product of the coefficients of P, Q. A fan-in two circuit computing this function could be built as follows:

- two parallel FFT subcircuits computing the coefficient representation of P, Q ($\log n$-depth and $n \log n$) size total for the 2 circuits);
- a subcircuit where at the first level the i-degree coefficients are multiplied with each other, and then all these products are added by a binary tree of additions $O(\log n)$-depth and $O(n)$ size);

Note that this circuit is regular, and has depth $2 \log n + 1$ and size $n \log n + n + 1$.

Consider a prover $\tilde{P}$ who pays $\tilde{C} < n \log n$ effort. Then, since the BFS strategy is optimal, the probability of convincing the Verifier of a wrong result of the FFT is $(1 - 2^{-\tilde{d}})^r$ where $\tilde{d} = c \log n$ with $c \leq 1$. Note also that $\frac{\tilde{C}}{C} < 1$. Therefore with $r = O(n^c)$ repetitions, the probability of success can be made smaller than $\frac{\tilde{C}}{C}$. The Verifier's complexity is $O(n^c \log n) = o(n \log n)$.

If $\tilde{C} \geq n \log n$ then the analysis above fails since $\tilde{d} > \log n$. Here we observe that in order for $\tilde{P}$ to earn a larger reward than P, it must be that P has run at least $k = O(\log n)$ executions (since it is possible to find $k + 1$ inputs such that $(k + 1)\tilde{C} \leq kC$ only if $k > \log n$).

Assume for a moment that the prover always executes the same strategy with the same running time. In this case we can use a "mixed" strategy for verification:

- The Verifier pays the Prover only after k executions. Each execution is verified as above (with n^c repetitions);
- Additionally the Verifier uses the "check by re-execution" (from Sect. 4.4) strategy every k executions (verifiying one execution by recomputing it);
- The Verifier pays R if all the checks are satisfied, 0 otherwise;
- The Verifier's complexity is $O(kn^c \log n + n \log n) = o(kn \log n)$ – the latter being the complexity of computing k instances.

Notice that there are many plausible ways to assume that the expected cost $\tilde{C}$ remains the same through the $k + 1$ proofs, for example by assuming that the Prover can be "resetted" at the beginning of each execution and made oblivious of the previous interactions.

7 Conclusion

Rational Proofs are a promising approach to the problem of verifying computations in a rational model, where the prover is not malicious, but only motivated by the goal of maximizing its utility function. We showed that Rational Proofs do not satisfy basic compositional properties in the case where a large number of "computation problems" are outsourced, e.g. volunteered computations. We showed that a "fast" incorrect answer is more remunerable for the prover, by allowing him to solve more problems and collect more rewards. We presented an enhanced definition of Rational Proofs that removes the economic incentive for this strategy and we presented a protocol that achieves it for some uniform bounded-depth circuits.

One thing to point out is that our protocol has two additional advantages:

- the honest Prover is always guaranteed a fixed reward R, as opposed to some of the protocols in [1,2] where the reward is a random variable even for the honest prover;
- Our protocol is the first example of a rational proof for arithmetic circuits.

Our work leaves many interesting research directions to explore:

- Is it possible to come up with a protocol that works for any bounded-depth circuit, and not circuits with special "topological" conditions such as the ones imposed by our results?
- Our results hold for "non-adaptive" prover strategies, though that seems more a proof artifact to simplify the analysis, than a technical requirement. Is it possible to lift that restriction?
- Are there other circuits which, like the FFT one, satisfy our notions and requirements?
- What about rational proofs for arbitrary poly-time computations? Even if the simpler stand-alone case?

References

1. Azar, P.D., Micali, S.: Rational proofs. In: 2012 ACM Symposium on Theory of Computing, pp. 1017–1028 (2012)
2. Azar, P.D., Micali, S.: Super-efficient rational proofs. In: 2013 ACM Conference on Electronic Commerce, pp. 29–30 (2013)
3. Belenkiy, M., Chase, M., Erway, C.C., Jannotti, J., Küpçü, A., Lysyanskaya, A.: Incentivizing outsourced computation. In: NetEcon 2008, pp. 85–90 (2008)
4. Cormen, T., Leiserson, C., Rivest, R., Stein, C.: Introduction to Algorithms. MIT Press (2001)
5. Dwork, C., Naor, M., Sahai, A.: Concurrent zero-knowledge. J. ACM **51**(6), 851–898 (2004)
6. Guo, S., Hubacek, P., Rosen, A., Vald, M.: Rational arguments: single round delegation with sublinear verification. In: 2014 Innovations in Theoretical Computer Science Conference (2014)
7. Goldwasser, S., Micali, S., Rackoff, C.: The knowledge complexity of interactive proof-systems. In: Proceedings of the seventeenth Annual ACM Symposium on Theory of computing. ACM (1985)
8. Walfish, M., Blumberg, A.J.: Verifying computations without reexecuting them. Commun. ACM **58**(2), 74–84 (2015)

Flip the Cloud: Cyber-Physical Signaling Games in the Presence of Advanced Persistent Threats

Jeffrey Pawlick(✉), Sadegh Farhang, and Quanyan Zhu

Department of Electrical and Computer Engineering,
Polytechnic School of Engineering, New York University, New York, USA
{jpawlick,farhang,quanyan.zhu}@nyu.edu

Abstract. Access to the cloud has the potential to provide scalable and cost effective enhancements of physical devices through the use of advanced computational processes run on apparently limitless cyber infrastructure. On the other hand, cyber-physical systems and cloud-controlled devices are subject to numerous design challenges; among them is that of security. In particular, recent advances in adversary technology pose Advanced Persistent Threats (APTs) which may stealthily and completely compromise a cyber system. In this paper, we design a framework for the security of cloud-based systems that specifies when a device should trust commands from the cloud which may be compromised. This interaction can be considered as a game between three players: a cloud defender/administrator, an attacker, and a device. We use traditional signaling games to model the interaction between the cloud and the device, and we use the recently proposed `FlipIt` game to model the struggle between the defender and attacker for control of the cloud. Because attacks upon the cloud can occur without knowledge of the defender, we assume that strategies in both games are picked according to prior commitment. This framework requires a new equilibrium concept, which we call *Gestalt Equilibrium*, a fixed-point that expresses the interdependence of the signaling and `FlipIt` games. We present the solution to this fixed-point problem under certain parameter cases, and illustrate an example application of cloud control of an unmanned vehicle. Our results contribute to the growing understanding of cloud-controlled systems.

1 Introduction

Advances in computation and information analysis have expanded the capabilities of the physical plants and devices in cyber-physical systems (CPS)[4,13]. Fostered by advances in cloud computing, CPS have garnered significant attention from both industry and academia. Access to the cloud gives administrators the opportunity to build virtual machines that provide to computational resources with precision, scalability, and accessibility.

Despite the advantages that cloud computing provides, it also has some drawbacks. They include - but are not limited to - accountability, virtualization, and security and privacy concerns. In this paper, we focus especially on providing

© Springer International Publishing Switzerland 2015
MHR Khouzani et al. (Eds.): GameSec 2015, LNCS 9406, pp. 289–308, 2015.
DOI: 10.1007/978-3-319-25594-1_16

accurate signals to a cloud-connected device and deciding whether to accept those signals in the face of security challenges.

Recently, system designers face security challenges in the form of *Advanced Persistent Threats (APTs)* [19]. APTs arise from sophisticated attackers who can infer a user's cryptographic key or leverage zero-day vulnerabilities in order to completely compromise a system without detection by the system administrator [16]. This type of stealthy and complete compromise has demanded new types of models [6,20] for prediction and design.

In this paper, we propose a model in which a device decides whether to trust commands from a cloud which is vulnerable to APTs and may fall under adversarial control. We synthesize a mathematical framework that enables devices controlled by the cloud to intelligently decide whether to obey commands from the possibly-compromised cloud or to rely on their own lower-level control.

We model the cyber layer of the cloud-based system using the recently proposed `FlipIt` game [6,20]. This game is especially suited for studying systems under APTs. We model the interaction between the cloud and the connected device using a signaling game, which provides a framework for modeling dynamic interactions in which one player operates based on a belief about the private information of the other. A significant body of research has utilized this framework for security [7–9,15,21]. The signaling and `FlipIt` games are coupled, because the outcome of the `FlipIt` game determines the likelihood of benign and malicious attackers in the robotic signaling game. Because the attacker is able to compromise the cloud without detection by the defender, we consider the strategies of the attacker and defender to be chosen with *prior commitment.* The circular dependence in our game requires a new equilibrium concept which we call a *Gestalt equilibrium*[1]. We specify the parameter cases under which the Gestalt equilibrium varies, and solve a case study of the game to give an idea of how the Gestalt equilibrium can be found in general. Our proposed framework has versatile applications to different cloud-connected systems such as urban traffic control, drone delivery, design of smart homes, etc. We study one particular application in this paper:ef control of an unmanned vehicle under the threat of a compromised cloud.

Our contributions are summarized as follows:

(i) We model the interaction of the attacker, defender/cloud administrator, and cloud-connected device by introducing a novel game consisting of two coupled games: a traditional signaling game and the recently proposed `FlipIt` game.

(ii) We provide a general framework by which a device connected to a cloud can decide whether to follow its own limited control ability or to trust the signal of a possibly-malicious cloud.

(iii) We propose a new equilibrium definition for this combined game: Gestalt equilibrium, which involves a fixed-point in the mappings between the two component games.

(iv) Finally, we apply our framework to the problem of unmanned vehicle control.

---

[1] Gestalt is a noun which means something that is composed of multiple arts and yet is different from the combination of the parts [2].

In the sections that follow, we first outline the system model, then describe the equilibrium concept. Next, we use this concept to find the equilibria of the game under selected parameter regimes. Finally, we apply our results to the control of an unmanned vehicle. In each of these sections, we first consider the signaling game, then consider the FlipIt game, and last discuss the synthesis of the two games. Finally, we conclude the paper and suggest areas for future research.

2   System Model

We model a cloud-based system in which a cloud is subject to APTs. In this model, an *attacker*, denoted by $\mathcal{A}$, capable of APTs can pay an attack cost to completely compromise the cloud without knowledge of the cloud defender. The *defender*, or cloud administrator, denoted by $\mathcal{D}$, does not observe these attacks, but has the capability to pay a cost to reclaim control of the cloud. The cloud transmits a message to a *robot* or other device, denoted by $\mathcal{R}$. The device may follow this command, but it is also equipped with an on-board control system for autonomous operation. It may elect to use its autonomous operation system rather than obey commands from the cloud.

This scenario involves two games: the FlipIt game introduced in [20], and the well-known signaling game. The FlipIt game takes place between the attacker and cloud defender, while the signaling game takes place between the possibly-compromised cloud and the device. For brevity, denote the FlipIt game by $\mathbf{G_F}$, the signaling game by $\mathbf{G_S}$, and the combined game - call it CloudControl - by $\mathbf{G_{CC}}$ as shown in Fig. 1. In the next subsections, we formalize this game model.

2.1   Cloud-Device Signaling Game

Let θ denote the type of the cloud. Denote *compromised* and *safe* types of clouds by θ_A and θ_D in the set Θ. Denote the probabilities that $\theta = \theta_A$ and that $\theta = \theta_D$ by p and $1 - p$. Signaling games typically give these probabilities *apriori*, but in CloudControl they are determined by the equilibrium of the FlipIt game $\mathbf{G_F}$.

Let m_H and m_L denote messages of high and low risk, respectively, and let $m \in M = \{m_H, m_L\}$ represent a message in general. After $\mathcal{R}$ receives the message, it chooses an action, $a \in A = \{a_T, a_N\}$, where a_T represents *trusting the cloud* and a_N represents *not trusting the cloud*.

For the device $\mathcal{R}$, let $u_{\mathcal{R}}^S : \Theta \times M \times A \to \mathscr{U}_{\mathcal{R}}$, where $\mathscr{U}_{\mathcal{R}} \subset \mathbb{R}$. $u_{\mathcal{R}}^S$ is a utility function such that $u_{\mathcal{R}}^S(\theta, m, a)$ gives the device's utility when the type is θ, the message is m, and the action is a. Let $u_{\mathcal{A}}^S : M \times A \to \mathscr{U}_{\mathcal{A}} \subset \mathbb{R}$ and $u_{\mathcal{D}}^S : M \times A \to \mathscr{U}_{\mathcal{D}} \subset \mathbb{R}$ be utility functions for the attacker and defender. Note that these players only receive utility in $\mathbf{G_S}$ if their own type controls the cloud in $\mathbf{G_F}$, so that type is not longer a necessary argument for $u_{\mathcal{A}}^S$ and $u_{\mathcal{D}}^S$.

Denote the strategy of $\mathcal{R}$ by $\sigma_{\mathcal{R}}^S : A \to [0,1]$, such that $\sigma_{\mathcal{R}}^S(a \mid m)$ gives the mixed-strategy probability that $\mathcal{R}$ plays action a when the message is m. The role of the sender may be played by $\mathcal{A}$ or $\mathcal{D}$ depending on the state of the cloud, determined by $\mathbf{G_F}$. Let $\sigma_{\mathcal{A}}^S : M \to [0,1]$ denote the strategy that $\mathcal{A}$ plays when she controls the cloud, so that $\sigma_{\mathcal{A}}^S(m)$ gives the probability that $\mathcal{A}$ sends message m. (The superscript S specifies that this strategy concerns the signaling game.) Similarly, let $\sigma_{\mathcal{D}}^S : M \to [0,1]$ denote the strategy played by $\mathcal{D}$ when he controls the cloud. Then $\sigma_{\mathcal{D}}^S(m)$ gives the probability that $\mathcal{D}$ sends message m. Let $\Gamma_{\mathcal{R}}^S$, $\Gamma_{\mathcal{A}}^S$, and $\Gamma_{\mathcal{D}}^S$ denote the sets of mixed strategies for each player.

For $\mathcal{X} \in \{\mathcal{D}, \mathcal{A}\}$, define functions $\bar{u}_{\mathcal{X}}^S : \Gamma_{\mathcal{R}}^S \times \Gamma_{\mathcal{X}}^S \to \mathcal{U}_{\mathcal{X}}$, such that $\bar{u}_{\mathcal{X}}^S\left(\sigma_{\mathcal{R}}^S, \sigma_{\mathcal{X}}^S\right)$ gives the expected utility to sender $\mathcal{X}$ when he or she plays mixed-strategy $\sigma_{\mathcal{X}}^S$ and the receiver plays mixed-strategy $\sigma_{\mathcal{R}}^S$. Equation (1) gives $\bar{u}_{\mathcal{X}}^S$.

$$\bar{u}_{\mathcal{X}}^S\left(\sigma_{\mathcal{R}}^S, \sigma_{\mathcal{X}}^S\right) = \sum_{a \in A} \sum_{m \in M} u_{\mathcal{X}}^S(m,a)\,\sigma_{\mathcal{R}}^S(a \mid m)\,\sigma_{\mathcal{X}}^S(m), \quad \mathcal{X} \in \{\mathcal{A}, \mathcal{D}\} \tag{1}$$

Next, let $\mu : \Theta \to [0,1]$ represent the belief of $\mathcal{R}$, such that $\mu(\theta \mid m)$ gives the likelihood with which $\mathcal{R}$ believes that a sender who issues message m is of type θ. Then define $\bar{u}_{\mathcal{R}}^S : \Gamma_{\mathcal{R}}^S \to \mathcal{U}_{\mathcal{R}}$ such that $\bar{u}_{\mathcal{R}}^S\left(\sigma_{\mathcal{R}}^S \mid m, \mu(\bullet \mid m)\right)$ gives the expected utility for $\mathcal{R}$ when it has belief μ, the message is m, and it plays strategy $\sigma_{\mathcal{R}}^S$. $\bar{u}_{\mathcal{R}}^S$ is given by

$$\bar{u}_{\mathcal{R}}^S\left(\sigma_{\mathcal{R}}^S \mid m, \mu\right) = \sum_{\theta \in \Theta} \sum_{a \in A} u_{\mathcal{R}}^S(\theta, m, a)\,\mu_R(\theta \mid m)\,\sigma_{\mathcal{R}}^S(a \mid m). \tag{2}$$

The expected utilities to the sender and receiver will determine their incentives to control the cloud in the game $\mathbf{G_F}$ described in the next subsection.

2.2   FlipIt Game for Cloud Control

The basic version of FlipIt [20][2] is played in continuous time. Assume that the defender controls the resource - here, the cloud - at $t = 0$. Moves for both players obtain control of the cloud if it is under the other player's control. In this paper, we limit our analysis to *periodic* strategies, in which the moves of the attacker and the moves of the defender are both spaced equally apart, and their phases are chosen randomly from a uniform distribution. Let $f_{\mathcal{A}} \in \mathbb{R}_+$ and $f_{\mathcal{D}} \in \mathbb{R}_+$ (where $\mathbb{R}_+$ represents non-negative real numbers) denote the attack and renewal frequencies, respectively.

Players benefit from controlling the cloud, and incur costs from moving. Let $w_{\mathcal{X}}(t)$ denote the average proportion of the time that player $\mathcal{X} \in \{\mathcal{D}, \mathcal{A}\}$ has controlled the cloud up to time t. Denote the number of moves up to t per unit time of player $\mathcal{X}$ by $z_{\mathcal{X}}(t)$. Let $\alpha_{\mathcal{D}}$ and $\alpha_{\mathcal{A}}$ represent the costs of each defender and attacker move. In the original formulation of FlipIt, the authors consider a fixed benefit for controlling the cloud. In our formulation, the benefit depends on the equilibrium outcomes of the signaling game $\mathbf{G_S}$. Denote these

---

[2] See [20] for a more comprehensive definition of the players, time, game state, and moves in FlipIt. Here, we move on to describing aspects of our game important for analyzing $\mathbf{G_{CC}}$.

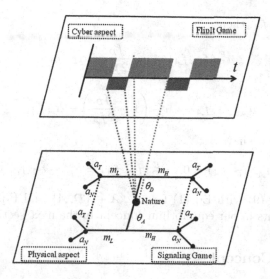

Fig. 1. The `CloudControl` game. The `FlipIt` game models the interaction between an attacker and a cloud administrator for control of the cloud. The outcome of this game determines the type of the cloud in a signaling game in which the cloud conveys commands to the robot or device. The device then decides whether to accept these commands or rely on its own lower-level control. The `FlipIt` and signaling games are played concurrently.

equilibrium utilities of $\mathcal{D}$ and $\mathcal{A}$ by $\bar{u}_{\mathcal{D}}^{S*}$ and $\bar{u}_{\mathcal{A}}^{S*}$. These give the expected benefit of controlling the cloud. Finally, let $u_{\mathcal{D}}^{F}(t)$ and $u_{\mathcal{A}}^{F}(t)$ denote the time-averaged benefit of $\mathcal{D}$ and $\mathcal{A}$ up to time t in $\mathbf{G_F}$. Then

$$u_{\mathcal{X}}^{F}(t) = \bar{u}_{\mathcal{X}}^{S*} w_{\mathcal{X}}(t) - \alpha_{\mathcal{X}} z_{\mathcal{X}}(t), \ \mathcal{X} \in \{\mathcal{D}, \mathcal{A}\}, \tag{3}$$

and, as time continues to evolve, the average benefits over all time become

$$\liminf_{t \to \infty} \bar{u}_{\mathcal{X}}^{S*} w_{\mathcal{X}}(t) - \alpha_{\mathcal{X}} z_{\mathcal{X}}(t), \ \mathcal{X} \in \{\mathcal{D}, \mathcal{A}\}. \tag{4}$$

We next express these expected utilities over all time as a function of periodic strategies that $\mathcal{D}$ and $\mathcal{A}$ employ. Let $\bar{u}_{\mathcal{X}}^{F} : \mathbb{R}_+ \times \mathbb{R}_+ \to \mathbb{R}$, $\mathcal{X} \in \{\mathcal{D}, \mathcal{A}\}$ be expected utility functions such that $\bar{u}_{\mathcal{D}}^{F}(f_{\mathcal{D}}, f_{\mathcal{A}})$ and $\bar{u}_{\mathcal{A}}^{F}(f_{\mathcal{D}}, f_{\mathcal{A}})$ give the average utility to $\mathcal{D}$ and $\mathcal{A}$, respectively, when they play with frequencies $f_{\mathcal{D}}$ and $f_{\mathcal{A}}$. If $f_{\mathcal{D}} \geq f_{\mathcal{A}} > 0$, it can be shown that

$$\bar{u}_{\mathcal{D}}^{F}(f_{\mathcal{D}}, f_{\mathcal{A}}) = \bar{u}_{\mathcal{D}}^{S*}\left(1 - \frac{f_{\mathcal{A}}}{2f_{\mathcal{D}}}\right) - \alpha_{\mathcal{D}} f_{\mathcal{D}}, \tag{5}$$

$$\bar{u}_{\mathcal{A}}^{F}(f_{\mathcal{D}}, f_{\mathcal{A}}) = \bar{u}_{\mathcal{A}}^{S*}\frac{f_{\mathcal{A}}}{2f_{\mathcal{D}}} - \alpha_{\mathcal{A}} f_{\mathcal{A}}, \tag{6}$$

while if $0 \leq f_\mathcal{D} < f_\mathcal{A}$, then

$$\bar{u}_\mathcal{D}^F (f_\mathcal{D}, f_\mathcal{A}) = \bar{u}_\mathcal{D}^{S*} \frac{f_\mathcal{D}}{2 f_\mathcal{A}} - \alpha_\mathcal{D} f_\mathcal{D}, \tag{7}$$

$$\bar{u}_\mathcal{A}^F (f_\mathcal{D}, f_\mathcal{A}) = \bar{u}_\mathcal{A}^{S*} \left(1 - \frac{f_\mathcal{D}}{2 f_\mathcal{A}} \right) - \alpha_\mathcal{A} f_\mathcal{A}, \tag{8}$$

and if $f_\mathcal{A} = 0$, we have

$$\bar{u}_\mathcal{A}^F (f_\mathcal{D}, f_\mathcal{A}) = 0, \quad \bar{u}_\mathcal{D}^F (f_\mathcal{D}, f_\mathcal{A}) = \bar{u}_\mathcal{D}^{S*} - \alpha_\mathcal{D} f_\mathcal{D}. \tag{9}$$

Equations (5)–(9) with Eq. (1) for $\bar{u}_\mathcal{X}^S$, $\mathcal{X} \in \{\mathcal{D}, \mathcal{A}\}$ and Eq. (2) for $\bar{u}_\mathcal{R}^S$ will be main ingredients in our equilibrium concept in the next section.

3   Solution Concept

In this section, we develop a new equilibrium concept for our `CloudControl` game $\mathbf{G_{CC}}$. We study the equilibria of the `FlipIt` and signaling games individually, and then show how they can be related through a fixed-point equation in order to obtain an overall equilibrium for $\mathbf{G_{CC}}$.

3.1   Signaling Game Equilibrium

Signaling games are a class of dynamic Bayesian games. Applying the concept of *perfect Bayesian equilibrium* (as it *e.g.*, [10]) to $\mathbf{G_S}$, we have Definition 1.

Definition 1. *Let the functions $\bar{u}_\mathcal{X}^S (\sigma_\mathcal{R}^S, \sigma_\mathcal{X}^S)$, $\mathcal{X} \in \{\mathcal{D}, \mathcal{A}\}$ and $\bar{u}_\mathcal{R}^S (\sigma_\mathcal{R}^S)$ be formulated according to Eqs. (1) and (2), respectively. Then a* perfect Bayesian equilibrium *of the signaling game $\mathbf{G_S}$ is a strategy profile $(\sigma_\mathcal{D}^{S*}, \sigma_\mathcal{A}^{S*}, \sigma_\mathcal{R}^{S*})$ and posterior beliefs $\mu (\bullet \,|\, m)$ such that*

$$\forall \mathcal{X} \in \{\mathcal{D}, \mathcal{A}\}, \; \sigma_\mathcal{X}^{S*} (\bullet) \in \arg\max_{\sigma_\mathcal{X}^S} \bar{u}_\mathcal{X}^S (\sigma_\mathcal{R}^{S*}, \sigma_\mathcal{X}^S), \tag{10}$$

$$\forall m \in M, \; \sigma_\mathcal{R}^{S*} (\bullet \,|\, m) \in \arg\max_{\sigma_\mathcal{R}^S} \bar{u}_\mathcal{R}^S (\sigma_\mathcal{R}^S \,|\, m, \mu (\bullet \,|\, m)), \tag{11}$$

$$\mu (\theta \,|\, m) = \frac{1 \{\theta = \theta_\mathcal{A}\} \sigma_\mathcal{A}^{S*} (m) \, p + 1 \{\theta = \theta_\mathcal{D}\} \sigma_\mathcal{D}^{S*} (m) \, (1 - p)}{\sigma_\mathcal{A}^{S*} (m) \, p + \sigma_\mathcal{D}^{S*} (m) \, (1 - p)}, \tag{12}$$

if $\sigma_\mathcal{A}^{S*} (m) \, p + \sigma_\mathcal{D}^{S*} (m) \, (1 - p) \neq 0$, and

$$\mu (\theta \,|\, m) = any\ distribution\ on\ \Theta, \tag{13}$$

if $\sigma_\mathcal{A}^{S*} (m) \, p + \sigma_\mathcal{D}^{S*} (m) \, (1 - p) = 0$.

Next, let $\bar{u}_{\mathcal{D}}^{S*}$, $\bar{u}_{\mathcal{A}}^{S*}$, and $\bar{u}_{\mathcal{R}}^{S*}$ be the utilities for the defender, attacker, and device, respectively, when they play according to a strategy profile $(\sigma_{\mathcal{D}}^{S*}, \sigma_{\mathcal{A}}^{S*}, \sigma_{\mathcal{R}}^{S*})$ and belief $\mu(\bullet \mid m)$ that satisfy the conditions for a perfect Bayesian equilibrium. Define a set-valued mapping $T^S : [0,1] \to 2^{\mathcal{U}_{\mathcal{D}} \times \mathcal{U}_{\mathcal{A}}}$ such that $T^S(p; G_S)$ gives the set of equilibrium utilities of the defender and attacker when the prior probabilities are p and $1-p$ and the signaling game utilities are parameterized by G_S[3]. We have

$$\{(\bar{u}_{\mathcal{D}}^{S*}, \bar{u}_{\mathcal{A}}^{S*})\} = T^S(p; G_S). \tag{14}$$

We will employ T^S as part of the definition of an overall equilibrium for $\mathbf{G_{CC}}$ after examining the equilibrium of the FlipIt game.

3.2 FlipIt Game Equilibrium

The appropriate equilibrium concept for the FlipIt game, when $\mathcal{A}$ and $\mathcal{D}$ are restricted to periodic strategies, is *Nash equilibrium* [14]. Definition 2 applies the concept of Nash Equilibrim to $\mathbf{G_F}$.

Definition 2. *A Nash equilibrium of the game* $\mathbf{G_F}$ *is a strategy profile* $(f_{\mathcal{D}}^*, f_{\mathcal{A}}^*)$ *such that*

$$f_{\mathcal{D}}^* \in \arg\max_{f_{\mathcal{D}}} \bar{u}_{\mathcal{D}}^F(f_{\mathcal{D}}, f_{\mathcal{A}}^*), \tag{15}$$

$$f_{\mathcal{A}}^* \in \arg\max_{f_{\mathcal{A}}} \bar{u}_{\mathcal{D}}^F(f_{\mathcal{D}}^*, f_{\mathcal{A}}), \tag{16}$$

where $\bar{u}_{\mathcal{D}}^F$ *and* $\bar{u}_{\mathcal{A}}^F$ *are computed by Eqs. (5) and (6) if* $f_{\mathcal{D}} \geq f_{\mathcal{A}}$ *and Eqs. (7) and (8) if* $f_{\mathcal{D}} \leq f_{\mathcal{A}}$.

To find an overall equilibrium of $\mathbf{G_{CC}}$, we are interested in the proportion of time that $\mathcal{A}$ and $\mathcal{D}$ control the cloud. As before, denote these proportions by p and $1-p$, respectively. These proportions (as in [6]) can be found from the equilibrium frequencies by

$$p = \begin{cases} 0, & \text{if} \quad f_{\mathcal{A}} = 0 \\ \frac{f_{\mathcal{A}}}{2f_{\mathcal{D}}}, & \text{if} \quad f_{\mathcal{D}} \geq f_{\mathcal{A}} > 0 \\ 1 - \frac{f_{\mathcal{D}}}{2f_{\mathcal{A}}}, & \text{if} \quad f_{\mathcal{A}} > f_{\mathcal{D}} \geq 0 \end{cases} \tag{17}$$

Let G_F parameterize the FlipIt game. Now, we can define a mapping $T^F : \mathcal{U}_{\mathcal{D}} \times \mathcal{U}_{\mathcal{A}} \to [0,1]$ such that the expression $T^F(\bar{u}_{\mathcal{D}}^{S*}, \bar{u}_{\mathcal{A}}^{S*}; G_F)$ gives the proportion of time that the attacker controls the cloud in equilibrium from the values of controlling the cloud for the defender and the attacker. This mapping gives

$$p = T^F(\bar{u}_{\mathcal{D}}^{S*}, \bar{u}_{\mathcal{A}}^{S*}; G_F). \tag{18}$$

---

[3] Since $\mathcal{R}$ does not take part in $\mathbf{G_S}$, it is not necessary to include $\bar{u}_{\mathcal{R}}^{S*}$ as an output of the mapping.

In addition to interpreting p as the proportion of time that the attacker controls the cloud, we can view it as the likelihood that, at any random time, the cloud will be controlled by the attacker. Of course, this is precisely the value p of interest in $\mathbf{G_S}$. Clearly, $\mathbf{G_F}$ and $\mathbf{G_S}$ are coupled by Eqs. (14) and (18). These two equations specify the overall equilibrium for the CloudControl game $\mathbf{G_{CC}}$ through a fixed-point equation, which we describe next.

3.3 Gestalt Equilibrium of $\mathbf{G_{CC}}$

When the CloudControl game $\mathbf{G_{CC}}$ is in equilibrium the mapping from the parameters of $\mathbf{G_S}$ to that game's equilibrium and the mapping from the parameters of $\mathbf{G_F}$ to that game's equilibrium are simultaneously satisfied as shown in Fig. 2. Definition 3 formalizes this equilibrium, which we call *Gestalt equilibrium*.

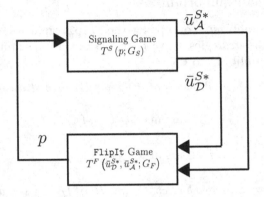

Fig. 2. $\mathbf{G_S}$ and $\mathbf{G_F}$ interact because the utilities in the FlipIt game are derived from the output of the signaling game, and the output of the FlipIt game is used to define prior probabilities in the signaling game. We call the fixed-point of the composition of these two relationships a Gestalt equilibrium.

Definition 3 (Gestalt Equilibrium). *The cloud control ratio* $p^\dagger \in [0,1]$ *and equilibrium signaling game utilities* $\bar{u}_{\mathcal{D}}^{S\dagger}$ *and* $\bar{u}_{\mathcal{A}}^{S\dagger}$ *constitute a Gestalt equilibrium of the game* $\mathbf{G_{CC}}$ *composed of coupled games* $\mathbf{G_S}$ *and* $\mathbf{G_F}$ *if the two components of Eq. (19) are simultaneously satisfied.*

$$\left(\bar{u}_{\mathcal{D}}^{S\dagger}, \bar{u}_{\mathcal{A}}^{S\dagger}\right) \in T^S\left(p^\dagger; G_S\right), \quad p^\dagger = T^F\left(\bar{u}_{\mathcal{D}}^{S\dagger}, \bar{u}_{\mathcal{A}}^{S\dagger}; G_F\right) \tag{19}$$

In short, the signaling game utilities $\left(\bar{u}_{\mathcal{D}}^{S\dagger}, \bar{u}_{\mathcal{A}}^{S\dagger}\right)$ *must satisfy the fixed-point equation*

$$\left(\bar{u}_{\mathcal{D}}^{S\dagger}, \bar{u}_{\mathcal{A}}^{S\dagger}\right) \in T^S\left(T^F\left(\bar{u}_{\mathcal{D}}^{S\dagger}, \bar{u}_{\mathcal{A}}^{S\dagger}; G_F\right); G_S\right) \tag{20}$$

In this equilibrium, $\mathcal{A}$ *receives* $\bar{u}_{\mathcal{A}}^F$ *according to Eq. (6), Eq. (8), or Eq. (9),* $\mathcal{D}$ *receives* $\bar{u}_{\mathcal{D}}^F$ *according to Eq. (5), Eq. (7), or Eq. (9), and* $\mathcal{R}$ *receives* $\bar{u}_{\mathcal{R}}^S$ *according to Eq. (2).*

Solving for the equilibrium of $\mathbf{G_{CC}}$ requires a fixed-point equation essentially because the games $\mathbf{G_F}$ and $\mathbf{G_S}$ are played according to *prior committment*. Prior commitment specifies that players in $\mathbf{G_S}$ do not know the outcome of $\mathbf{G_F}$. This structure prohibits us from using a sequential concept such as sub-game perfection and suggests instead a fixed-point equation.

4 Analysis

In this section, we analyze the game proposed in Sect. 2 based on our solution concept in Sect. 3. First, we analyze the signaling game and calculate the corresponding equilibria. Then, we solve the FlipIt game for different values of expected payoffs resulting from signaling game. Finally, we describe the solution of the combined game.

4.1 Signaling Game Analysis

The premise of $\mathbf{G_{CC}}$ allows us to make some basic assumptions about the utility parameters that simplifies the search for equilibria. We expect these assumptions to be true across many different contexts.

(A1) $u_{\mathcal{R}}(\theta_D, m_L, a_T) > u_{\mathcal{R}}(\theta_D, m_L, a_N)$: It is beneficial for the receiver to trust a low risk message from the defender.

(A2) $u_{\mathcal{R}}(\theta_A, m_H, a_T) < u_{\mathcal{R}}(\theta_A, m_H, a_N)$: It is harmful for the receiver to trust a high risk message from the attacker.

(A3) $\forall m, m' \in M, \ u_{\mathcal{A}}(m, a_T) > u_{\mathcal{A}}(m', a_N)$ and $\forall m, m' \in M, u_{\mathcal{D}}(m, a_T) > u_{\mathcal{D}}(m', a_N)$: Both types of sender prefer that either of their messages is trusted rather than that either of their messages is rejected.

(A4) $u_{\mathcal{A}}(m_H, a_T) > u_{\mathcal{A}}(m_L, a_T)$: The attacker prefers an outcome in which the receiver trusts his high risk message to an outcome in which the receiver trusts his low risk message.

Pooling equilibria of the signaling game differ depending on the prior probabilities p and $1 - p$. Specifically, the messages on which $\mathcal{A}$ and $\mathcal{D}$ pool and the equilibrium action of $\mathcal{R}$ depend on quantities in Eqs. (21) and (22) which we call *trust benefits*.

$$TB_H(p) = \begin{array}{l} p\left[u_{\mathcal{R}}(\theta_A, m_H, a_T) - u_{\mathcal{R}}(\theta_A, m_H, a_N)\right] \\ + (1 - p)\left[u_{\mathcal{R}}(\theta_D, m_H, a_T) - u_{\mathcal{R}}(\theta_D, m_H, a_N)\right] \end{array} \tag{21}$$

$$TB_L(p) = \begin{array}{l} p\left[u_{\mathcal{R}}(\theta_A, m_L, a_T) - u_{\mathcal{R}}(\theta_A, m_L, a_N)\right] \\ + (1 - p)\left[u_{\mathcal{R}}(\theta_D, m_L, a_T) - u_{\mathcal{R}}(\theta_D, m_L, a_N)\right] \end{array} \tag{22}$$

$TB_H(p)$ and $TB_L(p)$ give the benefit of trusting (compared to not trusting) high and low messages, respectively, when the prior probability is p. These quantities specify whether $\mathcal{R}$ will trust a message that it receives in a pooling equilibrium. If $TB_H(p)$ (respectively, $TB_L(p)$) is positive, then, in equilibrium, $\mathcal{R}$ will trust all messages when the senders pool on m_H (respectively, m_L).

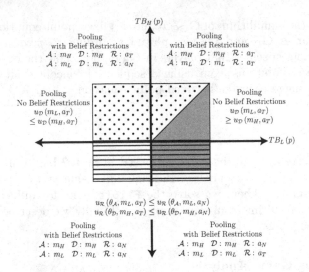

Fig. 3. The four quadrants represent parameter regions of **G$_S$**. The regions vary based on the types of pooling equilibria that they support. For instance, quadrant IV supports a pooling equilibrium in which $\mathcal{A}$ and $\mathcal{D}$ both send m_H and $\mathcal{R}$ plays a_N, as well as a pooling equilibrium in which $\mathcal{A}$ and $\mathcal{D}$ both send m_L and $\mathcal{R}$ plays a_T. The shaded regions denote special equilibria that occur under further parameter restrictions.

We illustrate the different possible combinations of $TB_H(p)$ and $TB_L(p)$ in the quadrants of Fig. 3. The labeled messages and actions for the sender and receiver, respectively, in each quadrant denote these pooling equilibria. These pooling equilibria apply throughout each entire quadrant. Note that we have not listed the requirements on belief μ here. These are addressed in the Appendix A.2, and become especially important for various equilibrium refinement procedures.

The shaded regions of Fig. 3 denote additional special equilibria which only occur under the additional parameter constraints listed within the regions. (The geometrical shapes of the shaded regions are not meaningful, but their overlap and location relative to the four quadrants are accurate.) The dotted and uniformly shaded zones contain equilibria similar to those already denoted in the equilibria for each quadrant, except that they do not require restrictions on μ. The zone with horizontal bars denotes the game's only separating equilibrium. It is a rather unproductive one for $\mathcal{D}$ and $\mathcal{A}$, since their messages are not trusted. (See the derivation in Appendix A.1.) The equilibria depicted in Fig. 3 will become the basis of analyzing the mapping $T^S(p; G_S)$, which will be crucial for forming our fixed-point equation that defines the Gestalt equilibrium. Before studying this mapping, however, we first analyze the equilibria of the FlipIt game on its own.

4.2 FlipIt Analysis

In this subsection, we calculate the Nash equilibrium in the FlipIt game. Equations (5)–(9) represent both players' utilities in FlipIt game. The solution of this game is similar to what has presented in [6,20], except that the reward of controlling the resource may vary. To calculate Nash equilibrium, we normalize both players' benefit with respect to the reward of controlling the resource. For different cases, the frequencies of move at Nash equilibrium are:

- $\dfrac{\alpha_D}{\bar{u}_D^{S*}} < \dfrac{\alpha_A}{\bar{u}_A^{S*}}$ and $\bar{u}_A^{S*}, \bar{u}_D^{S*} > 0$:

$$f_D^* = \frac{\bar{u}_A^{S*}}{2\alpha_A}, \quad f_A^* = \frac{\alpha_D}{2\alpha_A^2} \times \frac{(\bar{u}_A^{S*})^2}{\bar{u}_D^{S*}}, \tag{23}$$

- $\dfrac{\alpha_D}{\bar{u}_D^{S*}} > \dfrac{\alpha_A}{\bar{u}_A^{S*}}$ and $\bar{u}_A^{S*}, \bar{u}_D^{S*} > 0$:

$$f_D^* = \frac{\alpha_A}{2\alpha_D^2} \times \frac{(\bar{u}_D^{S*})^2}{\bar{u}_A^{S*}}, \quad f_A^* = \frac{\bar{u}_D^{S*}}{2\alpha_D}, \tag{24}$$

- $\dfrac{\alpha_D}{\bar{u}_D^{S*}} = \dfrac{\alpha_A}{\bar{u}_A^{S*}}$ and $\bar{u}_A^{S*}, \bar{u}_D^{S*} > 0$:

$$f_D^* = \frac{\bar{u}_A^{S*}}{2\alpha_A}, \quad f_A^* = \frac{\bar{u}_D^{S*}}{2\alpha_D}, \tag{25}$$

- $\bar{u}_A^{S*} \leq 0$:

$$f_D^* = f_A^* = 0, \tag{26}$$

- $\bar{u}_A^{S*} > 0$ and $\bar{u}_D^{S*} \leq 0$:

$$f_D^* = 0 \quad f_A^* = 0^+. \tag{27}$$

In the case that $\bar{u}_A^{S*} \leq 0$, the attacker has no incentive to attack the cloud. In this case, the defender need not move since we assume that she controls the cloud initially. In the case that $\bar{u}_A^{S*} > 0$ and $\bar{u}_D^{S*} \leq 0$, only the attacker has an incentive to control the cloud. We use $f_A^* = 0^+$ to signify that the attacker moves only once. Since the defender never moves, the attacker's single move is enough to retain control of the cloud at all times.

Next, we put together the analysis of $\mathbf{G_S}$ and $\mathbf{G_F}$ in order to study the Gestalt equilibria of the entire game.

4.3 $\mathbf{G_{CC}}$ Analysis

To identify the Gestalt Equilibrium of $\mathbf{G_{CC}}$, it is necessary to examine the mapping $T^S(p; G_S)$ for all $p \in [0,1]$. As noted in Sect. 4.1, this mapping depends on $TB_H(p)$ and $TB_L(p)$. From assumptions A1-A4, it is possible to verify that $(TB_L(0), TB_H(0))$ must fall in Quadrant I or Quadrant IV and that $(TB_L(1), TB_H(1))$ must lie in Quadrant III or Quadrant IV. There are numerous ways in which the set $(TB_L(p), TB_H(p))$, $p \in [0,1]$ can transverse different parameter regions. Rather than enumerating all of them, we consider one here.

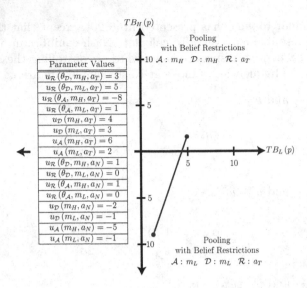

Fig. 4. For the parameter values overlayed on the figure, as p ranges from 0 to 1, $TB_H(p)$ and $TB_L(p)$ move from Quadrant I to Quadrant IV. The equilibria supported in each of these quadrants, as well as the equilibria supported on the interface between them, are presented in Table 1.

Consider parameters such that $TB_L(0), TB_H(0) > 0$ and $TB_L(1) > 0$ but $TB_H(1) < 0$[4]. This leads to an $\mathscr{L}$ that will traverse from Quadrant I to Quadrant IV. Let us also assume that $u_D(m_L, a_T) < u_D(m_H, a_T)$, so that Equilibrium 5 is not feasible. In Fig. 4, we give specific values of parameters that satisfy these conditions, and we plot $(TB_L(p), TB_H(p))$ for $p \in [0,1]$. Then, in Table 1, we give the equilibria in each region that the line segment traverses. The equilibrium numbers refer to the derivations in the Appendix A.2.

If p is such that the signaling game is played in Quadrant I, then both senders prefer pooling on m_H. By the *first mover advantage*, they will select Equilibrium 8. On the border between Quadrants I and IV, $\mathcal{A}$ and $\mathcal{D}$ both prefer an equilibrium in which $\mathcal{R}$ plays a_T. If they pool on m_L, this is guaranteed. If they pool on m_H, however, $\mathcal{R}$ receives equal utility for playing a_T and a_N; thus, the senders cannot guarantee that the receiver will play a_T. Here, we assume that the senders maximize their worst-case utility, and thus pool on m_L. This is Equilibrium 3. Finally, in Quadrant IV, both senders prefer to be trusted, and so select Equilibrium 3. From the table, we can see that the utilities will have a jump at the border between Quadrants I and IV. The solid line in Fig. 5 plots the ratio $\bar{u}_{\mathcal{A}}^{S*}/\bar{u}_{\mathcal{D}}^{S*}$ of the utilities as a function of p.

---

[4] These parameters must satisfy $u_{\mathcal{R}}(\theta_D, m_H, a_T) > u_{\mathcal{R}}(\theta_D, m_H, a_N)$ and $u_{\mathcal{R}}(\theta_A, m_L, a_T) > u_{\mathcal{R}}(\theta_A, m_L, a_N)$. Here, we give them specific values in order to plot the data.

Table 1. Signaling game equilibria by region for a game that traverses between Quadrant I and Quadrant IV. Some of the equilibria are feasible only for constrained beliefs μ, specified in Appendix A.2. We argue that the equilibria in each region marked by (*) will be selected.

Region	Equilibria
Quadrant I	Equilibrium 3: Pool on m_L; μ constrained; $\mathcal{R}$ plays a_T
	*Equilibrium 8: Pool on m_H; μ unconstrained; $\mathcal{R}$ plays a_T
$TB_H\,(p) = 0$ Axis	*Equilibrium 3: Pool on m_L; μ constrained; $\mathcal{R}$ plays a_T
	Equilibrium 8: Pool on m_H; μ unconstrained; $\mathcal{R}$ plays a_T
	Equilibrium 6: Pool on m_H; μ constrained; $\mathcal{R}$ plays a_N
Quadrant IV	*Equilibrium 3: Pool on m_L; μ constrained; $\mathcal{R}$ plays a_T
	Equilibrium 6: Pool on m_H; μ constrained; $\mathcal{R}$ plays a_N

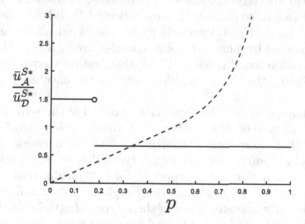

Fig. 5. T^F and T^S are combined on a single set of axis. In T^S (the solid line), the independent variable is on the horizontal axis. In T^F (the dashed line), the independent variable is on the vertical axis. The intersection of the two curves represents the Gestalt equilibrium.

Next, consider the mapping $p = T^F\left(\bar{u}_{\mathcal{D}}^{S*}, \bar{u}_{\mathcal{A}}^{S*}\right)$. As we have noted, p depends only on the ratio $\bar{u}_{\mathcal{A}}^{S*}/\bar{u}_{\mathcal{D}}^{S*}$[5]. Indeed, it is continuous in that ratio when the outcome at the endpoints is appropriately defined. This mapping is represented by the dashed line in Fig. 5, with the independent variable on the vertical axis.

We seek a fixed-point, in which $p = T^F\left(\bar{u}_{\mathcal{D}}^{S*}, \bar{u}_{\mathcal{A}}^{S*}\right)$ and $\left(\bar{u}_{\mathcal{D}}^{S*}, \bar{u}_{\mathcal{A}}^{S*}\right) = T^S\,(p)$. This shown by the intersection of the solid and dashed curves plotted in Fig. 5.

---

[5] When $\bar{u}_{\mathcal{A}}^{S*} = \bar{u}_{\mathcal{D}}^{S*} = 0$, we define that ratio to be equal to zero, since this will yield $f_{\mathcal{A}} = 0$ and $p = 0$, as in Eqs. (9) and (17). When $\bar{u}_{\mathcal{D}}^{S*} = 0$ and $\bar{u}_{\mathcal{A}}^{S*} > 0$, it is convenient to consider the ratio to be positively infinite since this is consistent with $p \rightarrow 1$.

At these points, the mappings between the signaling and the FlipIt games are mutually satisfied, and we have a Gestalt equilibrium.[6]

5   Cloud Control Application

In this section, we describe one possible application of our model: a cyber-physical system composed of autonomous vehicles with some on-board control but also with the ability to trust commands from the cloud. Access to the cloud can offer automated vehicles several benefits [12]. First, it allows access to massive computational resources - *i.e., infrastructure as a service (IaaS)*. (See [5].) Second, it allows access to large datasets. These datasets can offer super-additive benefits to the sensing capabilities of the vehicle itself, as in the case of the detailed road and terrain maps that automated cars such as those created by Google and Delphi combine with data collected by lidar, radar and vision-based cameras [1,11]. Third, interfacing with the cloud allows access to data collected or processed by humans through crowd-sourcing applications; consider, for instance, location-based services [17,18] that feature recommendations from other users. Finally, the cloud can allow vehicles to collectively learn through experience [12].

Attackers may attempt to influence cloud control of the vehicle through several means. In one type of attack, adversaries may be able to *steal or infer cryptographic keys* that allow them authorization into the network. These attacks are of the *complete compromise* and *stealth* types that are studied in the FlipIt framework [6,20] and thus are appropriate for a CloudControl game. FlipIt also provides the ability to model *zero-day exploits*, vulnerabilities for which a patch is not currently available. Each of these types of attacks on the cloud pose threats to unmanned vehicle security and involve the complete compromise and steathiness that motivate the FlipIt framework.

5.1   Dynamic Model for Cloud Controlled Unmanned Vehicles

In this subsection, we use a dynamic model of an autonomous car to illustrate one specific context in which a cloud-connected device could be making a decision of whether to trust the commands that it would receive or to follow its own on-board control.

We consider a car moving in two-dimensional space with a fixed speed v_0 but with steering that can be controlled. (See Fig. 6, which illustrates the "bicycle model" of steering control from [3].) For simplicity, assume that we are interested in the car's deviation from a straight line. (This line might, *e.g.*, run along the

---

[6] Note that this example featured a discontinuity in signaling game utilities on the border between equilibrium regions. Interestingly, even when the pooling equilibria differ between regions, it is possible that the equilibrium on the border admits a mixed strategy that provides continuity between the different equilibria in the two regions, and thus makes T^S continuous. This could allow $\mathbf{G_{CC}}$ to have multiple Gestalt equilibria.

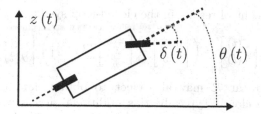

Fig. 6. A bicycle model is a type of representation of vehicle steering control. Here, $\delta(t)$ is used to denote the angle between the orientation of the front wheel and the heading $\theta(t)$. The deviation of the vehicle from a straight line is given by $z(t)$

center of the proper driving lane.) Let $z(t)$ denote the car's vertical distance from the horizontal line, and let $\theta(t)$ denote the heading of the car at time t. The state of the car can be represented by a two-dimensional vector $w(t) \triangleq \begin{bmatrix} z(t) & \theta(t) \end{bmatrix}^T$. Let $\delta(t)$ denote the angle between the orientation of the front wheel - which implements steering - and the orientation of the length of the car. We can consider $\delta(t)$ to be the input to the system. Finally, let $y(t)$ represent a vector of outputs available to the car's control system. The self-driving cars of both Google and Delphi employ radar, lidar, and vision-based cameras for localization. Assume that these allow accurate measurement of both states, such that $y_1(t) = z(t)$ and $y_2(t) = \theta(t)$. If the car stays near $w(t) = \begin{bmatrix} 0 & 0 \end{bmatrix}^T$, then we can approximate the system with a linear model. Let a and b denote the distances from the rear wheel to the center of gravity and the rear wheel to the front wheel of the car, respectively. Then the linearized system is given in [3] by the equations:

$$\frac{d}{dt} \begin{bmatrix} z(t) \\ \theta(t) \end{bmatrix} = \begin{bmatrix} 0 & v_0 \\ 0 & 0 \end{bmatrix} \begin{bmatrix} z(t) \\ \theta(t) \end{bmatrix} + \begin{bmatrix} \frac{av_0}{b} \\ \frac{v_0}{b} \end{bmatrix} \delta(t), \qquad (28)$$

$$\begin{bmatrix} y_1(t) \\ y_2(t) \end{bmatrix} = \begin{bmatrix} 1 & 0 \\ 0 & 1 \end{bmatrix} \begin{bmatrix} z(t) \\ \theta(t) \end{bmatrix} \qquad (29)$$

5.2   Control of Unmanned Vehicle

Assume that the unmanned car has some capacity for automatic control without the help of the cloud, but that the cloud typically provides more advanced navigation.

Specifically, consider a control system onboard the unmanned vehicle designed to return it to the equilibrium $w(t) = \begin{bmatrix} 0 & 0 \end{bmatrix}^T$. Because the car has access to both of the states, it can implement a state-feedback control. Consider a linear, time-invariant control of the form

$$\delta_{car}(t) = - \begin{bmatrix} k_1 & k_2 \end{bmatrix} \begin{bmatrix} z(t) \\ \theta(t) \end{bmatrix}. \qquad (30)$$

This proportional control results in the closed-loop system

$$\frac{d}{dt}\begin{bmatrix} z(t) \\ \theta(t) \end{bmatrix} = \left(\begin{bmatrix} 0 & v_0 \\ 0 & 0 \end{bmatrix} - \begin{bmatrix} \frac{av_0}{b} \\ \frac{v_0}{b} \end{bmatrix} \begin{bmatrix} k_1 & k_2 \end{bmatrix} \right) \begin{bmatrix} z(t) \\ \theta(t) \end{bmatrix} \tag{31}$$

The unmanned car $\mathcal{R}$ may also elect to obtain data or computational resources from the cloud. Typically, this additional access would improve the control of the car. The cloud administrator (defender $\mathcal{D}$), however, may issue faulty commands or there may be a breakdown in communication of the desired signals. In addition, the cloud may be compromised by $\mathcal{A}$ in a way that is stealthy. Because of these factors, $\mathcal{R}$ sometimes benefits from rejecting the cloud's command and relying on its own navigational abilities. Denote the command issued by the cloud at time t by $\delta_{cloud}(t) \in \delta_{\mathcal{A}}(t), \delta_{\mathcal{D}}(t)$, depending on who controls the cloud. With this command, the system is given by

$$\frac{d}{dt}\begin{bmatrix} z(t) \\ \theta(t) \end{bmatrix} = \begin{bmatrix} 0 & v_0 \\ 0 & 0 \end{bmatrix} \begin{bmatrix} z(t) \\ \theta(t) \end{bmatrix} + \begin{bmatrix} \frac{av_0}{b} \\ \frac{v_0}{b} \end{bmatrix} \delta_{cloud}(t). \tag{32}$$

5.3   Filter for High Risk Cloud Commands

In cloud control of an unmanned vehicle, the self-navigation state feedback input given by $\delta_{car}(t)$ in Eq. (30) represents the control that is expected by the vehicle given its state. If the signal from the cloud differs significantly from the signal given by the self-navigation system, then the vehicle may classify the message as "high-risk." Specifically, define a *difference threshold* τ, and let

$$m = \begin{cases} m_H, & \text{if} \quad |\delta_{cloud}(t) - \delta_{car}(t)| > \tau \\ m_L, & \text{if} \quad |\delta_{cloud}(t) - \delta_{car}(t)| \leq \tau \end{cases} \tag{33}$$

Equation (33) translates the actual command from the cloud (controlled by $\mathcal{D}$ or $\mathcal{A}$) into a message in the cloud signaling game.

Equations (31) and (32) give the dynamics of the unmanned car electing to trust and not trust the cloud. Based on these equations, Fig. 7 illustrates the combined self-navigating and cloud controlled system for vehicle control.

6   Conclusion and Future Work

In this paper, we have proposed a general framework for the interaction between an attacker, cloud administrator/defender, and cloud-connected device. We have described the struggle for control of the cloud using the FlipIt game and the interaction between the cloud and the connected device using a traditional signaling game. Because these two games are played by prior commitment, they are coupled. We have defined a new equilibrium concept - *i.e., Gestalt equilibrium*, which defines a solution to the combined game using a fixed-point equation. After illustrating various parameter regions under which the game may be

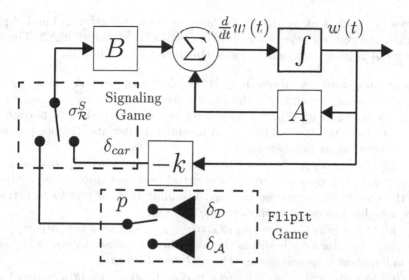

Fig. 7. Block-diagram model for unmanned vehicle navigation control. At any time, the vehicle uses strategy $\sigma_{\mathcal{R}}^{S}$ to decide whether to follow its own control or the control signal from the cloud, which may be $\delta_{\mathcal{A}}$ or $\delta_{\mathcal{D}}$, depending on the probabilities p, $1-p$ with which $\mathcal{A}$ and $\mathcal{D}$ control the cloud. Its own control signal, δ_{car}, is obtained via feedback control.

played, we solved the game in a sample parameter region. Finally, we showed how the framework may be applied to unmanned vehicle control.

Several directions remain open for future work. First, the physical component of the cyber-physical system can be further examined. Tools from optimal control such as the linear-quadratic regulator could offer a rigorous framework for defining the costs associated with the physical dynamic system, which in turn would define the payoffs of the signaling game. Second, future work could search for conditions under which a Gestalt equilibrium of the CloudControl game is guaranteed to exist. Finally, devices that use this framework should be equipped to learn online. Towards that end, a learning algorithm could be developed that is guaranteed to converge to the Gestalt equilibrium. Together with the framework developed in the present paper, these directions would help to advance our ability to secure cloud-connected and cyber-physical systems.

A   Derivation of Signaling Game Equilibria

In this appendix, we solve for the equilibria of $\mathbf{G_S}$.

A.1   Separating Equilibria

First, we search for separating equilibria of $\mathbf{G_S}$. In separating equilibria, $\mathcal{R}$ knows with certainty the type of the cloud.

$\mathcal{D}$ **plays** m_L **and** $\mathcal{A}$ **plays** m_H. If $\mathcal{D}$ plays m_L (as a pure strategy) and $\mathcal{A}$ plays m_H, then the receiver rejects any m_H according to assumption A2. The best action for $\mathcal{A}$ is to deviate to m_L. Thus, this is not an equilibrium.

$\mathcal{D}$ **plays** m_H **and** $\mathcal{A}$ **plays** m_L. If $\mathcal{D}$ plays m_H and $\mathcal{A}$ plays m_L, the $\mathcal{R}$'s best response depends on the utility parameters. If $u_{\mathcal{R}}^S(\theta_{\mathcal{A}}, m_L, a_T) \leq u_{\mathcal{R}}^S(\theta_{\mathcal{A}}, m_L, a_N)$ and $u_{\mathcal{R}}^S(\theta_{\mathcal{D}}, m_H, a_T) \leq u_{\mathcal{R}}^S(\theta_{\mathcal{D}}, m_H, a_N)$, then $\mathcal{R}$ plays a_N in response to both messages. There is no incentive to deviate. Denote this separating equilibrium as *Equilibrium #2*.

If $u_{\mathcal{R}}^S(\theta_{\mathcal{A}}, m_L, a_T) \leq u_{\mathcal{R}}^S(\theta_{\mathcal{A}}, m_L, a_N)$ and $u_{\mathcal{R}}^S(\theta_{\mathcal{D}}, m_H, a_T) > u_{\mathcal{R}}^S(\theta_{\mathcal{D}}, m_H, a_N)$, then a_N is within the set of best responses to m_L, whereas a_T is the unique best response to m_H. Assuming that he prefers to certainty receive a higher utility, $\mathcal{A}$ deviates to m_H.

If $u_{\mathcal{R}}^S(\theta_{\mathcal{A}}, m_L, a_T) > u_{\mathcal{R}}^S(\theta_{\mathcal{A}}, m_L, a_N)$ and $u_{\mathcal{R}}^S(\theta_{\mathcal{D}}, m_H, a_T) \leq u_{\mathcal{R}}^S(\theta_{\mathcal{D}}, m_H, a_N)$, then a_N is within the set of best responses to m_H, whereas a_T is the unique best response to m_L. Thus, $\mathcal{D}$ deviates to m_L.

If $u_{\mathcal{R}}^S(\theta_{\mathcal{A}}, m_L, a_T) > u_{\mathcal{R}}^S(\theta_{\mathcal{A}}, m_L, a_N)$ and $u_{\mathcal{R}}^S(\theta_{\mathcal{D}}, m_H, a_T) > u_{\mathcal{R}}^S(\theta_{\mathcal{D}}, m_H, a_N)$, then $\mathcal{R}$ plays a_T in response to both messages. We have assumed, however, that $\mathcal{A}$ prefers to be trusted on m_H compared to being trusted on m_L (A4), so $\mathcal{A}$ deviates and this is not an equilibrium.

A.2 Pooling Equilibria

Next, we search for pooling equilibria of $\mathbf{G_S}$. In pooling equilibria, $\mathcal{R}$ relies only on the prior probabilities p and $1 - p$ in order to form his belief about the type of the cloud. The existence of pooling equilibria depend essentially on the trust benefits $TB_H(p)$ and $TB_L(p)$.

Pooling on m_L. If $TB_L(p) < 0$, then $\mathcal{R}$'s best response is a_N. This will only be an equilibrium if his best response to m_H would also be a_N. This is the case only when the belief satisfies

$$\mu(\theta_{\mathcal{A}} \mid m_H) u_{\mathcal{R}}(\theta_{\mathcal{A}}, m_H, a_T) + (1 - \mu(\theta_{\mathcal{A}} \mid m_H)) u_{\mathcal{R}}(\theta_{\mathcal{D}}, m_H, a_T) \leq \mu(\theta_{\mathcal{A}} \mid m_H) u_{\mathcal{R}}(\theta_{\mathcal{A}}, m_H, a_N) + (1 - \mu(\theta_{\mathcal{A}} \mid m_H)) u_{\mathcal{R}}(\theta_{\mathcal{D}}, m_H, a_N) \tag{34}$$

Moreover, this can only be an equilibrium when neither $\mathcal{A}$ nor $\mathcal{D}$ have an incentive to deviate: *i.e.*, when

$$u_{\mathcal{A}}^S(m_H, a_N) \leq u_{\mathcal{A}}^S(m_L, a_N) \text{ and } u_{\mathcal{D}}^S(m_H, a_N) \leq u_{\mathcal{D}}^S(m_L, a_N) \tag{35}$$

If these conditions are satisfied, then denote this equilibrium by *Equilibrium #1*.

If $TB_L(p) \geq 0$, then $\mathcal{R}$'s best response us a_T. Whether this represents an equilibrium depends on if $\mathcal{A}$ or $\mathcal{D}$ have incentives to deviate from m_L. If $u_{\mathcal{D}}^S(m_H, a_T) \leq u_{\mathcal{D}}^S(m_L, a_T)$ and $u_{\mathcal{A}}^S(m_H, a_T) \leq u_{\mathcal{A}}^S(m_L, a_T)$, then neither has an incentive to deviate. This is *Equilibrium #5*. If one of these inequalities does *not* hold, then the player who prefers m_H to m_L will deviate if $\mathcal{R}$ would play a_T in response to the deviation. The equilibrium condition is narrowed to when the belief makes $\mathcal{R}$ not trust m_H; when Eq. (34) is satisfied. Call this *Equilibrium #3*.

Pooling on m_H. The pattern of equilibria for pooling on m_H follows a similar structure to the pattern of equilibria for pooling on m_L.

If $TB_H(p) < 0$, then $\mathcal{R}$'s best response is a_N. This will only be an equilibrium if his best response to m_L would also be a_N. This is the case only when the belief satisfies

$$\mu(\theta_\mathcal{A} \mid m_L) u_\mathcal{R}(\theta_\mathcal{A}, m_L, a_T) + (1 - \mu(\theta_\mathcal{A} \mid m_L)) u_\mathcal{R}(\theta_\mathcal{D}, m_L, a_T)$$
$$\leq \mu(\theta_\mathcal{A} \mid m_L) u_\mathcal{R}(\theta_\mathcal{A}, m_L, a_N) + (1 - \mu(\theta_\mathcal{A} \mid m_L)) u_\mathcal{R}(\theta_\mathcal{D}, m_L, a_N) \quad (36)$$

To guarantee that $\mathcal{A}$ and $\mathcal{D}$ do not deviate, we require

$$u_\mathcal{A}^S(m_H, a_N) \geq u_\mathcal{A}^S(m_L, a_N) \text{ and } u_\mathcal{D}^S(m_H, a_N) \geq u_\mathcal{D}^S(m_L, a_N) \quad (37)$$

If these conditions are satisfied, then we have *Equilibrium #6*.

If $TB_H \geq 0$, then $\mathcal{R}$'s best response is a_T. If $u_\mathcal{D}^S(m_H, a_T) \geq u_\mathcal{D}^S(m_L, a_T)$ and $u_\mathcal{A}^S(m_H, a_T) \geq u_\mathcal{A}^S(m_L, a_T)$, then neither $\mathcal{A}$ nor $\mathcal{D}$ have an incentive to deviate. Call this *Equilibrium #8*. If one of these inequalities does not hold, then the belief must satisfy Eq. (36) for an equilibrium to be sustained. Denote this equilibrium by *Equilibrium #7*.

References

1. Delphi drive, Delphi Automotive. http://www.delphi.com/delphi-drive
2. Gestalt, Mirium-Webster. http://www.merriam-webster.com/dictionary/gestalt
3. Aström, K.J., Murray, R.M.: Feedback Systems: An Introduction for Scientists and Engineers. Princeton University Press, Princeton (2010)
4. Baheti, R., Gill, H.: Cyber-physical systems. In: The Impact of Control Technology, vol. 12, pp. 161–166 (2011)
5. Bhardwaj, S., Jain, L., Jain, S.: Cloud computing: A study of infrastructure as a service (IAAS). Int. J. Eng. Inf. Technol. **2**(1), 60–63 (2010)
6. Bowers, K.D., van Dijk, M., Griffin, R., Juels, A., Oprea, A., Rivest, R.L., Triandopoulos, N.: Defending against the unknown enemy: applying FLIPIT to system security. In: Grossklags, J., Walrand, J. (eds.) GameSec 2012. LNCS, vol. 7638, pp. 248–263. Springer, Heidelberg (2012)
7. Carroll, T.E., Grosu, D.: A game theoretic investigation of deception in network security. In: Security and Communication, Networks vol. 4(10), pp. 1162–1172 (2011)
8. Casey, W., Morales, J.A., Nguyen, T., Spring, J., Weaver, R., Wright, E., Metcalf, L., Mishra, B.: Cyber security via signaling games: toward a science of cyber security. In: Natarajan, R. (ed.) ICDCIT 2014. LNCS, vol. 8337, pp. 34–42. Springer, Heidelberg (2014)

9. Farhang, S., Manshaei, M.H., Esfahani, M.N., Zhu, Q.: A dynamic bayesian security game framework for strategic defense mechanism design. In: Poovendran, R., Saad, W. (eds.) GameSec 2014. LNCS, vol. 8840, pp. 319–328. Springer, Heidelberg (2014)
10. Fudenberg, D., Tirole, J.: Game Theory, vol. 393. MIT press, Cambridge (1991)
11. Guizzo, E.: How googles self-driving car works. IEEE Spectrum Online, 18 October
12. Kehoe, B., Patil, S., Abbeel, P., Goldberg, K.: A survey of research on cloud robotics and automation. IEEE Trans. Autom. Sci. Eng. 12(2), 398–409 (2015)
13. Lee, E.A.: Cyber physical systems: design challenges. In: 2008 11th IEEE International Symposium on Object Oriented Real-Time Distributed Computing (ISORC), pp. 363–369. IEEE (2008)
14. Nash, J.F., et al.: Equilibrium points in n-person games. Proc. Nat. Acad. Sci. USA 36(1), 48–49 (1950)
15. Pawlick, J., Zhu, Q.: Deception by design: Evidence-based signaling games for network defense. arXiv preprint arXiv:1503.05458 (2015)
16. Portokalidis, G., Slowinska, A., Bos, H.: Argos: an emulator for fingerprinting zero-day attacks for advertised honeypots with automatic signature generation. ACM SIGOPS Operating Syst. Rev. 40(4), 15–27 (2006)
17. Sampigethaya, K., Huang, L., Li, M., Poovendran, R., Matsuura, K., Sezaki, K.: Caravan: Providing location privacy for vanet. Technical report, DTIC Document (2005)
18. Sampigethaya, K., Li, M., Huang, L., Poovendran, R.: Amoeba: Robust location privacy scheme for vanet. IEEE J. Sel. Areas Commun. 25(8), 1569–1589 (2007)
19. Tankard, C.: Advanced persistent threats and how to monitor and deter them. Netw. Secur. 2011(8), 16–19 (2011)
20. van Dijk, M., Juels, A., Oprea, A., Rivest, R.L.: Flipit: The game of "stealthy takeover". J. Cryptol. 26(4), 655–713 (2013)
21. Zhuang, J., Bier, V.M., Alagoz, O.: Modeling secrecy and deception in a multiple-period attacker-defender signaling game. Eur. J. Oper. Res. 203(2), 409–418 (2010)

Short Papers

Genetic Approximations for the Failure-Free Security Games

Aleksandr Lenin[1,2]([✉]), Jan Willemson[1], and Anton Charnamord[1,2]

[1] Cybernetica AS, Mäealuse 2/1, Tallinn, Estonia
{aleksandr.lenin,janwil,anton.charnamord}@cyber.ee
[2] Tallinn University of Technology, Ehitajate Tee 5, Tallinn, Estonia

Abstract. This paper deals with computational aspects of attack trees, more precisely, evaluating the expected adversarial utility in the failure-free game, where the adversary is allowed to re-run failed atomic attacks an unlimited number of times. It has been shown by Buldas and Lenin that exact evaluation of this utility is an NP-complete problem, so a computationally feasible approximation is needed. In this paper we consider a genetic approach for this challenge. Since genetic algorithms depend on a number of non-trivial parameters, we face a multi-objective optimization problem and we consider several heuristic criteria to solve it.

1 Introduction

Hierarchical methods for security assessment have been used for several decades already. Called *fault trees* and applied to analyze general security-critical systems in early 1980-s [1], they were adjusted for information systems and called *threat logic trees* by Weiss in 1991 [2]. In the late 1990-s, the method was popularized by Schneier under the name *attack trees* [3].

There are several ways attack trees can be used in security assessment. The simplest way is purely descriptive. Such an approach is limited only to qualitative assessment of security. Based on such an assessment, it is difficult to talk about optimal level of security or return of security investments. Already the first descriptions of attack trees introduced computational aspects [2,3]. The framework for a sound formal model for such computations was introduced in 2005 by Mauw and Oostdijk [4].

Most of the earlier studies focus on the analysis of a single parameter only. A substantial step forward was taken by Buldas et al. [5] who introduced the idea of game-theoretic modeling of the adversarial decision making process based on several interconnected parameters like the cost, risks and penalties associated

This research was supported by the European Regional Development Fund through Centre of Excellence in Computer Science (EXCS), the Estonian Research Council under Institutional Research Grant IUT27-1 and the European Union Seventh Framework Programme (FP7/2007–2013) under grant agreement ICT-318003 (TREsPASS). This publication reflects only the authors' views and the Union is not liable for any use that may be made of the information contained herein.

MHR Khouzani et al. (Eds.): GameSec 2015, LNCS 9406, pp. 311–321, 2015.
DOI: 10.1007/978-3-319-25594-1_17

with different atomic attacks. Their approach was later refined by Jürgenson and Willemson [6] to achieve compliance with Mauw-Oostdijk framework [7]. However, increase in the model precision was accompanied by significant drop in computational efficiency. To compensate for that, a genetic algorithm approach was proposed by Jürgenson and Willemson [8]. It was later shown by Lenin, Willemson and Sari that this approach is flexible enough to allow extensions like attacker models [9].

Buldas and Stepanenko [10] introduced the upper bound ideology by pointing out that in order to verify the security of the system, it is not necessary to compute the exact adversarial utility but only upper bounds. Buldas and Lenin further improved the fully adaptive model by eliminating the force failure states and suggested the new model called the failure-free model [11]. The model more closely followed the upper bounds ideology originally introduced by Buldas et al. [10] and turned out to be computationally somewhat easier to analyze. It has been shown that finding the optimal strategy is (still) an NP-complete problem, hence looking for a good heuristic approximation is an important goal. Additionally, one of the goals of the paper is to find empirical evidence for the rational choice of the parameters of the genetic algorithm.

The paper has the following structure. First, Sect. 2 defines the required terms. Section 3 presents and evaluates our genetic algorithms. These algorithms are improved with adaptiveness in Sect. 4. Finally, Sect. 5 draws some conclusions.

2   Definitions

Let $\mathcal{X} = \{\mathcal{X}_1, \mathcal{X}_2, \ldots, \mathcal{X}_n\}$ be the set of all possible atomic attacks and $\mathcal{F}$ be a monotone Boolean function corresponding to the considered attack tree.

Definition 1 (Attack Suite). *Attack suite $\sigma \subseteq \mathcal{X}$ is a set of atomic attacks which have been chosen by the adversary to be launched and used to try to achieve the attacker's goal. Also known as individual.*

Definition 2 (Satisfying Attack Suite). *A satisfying attack suite σ evaluates $\mathcal{F}$ to true when all the atomic attacks from the attack suite σ have been evaluated to true. Also known as live individual.*

Definition 3 (Satisfiability Game). *By a satisfiability game we mean a single-player game in which the player's goal is to satisfy a monotone Boolean function $\mathcal{F}(x_1, x_2, \ldots, x_k)$ by picking variables x_i one at a time and assigning $x_i = 1$. Each time the player picks the variable x_i he pays some amount of expenses $\mathcal{E}_i$, which is modeled as a random variable. With a certain probability p_i the move x_i succeeds. The game ends when the condition $\mathcal{F} \equiv 1$ is satisfied and the player wins the prize $\mathcal{P} \in \mathbb{R}$, or when the condition $\mathcal{F} \equiv 0$ is satisfied, meaning the loss of the game, or when the player stops playing. Thus we can define three common types of games:*

1. *SAT Game Without Repetitions* - the type of a game where a player can perform a move only once.
2. *SAT Game With Repetitions* - the type of a game where a player can re-run failed moves an arbitrary number of times.
3. *Failure-Free SAT Game* - the type of a game in which all success probabilities are equal to 1. It has been shown that any game with repetitions is equivalent to a failure-free game [11, Thm. 5].

3   Genetic Approximations for the Failure-Free Satisfiability Games

The whole family of satisfiability games tries to maximize expected adversarial profit by solving an optimization problem: given a monotone Boolean function $\mathcal{F}(x_1, x_2, \ldots, x_n)$ optimize the utility function $\mathcal{U}(x_{i_1}, x_{i_2}, \ldots, x_{i_n})$ over the set of all satisfying assignments fulfilling a set of model-specific conditions (in some specific cases). The models for the SAT games without move repetitions and the failure-free SAT games differ only by their corresponding utility functions, as in both cases the order in which atomic attacks are launched by an adversary is irrelevant. On the contrary, models for SAT games with repetitions (e.g. [12]) consider strategic adversarial behavior in the case of which the order in which the atomic attacks are launched does matter. In this paper we focus on the genetic approximations suitable to be applied to the SAT games without repetitions, as well as the failure-free SAT games. The suggested algorithm is practically validated by the example of the computational model for the failure-free SAT game.

3.1   Genetic Algorithm (GA)

A genetic algorithm is typically characterized by the set of the following parameters: a genetic representation of chromosomes or individuals (feasible solutions for the optimization problem), a population of encoded solutions, fitness function which evaluates the optimality of the solutions, genetic operators (selection, crossover, mutation) that generate a new population from the existing one, and control parameters (population size, crossover rate, mutation rate, condition under which the reproduction process terminates).

The reproduction process, as well as the condition, under which reproduction terminates is identical to the one described in [9]. We refer the readers to this paper for further details. An individual is any feasible solution to the considered optimization problem. Thus, for the SAT games a solution is any of the *satisfying attack suites*. We have chosen linear binary representation of individuals to facilitate the robustness of the crossover and mutation operations. The algorithm used to generate individuals is shown in Algorithm 1.

We allow duplicate entries to be present in the population for the sake of maintaining genetic variation and keep the population size constant throughout the reproduction process. It is well known in the field of genetic algorithms that genetic variation directly influences the chances of premature convergence – thus increasing genetic variation in the population is one of the design goals.

Algorithm 1: Recursive individual generation algorithm

Data: The root of a propositional directed acyclic graph (PDAG) representing
a monotone Boolean function. An empty individual with all bits set to 0.
Result: Live individual.

if *the root is a leaf* **then**
  get the index of the leaf;
  set corresponding individual's bit to 1;
end
else if *the root is an AND node* **then**
  forall *children of the root* **do**
    recursive call: child considered as root parameter;
  end
end
else if *the root is an OR node* **then**
  choose at least one child;
  forall *chosen children* **do**
    recursive call: child considered as root parameter;
  end
end

The choice of the population size is important – too small population does not contain enough genetic variation to maintain the exploration capabilities, too big population already contains enough genetic variation to efficiently explore the search space and only results in the performance overhead in the crossover operator. This means that there exists an optimal population size corresponding to the minimal population size capable of producing the best result. Thus the optimal size of the population sets the lower bound of reasonable choice for the population size and the upper bound is solely based on performance considerations – what is the reasonable time the analysts would agree to wait for the analysis to produce the result. If the population size is suboptimal, there is a high risk to converge to suboptimal solutions and if the population is bigger than the optimal size it does not add anything, except for the increase in the time required to run the analysis. If the optimal population in some certain case is $k\%$ of the size of the attack tree (the number of leaves in an attack tree), then any population size greater than $k\%$ and capable of producing the result in reasonable time, would suit to be used for analysis.

All the following computations were made with PC/Intel Core i5-4590 CPU @ 3.30 GHz, 8 GB RAM, Windows 8.1 (64 bit) operating system. Figure 1 on the left demonstrates the effect of the population size on the result in the case of a single attack tree. Measurements were taken for the attack tree with 100 leaves using uniform crossover operator and mutation rate 0.1.

We have conducted experiments on the set of attack trees of different sizes (ranging from 10 to 100 leaves with steps of size 3) and observed that there is no obvious relation between the size of the analyzed tree and the optimal population size. Apart from the size of the tree, the optimal population size might depend on, at the very least, the structure of the tree itself. Measurements were taken

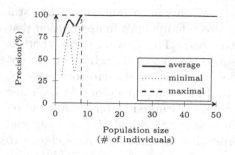

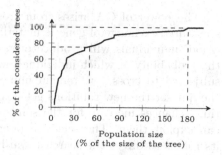

Fig. 1. Optimal population size

Fig. 2. Reasonable choice for population size

with the same crossover operator and mutation rate. Figure 2 shows how many trees (%) from the conducted experiment the considered population size would fit. It can be seen that, in general, the population size equal to 180 % of the size of the tree would fit every considered attack tree. The population size 200 %, chosen by Jürgenson and Willemson in [8] for their ApproxTree model, was a reasonable choice.

Lenin, Willemson and Sari have shown that the crossover operations take 90–99% of the time required to run the analysis [9]. Figure 3 shows the time measurement for the suggested GA, depending on the size of the population.

The *fitness function* is the model-specific utility function for the corresponding type of the security game. For further details we refer the reader to the detailed descriptions of the security games [7–11].

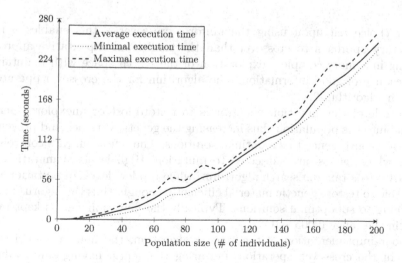

Fig. 3. Population size effect on GA execution time.

The power of GA arises from crossover which causes randomized but still structured exchange of genetic material between individuals in assumption that 'good' individuals will produce even better ones. The *crossover rate* controls the probability at which individuals are subjected to crossover. Individuals, not subjected to crossover, remain unmodified. The higher the crossover rate is, the quicker the new solutions get introduced into the population. At the same time, chances increase for the solutions to get disrupted faster than selection can exploit them. The selection operator selects individuals for crossing and its role is to direct the search towards promising solutions. We have chosen to disable parent selection entirely thus defaulting to crossing every individual with every other individual in the population (crossover rate equal to 1), as scalable selection pressure comes along with the selection mechanisms after reproduction.

Notable crossover techniques include the single-point, the two-point, and the uniform crossover types. Figures 4 and 5 demonstrate the differences between the convergence speeds resulting from using various crossover operators. It can be seen that the considered crossover operators do not have any major differences nor effect on the convergence speed of the GA.

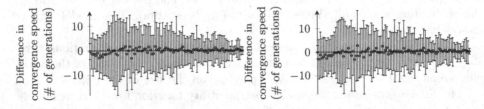

Fig. 4. Uniform crossover compared to single point crossover

Fig. 5. Uniform crossover compared to two point crossover

Our choice fell upon using the uniform crossover – this enables a more exploratory approach to crossover than the traditional exploitative approach, resulting in a more complete exploration of the search space with maintaining the exchange of good information. The algorithm for the crossover operator is shown in Algorithm 2.

The role of the mutation operator is to restore lost or unexplored genetic material into the population thus increasing the genetic variance and preventing premature convergence to suboptimal solutions. The mutation rate controls the rate at which 'genes' are subjected to mutation. High levels of mutation rate turn GA into a random search algorithm, while too low levels of mutation rates are unable to restore genetic material efficiently enough, thus the algorithm risks converging to suboptimal solutions. Typically the mutation rate is kept rather small, in the range $0.005 - 0.05$.

In our implementation of the genetic algorithm, the mutation operator is a part of the crossover operation, mutating the genes, having same value in the corresponding positions in both parent individuals. The uniform crossover randomly picks corresponding bits in the parent individuals to be used in the

---

Algorithm 2: The uniform crossover operation

Data: The population of individuals represented as a sorted set.
Result: The population with new added individuals, created during the
      crossover operation.

initialize a new set of individuals;
forall *individual* i *in the population* **do**
    forall *individual* j *different from* i **do**
        new individual := the result of cross operation between individuals i
        and j;
        if *new individual is alive* **then**
            add the new individual to the set of new individuals;
        end
    end
end
add the set of new individuals to the population;

---

new individual, and thus in the case bits are different, this already provides
sufficient genetic variation. However, in the case when bits have the same value
this yields just a single choice and in order to increase the genetic variation
(compared to its parents) we mutate just these bits. Figure 6 demonstrates the
mutation rate effect on the utility function for the case of a specific attack tree
with 100-leaves with initial population of 50 individuals. It shows that when
the mutation rate exceeds value 0.1 GA turns into a random search algorithm,
thus it is reasonable to keep the mutation rate rather small. We have conducted
similar experiments on a larger set of attack trees and the results have shown
that the optimal value for the mutation rate is not necessarily small – in some

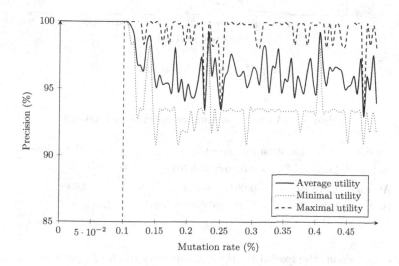

Fig. 6. GA mutation rate effect

cases the optimal mutation rate was 0.6 or even higher. This means that the optimal value for the mutation rate cannot be set from the very beginning – it highly depends on the structure of the fitness landscape. However, it is still reasonable to follow the general rule of thumb to keep the mutation rate small, assuming that this should work for the majority of the cases.

It is important to determine the practical applicability boundaries for the suggested method. By practical applicability we mean the maximal size of the attack tree, which the computational method is capable of analyzing in reasonable time set to two hours. Extrapolating the time consumption curve in Fig. 7 we have come to a conclusion that theoretically the suggested GA is capable of analyzing attack trees containing up to 800 leaves in reasonable time. This is a major advancement compared to the ApproxTree model [8] which would take more than 900 hours to complete such a task.

The execution time complexity estimations for GA are outlined in Table 1.

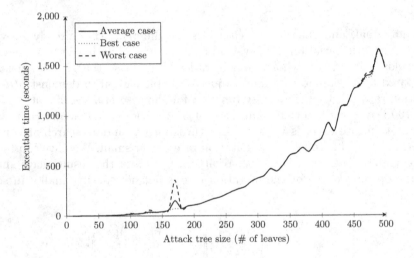

Fig. 7. GA execution time

Table 1. GA execution time complexity estimations

Case	Approximation polynomial	R^2 coefficient
Worst	$1.68 \cdot 10^{-5}n^3 - 0.003n^2 + 0.7015n - 23.03$	0.99
Average	$1.41 \cdot 10^{-5}n^3 - 0.001n^2 + 0.25n - 8.81$	0.99
Best	$1.26 \cdot 10^{-5}n^3 + 1.62 \cdot 10^{-5}n^2 + 0.047n - 2.55$	0.99

For comparison, the execution time complexity of the ApproxTree model [8] was estimated to be $\mathcal{O}(n^4)$, where n is the number of leaves in the attack tree. This difference comes from the fact that ApproxTree runs for a fixed number of generations, whereas the computations presented in this paper run until local convergence, as well as the fact that the utility function used in ApproxTree is considerably more complex compared to the corresponding utility function used in the Failure-Free model.

4   Adaptive Genetic Algorithm (AGA)

We compare the genetic algorithm suggested in Sect. 3 to the adaptive genetic approach described in [13]. The authors suggest to adaptively vary the values of crossover and mutation rates, depending on the fitness values of the solutions in the population. High fitness solutions are 'protected' and solutions with subaverage fitness are totally disrupted. It was suggested to detect whether the algorithm is converging to an optimum by evaluating the difference between the maximal and the average fitness values in the population $f_{max} - \bar{f}$ which is likely to be less for the population which is converging to an optimum solution than for a population scattered across the solution space. Thus the corresponding values of the mutation and crossover rates are increased when the algorithm is converging to an optimum and decreased when the population gets too scattered. The authors concluded that the performance of AGA is in general superior to the performance of GA but varies considerably from problem to problem. In this paper we apply the suggested method to the problem of the security games.

In the case of the adaptive genetic algorithm, the crossover and mutation rate parameters are assigned their initial values and are changed adaptively during the runtime of the algorithm and the only parameter which remains fixed is the population size. Similarly to the GA there exists an optimal population size corresponding to the minimal population size capable of producing the maximal result. Figure 8 shows the result corresponding to the computations using various population sizes in the experiment setup similar to the one for GA. In the case of GA the maximal value was stable with the increase in the population size, however in the case of AGA some fluctuations are present. Figure 9 shows how many trees (%) from the conducted experiment the considered population size would fit. It can be seen that, in general, the population size equal to 200 % of the size of the tree would fit every considered attack tree. Based on these observations we can say that AGA seems to be more robust, but less stable, compared to GA and requires bigger population sizes in order to produce optimal results for the majority of the cases.

Similarly to the GA, we estimate the maximal size of the attack tree which AGA is capable of analyzing within reasonable timeframe set to two hours. Extrapolating the time consumption curve with the most extreme values trimmed out in Fig. 10 we have come to a conclusion that theoretically AGA is capable of analyzing attack trees containing up to 26000 leaves in reasonable timeframe, which is approximately 32 times more efficient compared to GA.

The execution time complexity estimations for AGA are outlined in Table 2.

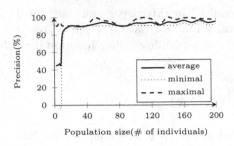

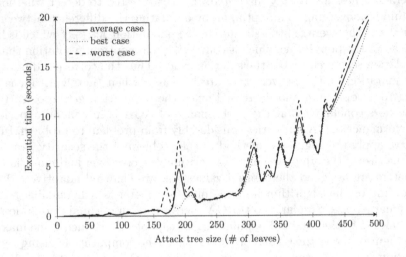

Fig. 8. Optimal population size

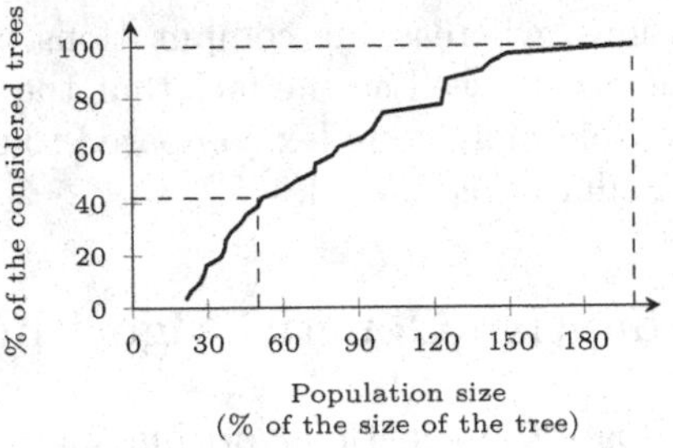

Fig. 9. Reasonable choice for population size

Fig. 10. AGA execution time

Table 2. AGA execution time complexity estimations

Case	Approximation polynomial	R^2 coefficient
Worst	$3.985x^3 - 0.0001x^2 + 0.0358x - 1.1970$	0.90
Average	$3.5731x^3 - 0.0001x^2 + 0.0267x - 0.8786$	0.94
Best	$3.1892x^3 - 0.0001x^2 + 0.0192x - 0.6115$	0.96

5   Conclusions

This paper addressed the problem of efficient approximation of attack tree evaluation of the failure-free game. We considered the genetic approach to approximation, since it is known to have worked on similar problems previously. However, genetic algorithms depend on various loosely connected parameters (e.g. crossover and mutation operators and their corresponding rates). Selecting them all simultaneously is a non-trivial task requiring a dedicated assessment effort for each particular problem type. The current paper presents the first systematic study of GA parameter optimization for the attack tree evaluation. We have conducted a series of experiments and collected heuristic evidence for optimal parameter selection.

The second contribution of the paper is the application of adaptive genetic algorithms (AGA) to the problem domain of attack tree computations. It turns out that AGA converges generally faster than GA and provides similar level of accuracy, but with the price of potentially larger population sizes. Since usually there are no major technical obstacles to increasing the population, we conclude that AGA should be preferred to plain GA in the considered application domain.

References

1. Vesely, W., Goldberg, F., Roberts, N., Haasl, D.: Fault tree handbook. US Government Printing Office: Systems and Reliability Research, Office of Nuclear Regulatory Research. U.S, Nuclear Regulatory Commission, January 1981
2. Weiss, J.D.: A system security engineering process. In: Proceedings of the 14th National Computer Security Conference, pp. 572–581 (1991)
3. Schneier, B.: Attack trees: modeling security threats. Dr. Dobb's J. 24(12), 21–29 (1999)
4. Mauw, S., Oostdijk, M.: Foundations of attack trees. In: Kim, S., Won, D.H. (eds.) ICISC 2005. LNCS, vol. 3935, pp. 186–198. Springer, Heidelberg (2006)
5. Buldas, A., Laud, P., Priisalu, J., Saarepera, M., Willemson, J.: Rational choice of security measures via multi-parameter attack trees. In: López, J. (ed.) CRITIS 2006. LNCS, vol. 4347, pp. 235–248. Springer, Heidelberg (2006)
6. Jürgenson, A., Willemson, J.: Serial model for attack tree computations. In: Lee, D., Hong, S. (eds.) ICISC 2009. LNCS, vol. 5984, pp. 118–128. Springer, Heidelberg (2010)
7. Jürgenson, A., Willemson, J.: Computing exact outcomes of multi-parameter attack trees. In: Meersman, R., Tari, Z. (eds.) OTM 2008, Part II. LNCS, vol. 5332, pp. 1036–1051. Springer, Heidelberg (2008)
8. Jürgenson, A., Willemson, J.: On fast and approximate attack tree computations. In: Wang, G., Deng, R.H., Won, Y., Kwak, J. (eds.) ISPEC 2010. LNCS, vol. 6047, pp. 56–66. Springer, Heidelberg (2010)
9. Lenin, A., Willemson, J., Sari, D.P.: Attacker profiling in quantitative security assessment based on attack trees. In: Bernsmed, K., Fischer-Hübner, S. (eds.) NordSec 2014. LNCS, vol. 8788, pp. 199–212. Springer, Heidelberg (2014)
10. Buldas, A., Stepanenko, R.: Upper bounds for adversaries' utility in attack trees. In: Walrand, J., Grossklags, J. (eds.) GameSec 2012. LNCS, vol. 7638, pp. 98–117. Springer, Heidelberg (2012)
11. Buldas, A., Lenin, A.: New efficient utility upper bounds for the fully adaptive model of attack trees. In: Das, S.K., Nita-Rotaru, C., Kantarcioglu, M. (eds.) GameSec 2013. LNCS, vol. 8252, pp. 192–205. Springer, Heidelberg (2013)
12. Lenin, A., Buldas, A.: Limiting adversarial budget in quantitative security assessment. In: Poovendran, R., Saad, W. (eds.) GameSec 2014. LNCS, vol. 8840, pp. 155–174. Springer, Heidelberg (2014)
13. Srinivas, M., Patnaik, L.M.: Adaptive probabilities of crossover and mutation in genetic algorithms. IEEE Trans. Syst. Man Cybern. 24(4), 656–667 (1994)

To Trust or Not: A Security Signaling Game Between Service Provider and Client

Monireh Mohebbi Moghaddam[1], Mohammad Hossein Manshaei[1($\boxtimes$)], and Quanyan Zhu[2]

[1] Department of Electrical and Computer Engineering,
Isfahan University of Technology (IUT), Isfahan, Iran
monireh.mohebbi@ec.iut.ac.ir, manshaei@cc.iut.ac.ir
[2] Department of Electrical and Computer Engineering,
New York University (NYU), New York, USA
quanyan.zhu@nyu.edu

Abstract. In this paper, we investigate the interactions between a service provider (SP) and a client, where the client does not have complete information about the security conditions of the service provider. The environment includes several resources of the service provider, a client who sends requests to the service provider, and the signal generated by the service provider and delivered to the client. By taking into account potential attacks on the service provider, we develop an extended signaling game model, where the prior probability of the signaling game is determined by the outcome of a normal form game between an attacker and the service provider as a defender. Our results show different equilibria of the game as well as conditions under which these equilibria can take place. This will eventually help the defender to select the best defense mechanism against potential attacks, given his knowledge about the type of the attacker.

Keywords: Network security · Computation outsourcing · Game theory · Signaling game

1 Introduction

Increasing the amount of generated data raises new challenges for processing data in large scale. The idea of outsourcing computational jobs is proposed to overcome the complexity and cost for applications that rely on big data processing. This trend is accelerated with the introduction of *Cloud Computing*. Cloud computing provides computational services that enable ubiquitous, inexpensive and on-demand access to vastly shared resources. This paradigm eliminates the requirements for setting up high-cost computing infrastructure and storage systems, making it more beneficial for clients with the limited storage and computational capacity [1]. Despite the huge benefits of this platform, it faces several challenges. Security issues are among the biggest challenges that hinder the ubiquitous adaption of cloud computing [2].

© Springer International Publishing Switzerland 2015
MHR Khouzani et al. (Eds.): GameSec 2015, LNCS 9406, pp. 322–333, 2015.
DOI: 10.1007/978-3-319-25594-1_18

Although SPs aim to secure their infrastructure against threats, but providing a full security has not been possible at least up to now [3]. As a result, it is likely that the providers face attackers. This issue becomes worse for the client who cannot detect whether SP is compromised or not. All of these entities including the service provider, the client, and the attacker are rational decision makers who aim to make the best decisions. When there are rational entities who face different choices, and the outcome of an entity's choice depends critically on the actions of other participants, we can use game theory to model their behaviors. As an example of applying this framework to the cloud security, authors in [4] used this tool for identifying untrustworthy cloud users. In [5], the effect of interdependency in a public cloud platform has been investigated with a game-theoretic framework. Also, as the providers have the economic incentive to return guessed results instead of performing the computation completely, auditing is an important issue in the data outsourcing. Nix et al. in [8] proposed a game theory-based approach to query verification on outsourced data. They proved that the incentive for cloud to cheat an outsourcing service could be reduced under their proposed structure. In [9], Pham et al. also addressed the auditing problem on the outsourced data and provided a general approach based on game theory for optimal contract design. Authors in these works considered the cheating or lazy behavior of the SPs, but ignored the probability of generating incorrect results because of compromised resources by attackers.

In contrast to the related works, in this paper, we model and analyze the behavior of the SP and his client in the presence of an attacker by using game-theoretic techniques. Our focus in this work is on the client side to make the best strategic decision. The client outsources his computation to a SP and receives the computational results as a service. This SP may be under attack. Subsequently, the returned results to the client may be incorrect. We model the interaction between the SP and the attacker with a normal form game. In this regard, the client faces with a safe or compromised SP. He should decide whether to rely on the provided service by SP or not. He should make this decision under incomplete information, as the type of the SP is not observable by the client. We are interested in modeling all these interactions by using a suitable game model. Signaling game, due to its special properties, is highly consistent with our purpose. In this game, the client receives signal (service) from a SP who has two types, e.g. legitimate and compromised. The prior probability of the signaling game can be derived from the game between the SP and the attacker. The composition of these games defines an extended form of signaling game. To the best of our knowledge, this is the first work that adopts a signaling game approach to model the interaction of the SP and the client in the presence of an attacker who attacks the SP. Although our model is proposed for cloud computing platform, it is generally applicable to other situations in which a client interacts with a SP exposed to the attacks.

The paper is structured as follows. In Sect. 2, we describe our system model. Then, we discuss and analyze our game models in Sect. 3. In Sect. 4, we analyze the proposed signaling game and derive all possible Nash equilibria. Finally, Sect. 5 summarizes the conclusion and proposes future works.

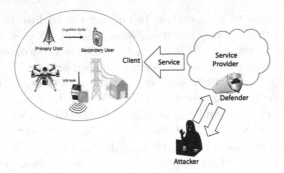

Fig. 1. System model: Service provider interacts with his client. In addition, the SP interacts with an attacker who tries to compromise the cloud resources.

2   System Model

In this paper, we focus on a system model depicted in Fig. 1. We assume a *service provider* (SP) provides the desired services to the clients. As we have mentioned in the previous section, the SP may be threatened by different attacks. To this end, we assume that there is an *attacker* in this system who tries to compromise the SP resources. The SP acts as a defender against him and employs defense mechanism. We model the interaction between the SP and the attacker as a two-player normal form game. The strategy space of these players consists of resources they select; the SP for processing the client requests and the attacker for compromising and maybe generating the wrong answer for the client. We will describe this game in the next section in more details for the purpose of deriving the probability of a successful attack. The *client* as another entity in this system participates in a separated game from the previous one. The client sends his computation to the SP and receives a service. This service is in the form of a signal. The client is aware of the probability of the attacks, but he does not know the selected SP is compromised by an attacker or not. Therefore, he participates in a dynamic game with either a compromised or a safe SP. The client should decide to trust the received signal from the SP or not. Different types of SP can play different strategies in these dynamic games. As a result of the above descriptions, we need a game to model the uncertainty of the client about the SP type as well as the dynamic nature of the SP and the client movements. Signaling game is a good choice for this purpose. In our game model which is an extended form of signaling game, the prior probability of the game is derived from the game between the SP and the attacker. Since signaling game is a game with incomplete information is useful for modeling the client uncertainty about the SP type. In addition, the received service by the client is in the form of a signal, which is compatible with this game model. The mentioned dynamic games between the client and the safe/compromised SP are merged to form different signals which are sent by different types of the SP.

The mentioned system with these entities and interactions can be found in many contexts. For example, when an unmanned aerial vehicle, also known as drone, outsources his computation to a SP, because of its limited computational resources. Drone acts according to the received service from the SP. But, it seems that relying on the received service is not the best action in all situations. Especially, when the SP is exposed to the various attacks and rely on the wrong signal has irreparable damage for drones. Cognitive Radio (CR) is another example. Dynamic spectrum assignment to the secondary users by CR should be done in a way that does not create harmful interference to primary licensed users. Similar to the drone example, CR devices can overcome their limited computational capacity in determining idle bands with computational outsourcing [6]. Despite these advantages, the cloud may be subject to attacks with the objective of creating interference. Given these risks, secondary user may face the same question for relying on the results obtained from the cloud. Sensor nodes in a Wireless Sensor Network (WSN) can utilize the vast computational capacity of cloud computing in the same way [7]. In this scenario as previous ones, it is probable that the computational results are incorrect and the user should be more careful about his decision whether to trust the cloud or not.

In the forementioned scenarios, clients have common concerns about the received outcomes. Can they trust the returned result? More importantly, in what situations, they should apply an auditing approach to investigate the results and under which conditions, it is better to trust? We will investigate and answer these questions along this paper.

3   Game Model

In this section, we will model the interactions between the mentioned individuals in the previous section using game theory. We employ two different game models for this purpose. The first game, which is a normal form one, models the scenario where a malicious agent attacks the SP. We will describe and analyze it in detail in Sect. 3.1. Since the SP can be under attack or not, the client faces two types of SPs. In Sect. 3.2, we will illustrate the relation between the client and each type of the SP using two different dynamic games. In Sect. 3.3, we will describe how we can merge these three games to form an extended form of a signaling game.

3.1   Attacker-Defender (AD) Game Model

An SP has several resources which can be used for processing the client requests. The SP should decide about the selected resources for computation. Simultaneously, the attacker determines his attack target. We model this interaction using a parametric normal form game. We start with a cloud with two resources and extend our analysis to $\mathcal{N}$ resources. Under the assumption of two resources, the strategy sets of both the attacker and defender as well as payoff values is shown in Table 1, we named *AD-game*.

Table 1. Strategic form of the security game between an attacker and the SP as a defender in a cloud computing with two resources.

D↓ A→	R_1	R_2
R_1	D_{11},A_{11}	D_{12},A_{12}
R_2	D_{21},A_{21}	D_{22},A_{22}

We now analyze *AD-game* for finding its Nash Equilibria (NE). This calculation is valuable as we use it for finding the probability of the successful attacks. If the SP selects R_i, the best response for the attacker is to attack the same resource. It is obvious that the defender prefers to select the other resource, except that one has been chosen by the attacker. Given this, we conclude that: $D_{21} \geq D_{11}$, $D_{12} \geq D_{22}$, $A_{11} \geq A_{12}$ and $A_{22} \geq A_{21}$. Therefore, there is no pure strategy Nash equilibrium for this game.

We now compute the mixed Nash equilibrium of this game. For simplicity in the mixed NE closed form representation, we define the following notations, δ_{iD} and δ_{iA}, in which $\delta_{iD} = D_{ji} - D_{ii}$ and $\delta_{iA} = A_{ii} - A_{ij}$, where $i, j = \{1, 2\}$. Suppose the attacker randomizes over R_i resource with P_{iA} and the probability of defending the same resource is equal to P_{iD}. The closed form of these probabilities are equal to the following equations at the mixed-NE:

$$P_{iA} = \frac{\delta_{jD}}{\sum_{k=1}^{2} \delta_{kD}}, \text{ where } i \in \{1, 2\}; j \neq i \tag{1}$$

$$P_{iD} = \frac{\delta_{jA}}{\sum_{k=1}^{2} \delta_{kA}}, \text{ where } i \in \{1, 2\}; j \neq i \tag{2}$$

The client will get a service from a compromised SP when both the attacker and the defender select the same resource. We represent the probability of this event by θ, which is calculated in (3). The value of θ is important for both the attacker and the defender, and is more valuable for a client who gets the service from the SP. The larger values of θ indicate a risky cloud.

$$\theta = \sum_{i=1}^{2}(P_{iD} \times P_{iA}) = \frac{\sum_{i=1}^{2}(\delta_{iD} \times \delta_{iA})}{(\sum_{i=1}^{2} \delta_{iD}) \times (\sum_{i=1}^{2} \delta_{iA})} \tag{3}$$

The above calculations can be extended to a cloud with $\mathcal{N}$ resources. We call this game as *Extended-AD game*. For simplicity, we assume that the attacker just selects one resource for the attack, but the following calculations can be extended to include any $m < \mathcal{N}$ resources. Since the attacker does not know which resource consists of the result, there is no difference between them for him. Additionally, the SP doesn't have any information about the target of the attacker. In this regard, we assume the same values for all $D_{ji}, \forall j \neq i$, and in the same way, the equal values for all $A_{ij}, \forall j \neq i$, where $i, j \in \{1, 2, ..., \mathcal{N}\}$. Under these assumptions, we can conclude that there is no pure strategy NE in this game. Values of the P_{iD} and P_{iA} in the mixed NE as well as θ can be calculated by the following equations:

$$P_{iA} - \frac{\prod_{j=1, j\neq i}^{\mathcal{N}} \delta_{jD}}{\sum_{\{i_1, i_2, \cdots, i_{\mathcal{N}-1}\} \in S} (\delta_{i_1 D} \times \cdots \times \delta_{i_{(\mathcal{N}-1)D}})} \qquad (4)$$

$$P_{iD} = \frac{\prod_{j=1, j\neq i}^{\mathcal{N}} \delta_{jA}}{\sum_{\{i_1, i_2, \cdots, i_{\mathcal{N}-1}\} \in S} (\delta_{i_1 A} \times \cdots \times \delta_{i_{(\mathcal{N}-1)A}})} \qquad (5)$$

$$\theta = \sum_{i=1}^{\mathcal{N}} (P_{iD} \times P_{iA}) \qquad (6)$$

where S is the all $(\mathcal{N} - 1)$-combinations of $\lfloor \mathcal{N} \rfloor$, $\lfloor \mathcal{N} \rfloor = \{1, 2, \cdots, \mathcal{N}\}$.

3.2 Service Provider-Client (SPC) Game Model

As discussed before, cloud resources may be compromised with probability θ. It is noticeable that the client does not know the SP is compromised or not. Therefore, the client may face a compromised cloud or a safe one. In this subsection, we examine the interactions between the client and each type of SP using dynamic games. In the next subsection, we will show how these interactions can merge to form a single game model. We begin with the game model between the client and the safe SP.

The SP first performs the required computation on the data. He can allocate enough and precise computational resources for these calculations. He has another choice in which uses the low precise computational resources, or maybe he is a *lazy* SP and returns the guessed result instead of the accurate one. It can cause the incorrect result generated. Therefor, in the first step, the SP takes one of the two choices: h or l as represented in Fig. 2. Depending on the chosen strategy by the provider, the returned result to the client can be correct or incorrect. We show the probabilities of data correctness with p_C and p'_C for h and l strategies, respectively. p_I and p'_I also stand for the probabilities of generating false results. Obviously, $p_I = 1 - p_C$, $p'_I = 1 - p'_C$ and $p_C \geq p'_C$. Following the *chance move*, the SP acts again as a player and sends the computed result as a service to the client. In our model, we suppose that the SP strategy space consists of two actions, but in general, the number of signals can be more than two. Once the client receives the signal, he should make the final decision to trust or not. These strategies are shown with T and N in the extended tree form. Depending on the SP and the client choices, the payoff values for both of them will determine.

Now, we investigate the game in the presence of an attacker. The attacker knows the possibility for generating false result by the SP. Also, he may ignore the changing the results, as this raises the probability of attacher detection by the defender. Additionally, his purpose could be the access to resources for eavesdropping, not changing the results. So, he should decide to change the computational results or leave them without any manipulation. Given this information, we define two actions for this attacker dilemma, that is shown with m and nm in Fig. 3. Note that the probability of chance move is equal to q_C and q'_C in this case. Similar to the safe cloud case, the client is the final player.

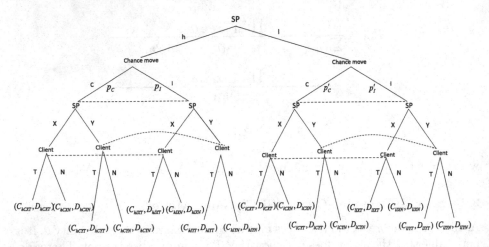

Fig. 2. A dynamic game model for a safe cloud.

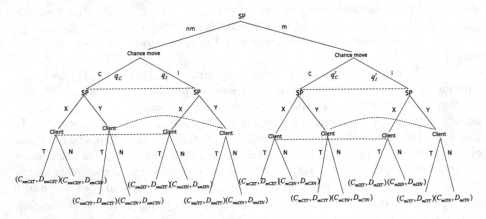

Fig. 3. A dynamic game model for a compromised cloud.

3.3 Service Provider-Client Signaling (S-SPC) Game

In the previous subsection, we have described the interaction between a safe
SP and the client in one hand and the interaction between the client and the
compromised SP, on the other hand. Now, we are interested in modeling all
of the interactions in the previous parts by a game. We require a two-player
game which is able to model the uncertainty of the client about the type of the
SP. In addition, this game must be a dynamic one, in which the client makes
the decision after receiving a signal from the SP. Given these requirements, it
seems signaling game can be a good choice. Signaling game is a non-cooperative
dynamic game of incomplete information with two players wherein one player
has private information [10].

Our model as a singling game is represented in Fig. 4. In this game, Nature
as the first mover chooses type of the SP. It chooses either type under attach (A)
or safe (NA) with probability θ and $1 - \theta$, respectively. This value is gained from

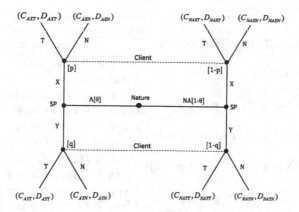

Fig. 4. A signaling game model with two different types of the service provider.

Table 2. Signaling game notations

Notation	Description
A	Cloud is under attack
NA	Cloud is safe
T	Trust to received signal
N	Do not trust to received signal
θ	The probability of the successful attack ($\theta \in [0,1]$)
$m(i)$	Sending signal by SP type i, $m(i) \in \{X, Y\}, i \in \{A, NA\}$
$a(j)$	The client's action in response to received signal j, $a(j) \in \{T, N\}$, $j \in \{X, Y\}$
p	Client's belief at the upper information set where the SP is under attack ($p \in [0,1]$)
q	Client's belief at the lower information set where the SP is under attack ($q \in [0,1]$)
D_{ijk}	Client's payoff, where SP's type is i and signal j is sent and client plays action k where $i \in \{A, NA\}, j \in \{X, Y\}$ and $k \in \{T, N\}$
C_{ijk}	SP's payoff, where SP type is i and signal j is sent and client plays action k, where $i \in \{A, NA\}, j \in \{X, Y\}$ and $k \in \{T, N\}$

the *Extended AD-game*. After that, selected SP type sends either X or Y signal. The other used notations in this figure are summarized in Table 2. Given the type of the SP, he, and the client choose their actions according to the extensive forms which is shown in Figs. 2 and 3. In this regard and after the action selected by the client, the payoff values for the SP and the client can be calculated. For instance, D_{NAXT} in the signaling game, is the client's payoff value when SP type NA sends the signal X and the client chooses the strategy T. The value of this payoff and other outcomes can be calculated using the payoff values of Figs. 2 and 3 as described in the following equations, where $j \in \{X, Y\}$ and $k \in \{T, N\}$:

Table 3. Separating equilibria and their conditions.

BNE	Separating BNE Profile	Conditions
SE1	$((X,Y),(T,T),p=1,q=0)$	$D_{AXT} \geq D_{AXN}, D_{NAYT} \geq D_{NAYN}$ $C_{AXT} \geq C_{AYT}, C_{NAYT} \geq C_{NAXT}$
SE2	$((X,Y),(T,N),p=1,q=0)$	$D_{AXT} \geq D_{AXN}, D_{NAYT} \leq D_{NAYN}$ $C_{AXT} \geq C_{AYN}, C_{NAYN} \geq C_{NAXT}$
SE3	$((X,Y),(N,T),p=1,q=0)$	$D_{AXT} \leq D_{AXN}, D_{NAYT} \geq D_{NAYN}$ $C_{AXN} \geq C_{AYT}, C_{NAYT} \geq C_{NAXN}$
SE4	$((X,Y),(N,N),p=1,q=0)$	$D_{AXT} \leq D_{AXN}, D_{NAYT} \leq D_{NAYN}$ $C_{AXN} \geq C_{AYN}, C_{NAYN} \geq C_{NAXN}$
SE5	$((Y,X),(T,T),p=0,q=1)$	$D_{AYT} \geq D_{AYN}, D_{NAXT} \geq D_{NAXN}$ $C_{AYT} \geq C_{AXT}, C_{NAXT} \geq C_{NAYT}$
SE6	$((Y,X),(N,T),p=0,q=1)$	$D_{AYT} \leq D_{AYN}, D_{NAXT} \geq D_{NAXN}$ $C_{AYT} \geq C_{AXN}, C_{NAXN} \geq C_{NAYT}$
SE7	$((Y,X),(T,N),p=0,q=1)$	$D_{AYT} \geq D_{AYN}, D_{NAXT} \leq D_{NAXN}$ $C_{AYN} \geq C_{AXT}, C_{NAXT} \geq C_{NAYN}$
SE8	$((Y,X),(N,N),p=0,q=1)$	$D_{AYT} \leq D_{AYN}, D_{NAXT} \leq D_{NAXN}$ $C_{AYN} \geq C_{AYN}, C_{NAXN} \geq C_{NAYN}$

$$C_{NAjk} = p_C(C_{hCjk} - C_{hIjk}) + p'_C(C_{lCjk} - C_{lIjk}) + C_{hIjk} + C_{lIjk} \quad (7)$$

$$D_{NAjk} = p_C(D_{hCjk} - D_{hIjk}) + p'_C(D_{lCjk} - D_{lIjk}) + D_{hIjk} + D_{lIjk} \quad (8)$$

$$C_{Ajk} = q_C(C_{nmCjk} - C_{nmIjk}) + q'_C(C_{mCjk} - C_{mIjk}) + C_{nmIjk} + C_{mIjk} \quad (9)$$

$$D_{Ajk} = q_C(D_{nmCjk} - D_{nmIjk}) + q'_C(D_{mCjk} - D_{mIjk}) + D_{nmIjk} + D_{mIjk} \quad (10)$$

4   Game Analysis

In this section, we analyze the signaling game for finding the pure strategy Bayesian Nash Equilibria (BNE).

4.1   Service Provider-Client Signaling Game's Equilibria

In the following, we examine the signaling game for the existence and properties of any pure strategy BNE. Basically, in non-Bayesian games, a strategy profile is a NE if every strategy in that profile is a best response to every other strategy. But, in Bayesian games, players are seeking to maximize their expected payoff, given their beliefs about the other players [10].

For the defined signaling game in Fig. 4, a pure strategy BNE profile is determined as tuple $((m(A), m(NA)), (a(X), a(Y)), p, q)$. It consists of the pair of strategies chosen by each type of the first player (SP), the actions taken by the second player (client) in response to the each signal and the client's beliefs. In each signaling game, two types of BNE are possible, called pooling and separating, which are defined as follows:

Definition 1 (Pooling equilibrium [10]): A BNE is a pooling equilibrium if the first player sends the same signal, regardless of his type.

Definition 2 (Separating equilibrium [10]): A BNE is a separating equilibrium if the first player sends different signals, depending on his type.

For example in our defined signaling game presented in Fig. 4, if $m(A) = m(NA)$ at a given BNE, it represents a pooling equilibrium. In contrast, a BNE in which $m(A) \neq m(NA)$ is a separating equilibrium.

Separating Equilibria. In this part, we examine the conditions under which our signaling game has separating equilibria. According to *Definition 2*, this occurs when $m(A) = X$ and $m(NA) = Y$ or $m(A) = Y$ and $m(NA) = X$ in Fig. 4. We first analyze the case in which equilibria contain strategy (X, Y). This means the SP type A chooses strategy X and other type selects Y. In this situation, the client chooses T in response to X if this choice is more beneficial for him rather than choosing strategy N. In other words, his gained payoff for action T, i.e., D_{AXT}, be greater than or equal to D_{AXN}. Otherwise, he plays action N. Similarly, the client will play T in response to signal Y sent by SP type NA, when $D_{NAYT} \geq D_{NAYN}$. In each case, we must check whether any type of the SP deviates from selected strategy, his payoff would increase or not. If deviation causes increasing the SP's payoffs, this strategy profile will not be an equilibrium. These conditions must be checked for other strategy profiles. All the separating equilibria and their conditions are given in Table 3. Recall that the values of payoffs can be calculated by Eqs. (7)–(10).

Pooling Equilibria. As *Definition 1* states, pooling equilibria occur either $m(A) = m(NA) = X$ or $m(A) = m(NA) = Y$. In both cases, the selected strategy by the client depends on the expected payoff he obtains by playing that strategy. In the earlier case, $m(A) = X$ and $m(NA) = X$, the client will play T if this action results in an expected payoff greater than or equal to expected payoff for playing N, i.e., $p \times D_{AXT} + (1-p) \times D_{NA1T} \geq p \times D_{AXN} + (1-p) \times D_{NAXN}$ Depending on the value of payoffs appeared in this inequality, the value of p will be determined. Since action X is on the equilibrium path, the value of p will be equal to θ.

Without loss of generality, we assume that the client's payoffs in Eqs. (8) and (10), have the following relations: $D_{iCjT} > D_{iIjN} > D_{iCjN} > D_{iIjT}, i \in \{nm, m, h, l\}$ and $j \in \{X, Y\}$. These relative relations can be true because of the following reasons. The client gains the greatest payoff value when the signal is correct and he trusts. The lowest payoff value belongs to the case he trusts the wrong signal. Additionally, he audits the received signal via another approach even when the signal is correct but he does not trust. Therefore, he should pay the auditing cost. In addition, assume that the sum of correctness probabilities in the case of the safe SP be greater than the same one in the presence of an attacker, i.e., $(p_C + p_C') > (q_C + q_C')$. With these assumptions, we can conclude that:

$$p \leq \frac{(D_{NAXT} - D_{NAXN})}{(D_{AXN} - D_{NAXN}) + (D_{NAXT} - D_{AXT})} =: F_1$$

We now should determine the beliefs and actions for the off-equilibrium path of sending signal Y by the SP. Similarly, the client will play T if the expected payoff for choosing this strategy is greater than or equal to expected payoff for playing N. Given the above assumptions, we will have:

$$q \leq \frac{(D_{NAYT} - D_{NAYN})}{(D_{AYN} - D_{NAYN}) + (D_{NAYT} - D_{AYT})} =: F_2.$$

Similar to separating equilibria, we perform the calculations to obtain the conditions in which there exist pooling equilibria. This is summarized in Table 4.

Table 4. Pooling equilibria and their conditions.

BNE	Pooling BNE Profile	Conditions
PE1	$((X,X),(T,T), p = \theta \leq F_1, q \leq F_2)$	$C_{AXT} \geq C_{AYT}, C_{NAXT} \geq C_{NAYT}$
PE2	$((X,X),(T,N), p = \theta \leq F_1, q \geq F_2)$	$C_{AXT} \geq C_{AYN}, C_{NAXT} \geq C_{NAYN}$
PE3	$((X,X),(N,T), p = \theta \geq F_1, q \leq F_2)$	$C_{AXN} \geq C_{AYT}, C_{NAXN} \geq C_{NAYT}$
PE4	$((X,X),(N,N), p = \theta \geq F_1, q \geq F_2)$	$C_{AXN} \geq C_{AYN}, C_{NAXN} \geq C_{NAYN}$
PE5	$((Y,Y),(T,T), p \leq F_2, q = \theta \leq F_2)$	$C_{AXT} \leq C_{AYT}, C_{NAXT} \leq C_{NAYT}$
PE6	$((Y,Y),(T,N), p \leq F_2, q = \theta \geq F_1)$	$C_{AXT} \leq C_{AYN}, C_{NAXT} \leq C_{NAYN}$
PE7	$((Y,Y),(N,T), p \geq F_2, q = \theta \leq F_1)$	$C_{AXN} \leq C_{AYT}, C_{NAXN} \leq C_{NAYT}$
PE8	$((Y,Y),(N,N), p \geq F_2, q = \theta \geq F_1)$	$C_{AXN} \leq C_{AYN}, C_{NAXN} \leq C_{NAYN}$
$F_1 := \frac{D_{NAXT}-D_{NAXN}}{D_{AXN}-D_{NAXN}+D_{NAXT}-D_{AXT}}, F_2 := \frac{D_{NAYT}-D_{NAYN}}{D_{A2N}-D_{NAYN}+D_{NAYT}-D_{AYT}}$		

5   Conclusion

In this paper, we have investigated a client's security concern where he outsources his computational tasks to a service provider. In this scenario, the SP might be compromised by different types of attackers. Consequently, the client involves in a dilemma whether to trust the received results from the service provider or not. To address this problem we have modeled the interactions between the service provider and the attacker as well as the relation between the client and the provider by two different game models. We have first defined and analyzed a normal form game for the interaction between the service provider and an attacker. Then, we employed the obtained NE of the first game to define the prior probability of the second game (i.e., the prior probability of the Nature), which is a signaling game. We have analyzed this extended signaling game and determined potential equilibria.

For future work, we will evaluate the mixed strategy BNE in the extended signaling game. We can also examine how updating the client's beliefs can affect his decision when the signaling game becomes repeated. Furthermore, we plan to investigate our game model using real case studies.

Acknowledgement. The authors would like to thank Amin Mohammadi for his constructive feedback and insightful suggestions on the primary version of the proposed model and notations.

References

1. Bhadauria, R., Sanyal, S.: Survey on Security Issues in Cloud Computing and Associated Mitigation Techniques (2012, arXiv preprint). arXiv:1204.0764
2. Chen, D., Zhao, H.: Data security and privacy protection issues in cloud computing. In: International Conference on Computer Science and Electronics Engineering (ICCSEE), pp. 647–651 (2012)
3. Toosi, A.N., Calheiros, R.N., Buyya, R.: Interconnected cloud computing environments: challenges, taxonomy, and survey. ACM Comput. Surv. (CSUR) **47**(1), 7–53 (2014)
4. Chen, Y.R., Tian, L.Q., Yang, Y.: Model and analysis of user behavior based on dynamic game theory in cloud computing. Dianzi Xuebao Acta Electronica Sinica **39**(8), 1818–1823 (2011)
5. Kamhoua, C., Kwiat, L., Kwiat, K., Park, J. S., Zhao, M., Rodriguez, M.: Game theoretic modeling of security and interdependency in a public cloud. In: IEEE 7th International Conference on Cloud Computing (CLOUD), pp. 514–521 (2014)
6. Gulbhile, A.S., Patil, M.P., Pawar, P.P., Mahajan, S.V., Barve, S.: Secure radio resource management in cloud computing based cognitive radio networks. Int. J. Sci. Technol. Res. **3**(5), 136–139 (2014)
7. Rajesh, V., Gnanasekar, J.M., Ponmagal, R.S., Anbalagan, P.: Integration of wireless sensor network with cloud. In: 2010 International Conference on Recent Trends in Information, Telecommunication and Computing, pp. 321–323 (2010)
8. Nix, R., Kantarcioglu, M.: Contractual agreement design for enforcing honesty in cloud outsourcing. In: Walrand, J., Grossklags, J. (eds.) GameSec 2012. LNCS, vol. 7638, pp. 296–308. Springer, Heidelberg (2012)
9. Pham, V., Khouzani, M.H.R., Cid, C.: Optimal Contracts for outsourced computation. In: Poovendran, R., Saad, W. (eds.) GameSec 2014. LNCS, vol. 8840, pp. 79–98. Springer, Heidelberg (2014)
10. Shoham, Y., Leyton-Brown, K.: Multiagent Systems: Algorithmic, Game-Theoretic, and Logical Foundations. Cambridge University Press, Cambridge (2008)

Game Theory and Security: Recent History and Future Directions

Jonathan S.A. Merlevede$^{(\boxtimes)}$ and Tom Holvoet

University of Leuven, Leuven, Belgium
{jonathan.s.a.merlevede,tom.holvoet}@cs.kuleuven.be

Abstract. Until twenty years ago, the application of game theory (GT) was mostly limited to toy examples. Today, as a result of major techno-logical and algorithmic advances, researchers use game-theoretical mod-els to motivate complex security decisions relating to real-life security problems. This requires models that are an accurate reflection of reality. This paper presents a biased bird's-eye view of the security-related GT research of the past decade. It presents this research as a move towards increasingly accurate and comprehensive models. We discuss the need for adversarial modeling as well as the internalization of externalities due to security interdependencies. Finally, we identify three promising directions for future research: relaxing common game-theoretical assump-tions, creating models that model interdependencies as well as a strategic adversary and modelling interdependencies between attackers.

1 Introduction

Ever since its inception by von Neumann and Morgenstern in 1944 [33], GT has been recognized as a tool with great potential for motivating security related decisions. While initial studies were by necessity restricted to highly stylized representations of e.g. warfare and disarmament inspections [4], the advent of (powerful) computers and computer networks has enabled us to examine more complex models as well as a plethora of new applications. Research at the inter-section of computer science (CS), security and GT is addressing previously non-existent security problems in cyberspace, and is leveraging the computational power of modern technology to apply GT concepts to security problems with an ever-increasing number of players, attacker types and strategy space sizes.

This process of increasing sophistication and applicability is one that all mathematized sciences go through [33]. A glance at recent literature shows that GT has mostly outgrown the stage where (elementary) applications only serve to corroborate its foundations. Current research seeks to apply GT to complicated situations that occur in real life, guiding decision making in a principled manner [43]. In the field of security, where decisions are still often made in a rather ad-hoc fashion and the amount of information and variables to take into consideration is becoming too much for domain experts to handle, such a formal guide can be of immense value [1].

© Springer International Publishing Switzerland 2015
MHR Khouzani et al. (Eds.): GameSec 2015, LNCS 9406, pp. 334–345, 2015.
DOI: 10.1007/978-3-319-25594-1_19

Correctly assessing model adequacy has proved itself a difficult task. Because of the quirks of human behavior, it is unlikely that we will ever be able to fully close the gap between model and reality of security-related interactions. Identifying useful abstractions is an ongoing effort that has captivated GT researchers for decades, and that now receives more and more attention of the security community. Where security-related GT research used to focus on the development of scalable algorithms and exploring the applicability of classic concepts from GT to novel problems, contemporary research increasingly covers ideas from behavioral GT and improvements to standard models.

This paper presents an opinionated view of the history of and motivation for using GT to make security decisions. It presents 'single-player' decision-theoretical models (Sect. 2) as the ancestor of current GT decision making. We divide and discuss GT decision models in three categories based on the number and nature of participating players. The first category contains attacker-defender games, which are games between a single malicious and a single benevolent player (Sect. 3). The second category contains interdependent games, which involve two or more non-malicious entities whose decisions influence each other's utility (Sect. 4). The third category contains all other security-related games (Sect. 5). We then discuss three interesting research directions that will lead to models of security problems as complex interdependent games with multiple behavioral attackers and defenders (Sect. 6).

Contribution. There already exist excellent recent review papers on the subject of security and GT. Manshaei et al. [31] provide a structured and comprehensive overview of game-theoretical models of security and privacy in communication networks, but do not really discuss security interdependencies. Laszka, Felegyhazi, and Buttyán [26] complements this work by giving a comprehensive overview of interdependent (investment) security games. Both papers offer significant technical detail. This paper is more introductory, is not comprehensive and does not offer comparable detail. Our main objective to provide the reader with a manageable high-level overview of the GT models that are being used to model security problems and to motivate their use. We focus on how models are changing, and seek to convince the reader of the importance of modeling interdependencies.

Remarks

– The way in which we present material is biased; non-game-theoretical models are hardly something of the past, and we do not know the future. Presenting modern risk management strategies as ancestral to attacker-defender games, some of which are over half a century old, is historically incorrect. Instead of being an objective or comprehensive reference work, we aim to present the recent history of game-theoretical security-related research in a way that should spark discussion and that may influence how its reader looks at past, current and future research.

– This paper focuses on cybersecurity, which is gaining more and more attention from policymakers, the general public and the scientific community.[1] However, the concepts presented are not limited to this application.

2 The Decision-Theoretical Approach

Economists have a long history of studying risk and reward, generally using decision-theoretic techniques that allow economic entities to maximize their expected gain. Per the law of the instrument, they often model the threat posed by external parties as an 'operational risk' [7]. This approach fails to incorporate the effect of the entity's policy on the behavior of the external parties. In game-theoretical terms, firms are playing a one-player game, and the external party is modelled as part of the game rules.

It is important to realize that non-game-theoretical models are still used and studied today, maybe even more so than game-theoretical approaches [9]. One popular application of single-player models seems to be the study of security investment decisions [18]. One-player models are also still used for motivating decisions that are more convincingly strategical in nature than security investment, such as when to deploy cyber resources like zero-day exploits [5].

3 Attacker-Defender Games

In contrast to decision-theoretical models, attacker-defender games (ADGs) explicitly include malevolent parties as players, able to make strategic decisions. This section contains a short motivation for this approach (Sect. 3.1), followed by an enumeration of some of the most popular ADG models (Sect. 3.2).

3.1 Motivation

The basic argument for ADGs is that attackers are people or organizations. They are capable of making strategic decisions in real life, which makes it inappropriate to model them as an unchanging presence. Regardless of the true strategic nature of an opponent, assuming random and constant attacker behavior might still be a reasonable assumption if nothing is known about how the attacker operates or what her preferences are. In most security-related interactions, this is not the case.

---

[1] The 2015 version of Panama Institute's yearly cost of data breach study shows (•) an increased number of data breaches resulting from attacks by malicious attackers (47 % versus 37 % in 2013), (•) a 23 % increase in total cost of data breach since 2013, and (•) shows that attacks have increased in frequency as well as in the cost to remediate the consequences [12,13]. ENISA, the European Union Agency for Network and Information Security, sees a 25 % increase in the number of data breaches in 2014 compared to 2013 and refers to 2014 as "the year of the data breach' [16]. They list nearly all cyber threats, such as denial of service attacks and cyber espionage, as increasing.

Firstly, cyberattackers are, in an important sense, becoming predictable [11]. In the past 15 years, there has been a shift from 'hacking for fame' and 'hacking for fun' to profit-driven attacks [28]. The cyber-crime economy has transformed from an unorganized reputation-based economy to an organized cash economy [17,32]. This new focus on financial gain is evidenced by increased financial damages associated with a successful cyberattack [13].

Secondly, many attackers seem to act approximately rational, even those that are not only motivated by financial gain. An example of this is given by Tambe [43], who convincingly argues that even terrorist groups such as Al-Qaeda are (approximately) rational in the game-theoretical sense (i.e. they act as if optimizing expected utility).

Thirdly, attacker rationality and our knowledge of attacker preferences do not have to be perfect for a GT analysis to be useful. Many GT models allow us to express uncertainty over the attackers' type. Even if we can give only a rough estimate of attacker preferences, stochastic game-theoretical models can be an improvement over one-player games. Concepts from behavioural GT can allow us to account for attacker irrationality.

3.2 Models and Applications

The class of ADGs comprises all two-player games between an attacker and a defender. The goal of the attacker varies but is always malicious; examples include preventing the legitimate use of a computer network and the successful execution of a terrorist attack. The goal of the defender is simply to minimize her expected damages by taking defensive measures. The single attacker and defender in ADGs do not have to represent individual people, and in fact typically represent multiple entities [1]. A wide variety of game models and solution concepts are employed, although the simultaneous-move game and Nash equilibrium (NE) remain very popular. While the ADG 'framework' might seem rather restrictive, a large fraction of the games considered in security-related GT falls within this category. The ADG model is being used to model a large variety of security-related topics including arms control, (anti-)jamming games, botnet defense, intrusion detection, (security) auditing and advanced persistent threats [14].

Stackelberg Security Games (SSGs). One very actively studied class of games within the ADG category that does not use the NE is the class of so-called security games or SSGs [24,43].[2] In SSGs, the defender has a fixed number of security resources (e.g. police patrols or blockades) that she needs to deploy to a (larger) number of targets. The defender's strategy consists of assigning resources to schedules, which are subsets of targets. The attacker observes the defender's strategy and then attacks a single target. There are always unprotected targets, but the attacker does not know which ones, because she only

---

[2] We think this is a confusing name, because there are a lot of security-related games that are not security games.

observes the defender's (stochastic) strategy, not her action. Both attacker and defender have two utilities for each target – one for the case where the target is attacked, and one where it is not.

More so than any other class of game-theoretical models, SSGs have repeatedly demonstrated their usefulness in real-world applications [43]. An early example of a deployed system is ARMOR, a software assistant agent that allocates checkpoints to the roadways entering the Los Angeles International Airport and canine patrols within the airport terminals to minimize the probability of a successful terrorist attack [37,38]. Since then, many more applications have been explored, including the allocation of air marshals to flights, scheduling inspection checkpoints on the roads of a large city and wildlife protection. Many of these have actually been deployed.

References. For a recent survey of network security and privacy ADGs we refer to Manshaei et al. [31], who classify the literature into six categories based on the problem that they model. Roy et al. [42] give a more concise overview of network security-related work; they classify games according to the underlying game model (cooperative, non-cooperative, static, . . .), which is perhaps shallow but nevertheless interesting. Alpcan and Başar [1] is a good book on GT and network security. We could not find a recent, comprehensive survey on the work on SSGs, although the book by Prof. Tambe [43] and the Teamcore website[3] are a good place to start.

4   Interdependent Games

Interdependent security gamess (ISGs) are games that explicitly model multiple players in the defending role. This approach has the advantage that it can include interdependencies between the decisions of multiple defenders.

4.1   Motivation

In real life, decisions are not made in a vacuum and can have influences that extend beyond the decision maker herself. Indeed, the difference between ADGs and single-player games is that they account for the strategic interdependency that exists between attacker and defender. This section shows that interdependencies also exist between defenders.

Central in this discussion is the concept of externalities [26], which result from interdependencies. In the context of game theory and security, the term externality refers to the impact that the decision of an individual or firm has on other actors, and for which there is no compensation. There exist both negative externalities (actions that harm the others, such as pollution) and positive externalities (actions that benefit the others). Externalities can sometimes be internalized by compensating for them (e. g. by making polluters liable for the harm that they inflict on others).

---

[3] http://teamcore.usc.edu

Connectivity. The connected nature of computerized systems is an important cause for interdependencies. The way in which networked systems are often set up has the effect that if an attacker is able to breach one defender, she is able to easily access or harm other connected defenders. Consequently, an investment in security by one defender increases the security of other defenders as a positive externality. Consider the following examples:

- Some viruses spread through email. As the people surrounding you become more conscious about security, the likelihood of you receiving and being infected by the virus decreases regardless of your own strategy.
- Computers infected by a virus often become part of a botnet that inflicts harm on third parties by sending spam or executing DDoS attacks. Removing the virus causes the attacks on the third parties to stop.
- Some of the material that people share with you on social networks may be considered private. Activating two-factor authentication decreases the chance of your account being compromised, decreasing the chance of someone violating your friends' privacy.

Logical Interdependencies. There exists a logical interdependence between otherwise unrelated systems because they run the same software [36]. One effect of this interdependency can be that using popular software makes you a popular target for attackers [30] (negative externality). Logical interdependencies also cause security risk correlation, because exploits targeting popular software can affect millions of computers (see e.g. Heartbleed[4] and Shellshock/Bashdoor). This can be important when modeling security insurance. Correlation of risk has been identified as one of the main causes of the unfortunate state of the security insurance market [8].

Strategic Interdependence. Attackers often do not just pick their targets randomly (Sect. 3). If one firm increases its security level to make it more resistant to attacks, this can cause other firms to be targeted instead (negative externality).

Poorly Aligned Incentives. It is well-known that positive externalities generally leads to social inefficiency (underinvestment) and free-riding (players avoid investing, expecting others to protect them). However, the case for cybersecurity is especially bad because the externalities are often large. The actors who are in the best position to prevent risk, are often not the ones who are liable or under attack [3].

Poorly aligned incentives are exemplified by ingress filtering. Ingress filtering is a denial-of-service (DoS) prevention technique where attack traffic gets discarded by routers close to the perpetrators, such as a router controlled by the perpetrator's ISP. Here, an investment by the ISP in filtering technology is of limited benefit to the ISP itself, but of great benefit to the target of the DoS

---

[4] http://heartbleed.com

attack. We say that the incentives of the victim and the ISP are 'poorly aligned'. It is a well-known result that poorly aligned incentives among defenders are at the root of security issues at least as often as technical factors [2,3]. The detriments associated with them are often exacerbated by information asymmetry.

Noncyber. Note that although the effect of externalities in cyberspace is larger than in most other domains, there also exist security externalities outside of cyberspace. E.g., Kunreuther and Heal [25] pointed out the interdependencies that used to exist between airports, because there was no time to check the bags of passengers transferring flights. Ayres and Levitt [6] provide another example by showing that unobservable precautions provide positive externalities, since criminals cannot determine a priori who is protected. In their investigation this precaution took the form of the Lojack, a hidden radio transmitter used for retrieving stolen vehicles.

4.2   Models

We consider interdependent security games to be games in which there are multiple selfish but nonmalicious players that can protect themselves from attack and whose decisions do not influence only their own utility, but also the utility of other players. These influences should be indirect, so we do not consider games where the players are adversarial to be interdependent games. This definition is more general than the common definition in literature, which is usually specific to security investment and risk minimization [25,26]. We refer to the latter as interdependent investment security games.

Interdependent investment security games are relatively new. They were introduced by economists Kunreuther and Heal [25] in 2003. In this work, Kunreuther and Heal show that the positive externalities associated with security investment generally lead to an underinvestment in security due to free riding. Similar findings are made by Varian [44], who argues that security should be treated as a common good.

An important aspect that is missing from the model of Kunreuther and Heal [25] and derivative models is that of the source of the risk. This has been observed by a few researchers. Chan, Ceyko, and Ortiz [10] introduce interdependent defense (IDD) games as an adaptation of Kunreuther's model that does account for the deliberate nature of attacker's actions, but they treat security investment as a binary choice that leads to perfect defense. Lou, Smith, and Vorobeychik [29] present a more general model, in which each defender has to protect multiple resources instead of just one resource that can be perfectly protected.

References. Laszka, Felegyhazi, and Buttyán [26] present a comprehensive survey of interdependent investment security games. They restrict themselves to those games where defenders' strategic decisions are related to security investments and risk minimization.

5   Other Games

Although many security-related games fit in one of the two categories discussed in the previous sections, a large body of security-related literature does not. We do not have the space for a comprehensive discussion, so instead make just a few observations.

- Some games do not consider parties to be malicious or benevolent, but instead consider a number of 'neutral' strategic entities that can take offensive or other actions based on whichever one benefits them most. E.g., Johnson, Laszka, and Grossklags [21] consider a two-player game between Bitcoin mining pools of varying size that have to make a binary choice between attacking the other party or investing in increased computing power.
- There is a substantial body of work considering self-organizing wireless ad-hoc networks, such as wireless sensor networks, mobile ad-hoc networks and vehicular ad-hoc networks (VANETs). There, every node is usually considered to be a player. The work in this area is often security-related because of the specifics of the scenario being modelled, e.g. because the purpose of the VANET is to communicate safety-related information.[5]
- Some games do model multiple attackers. However, in the absence of interdependencies between attackers these games generally reduce to ADGs. E.g., Yang et al. [46] model poaching using a single defender and a population of poachers. Poachers respond to the rangers' patrolling strategy independently; potential collaboration between poachers is explicitly not considered. The strategic considerations of the defender are therefore equivalent to those in a Bayesian SSG. As another example, Laszka, Johnson, and Grossklags [27] consider defense against covert cyberattacks by a targeted attacker and a large group of non-targeted attackers. Because every non-targeted attacker attacks a large number of targets with negligible probability, the influence of a single defender's strategy on the strategy of such an attacker is negligible. Their game reduces to an ADG played between the defender and the targetted attacker.
- Interesting work is being done at the intersection of cryptography and GT. One stream of research studies applications of cryptographic methods to GT; specifically, how secure cryptographic multi-party computation methods can eliminate the need for a trusted mediator in game-theoretical mechanism design [15,23]. Another stream of research studies how the game-theoretic concept or rationality and selfishness could replace or add to the assumption of benevolence or malevolence in in cryptographic protocols for secret sharing and multi-party computation [15,20,23]. Because of impossibility results or because the classical GT concepts are not always a good fit for the cryptographic setting, some new GT concepts such as the computationally robust (Nash) equilibrium [19] and the rational foresighted player (for secret sharing) [35] have been proposed. These new concepts may also prove useful for applications of GT that are not cryptography-related.

---

[5] These games do not fit our definition of interdependent game because there is no attack or attacker.

6 Future Directions

The previous sections have motivated why we should model attacks as originating from a strategic adversary (Sect. 3) and why we should consider interdependency effects between non-malicious entities (Sect. 4). This section identifies three promising directions for future research.

Relaxing Assumptions. Applying GT to real-life problems requires relaxing some of the assumptions that are common in game-theoretical models. There is already a lot of work being done in this area in the context of SSGs. E.g., the SSG model has been extended to account for imperfect execution and observation [49], to account for interdependencies and coordination between assets [45] and to be robust to deviations from optimal attacker behavior [39, 40]. Novel models try to explicitly account for the irrationality of attackers, e. g. by relaxing the expected utility assumption by accounting for risk aversion [41] or by using concepts from behavioral GT such as prospect theory [47, 48], the quantal response equilibrium [34, 47] and subjective expected utility [34]. There are even new human behavior models being proposed [22].

ISGs with Strategic Adversaries. Curiously, while the community surrounding ADGs advocates the importance of strategical adversaries, this concept is generally ignored in the context of ISGs. Conversely, ISGs take into account externalities, which are ignored by ADGs. We think that models such as those of Chan, Ceyko, and Ortiz [10] and Lou, Smith, and Vorobeychik [29] that combine ISGs with strategical adversaries could be a great improvement over both ISGs and ADGs. The current absence of such models may be due to the fact that ISGs are often studed in a more economical context, whereas ADGs are more popular in CS, even though both approaches are often used to model similar applications.

Interdependent Attackers. The success of a cyberattack usually depends not on the efforts of a single attacker but instead on the efforts of multiple parties. ADGs represent all of these parties as a single player. As is the case for the defending player, such an abstraction is only appropriate when the incentives of the attacking parties are perfectly aligned.

A classic motivation for defining a single abstract 'attacker' instead of multiple interdependent attackers, is that defenders have only vague, limited knowledge of attackers' preferences [1]. Ironically, the same argument was used against ADGs in favor of the single-player approach (Sect. 2). In our opinion ADG models are often used as a golden hammer even when enough information is available to warrant a more detailed approach.

For the case of cybersecurity, there is evidence that cybercrime has organized itself in criminal networks in which individual malefactors take on highly specialized roles [32]. Consider exploit kits like Blackhole, software that booby-traps hacked Web sites to serve malware and that is updated on a regular basis to abuse new vulnerabilities in browsers. Parties that want to construct a botnet or perform some other malicious activity can buy or rent such an exploit kit, which saves them the effort from looking for vulnerabilities themselves. The vendors of exploit kits in turn usually do not look for vulnerabilities themselves, but buy them from non-malicious but amoral hackers who sell them to the highest bidder.

The rich structure of attacker's markets seems hard to reconcile with the omnipresence of the ADG model, because valuable information and decision strategies are lost. By modeling the structure of criminal interaction, we can study the efficacy of strategies such as outbidding exploit kit vendors. If we also consider multiple defenders, we can model cooperation to take out the common adversary, the exploit vendor.

7 Conclusion

This paper has argued that the focus of current security-related GT research is on producing models that reflect reality more accurately, so that they can better guide us in making security-related decisions. Through a birds-eye view of current literature, it has shown that we do not shy away from increased model complexity when attempting to model different kinds of uncertainty, the quirks of human psychology or other aspects of real-life interaction that classical GT does not accurately reflect. We have seen that games with multiple strategic defenders and one or more strategic attacker have received almost no attention, and have contrasted this with evidence that shows that interdependencies between defenders and to a lesser extent attackers are necessary to explain the current state of security and cybersecurity.

Acknowledgements. This research is partially funded by the Research Fund KU Leuven.

References

1. Alpcan, T., Başar, T., et al.: Network Security: A Decision and Game-Theoretic Approach. Cambridge University Press, New York (2010)
2. Anderson, R.J.: Why information security is hard - an economic perspective. In: Proceedings of the ACSAC (2001)
3. Anderson, R.J., Moore, T., et al.: The economics of information security. Science **314**, 610–613 (2006)
4. Avenhaus, R., Canty, M.J., et al.: Inspection games. In: Meyers, R.A. (ed.) Computational Complexity, pp. 1605–1618. Springer, New York (2012)
5. Axelrod, R., Iliev, R., et al.: Timing of cyber conflict. Proc. Natl. Acad. Sci. U.S.A **111**, 1298–1303 (2014)
6. Ayres, I., Levitt, S.D., et al.: Measuring Positive Externalities from Unobservable Victim Precaution: An Empirical Analysis of Lojack. Working Paper. National Bureau of Economic Research, Cambridge (1997)
7. Böhme, R., Nowey, T.: Economic security metrics. In: Eusgeld, I., Freiling, F.C., Reussner, R. (eds.) Dependability Metrics. LNCS, vol. 4909, pp. 176–187. Springer, Heidelberg (2008)

8. Böhme, R., Schwartz, G., et al.: Modeling cyber-insurance: towards a unifying framework, June 2010
9. Cavusoglu, H., Raghunathan, S., et al.: Decision-theoretic and game-theoretic approaches to IT security investment. J. Manag. Inf. Syst. **25**, 281–304 (2008)
10. Chan, H., Ceyko, M., et al.: Interdependent defense games: Modeling interdependent security under deliberate attacks. ArXiv Prepr. http://www.ArXiv12104838 (2012)
11. Christin, N.: Network security games: combining game theory, behavioral economics, and network measurements. In: Baras, J.S., Katz, J., Altman, E. (eds.) GameSec 2011. LNCS, vol. 7037, pp. 4–6. Springer, Heidelberg (2011)
12. Cost of Data Breach Study: Global Analysis. Ponemon Institute (2013)
13. Cost of Data Breach Study: Global Analysis. Ponemon Institute (2015)
14. van Dijk, M., Juels, A., et al.: FlipIt: The game of "stealthy takeover". J. Cryptol. **26**(4), 655–713 (2012)
15. Dodis, Y., Rabin, T., et al.: Cryptography and game theory. In: Algorithmic Game Theory (2007)
16. ENISA Threat Landscape 2014 - Overview of current and emerging cyber-threats. Report/Study. ENISA (2015)
17. Franklin, J., Perrig, A., et al.: An inquiry into the nature and causes of the wealth of internet miscreants. In: Proceedings of the ACM CCS (2007)
18. Gordon, L.A., Loeb, M.P., et al.: The economics of information security investment. ACM Trans. Inf. Syst. Secur. **5**, 438–457 (2002)
19. Halpern, J.Y., Pass, R., et al.: Game theory with costly computation. In: Innovations in Computer Science, August 2010
20. Halpern, J., Teague, V., et al.: Rational secret sharing and multiparty computation: extended abstract. In: Proceedings of the ACM STOC (2004)
21. Johnson, B., Laszka, A., Grossklags, J., Vasek, M., Moore, T.: Game-theoretic analysisof DDoS attacks against bitcoin mining pools. In: Böhme, R., Brenner, M., Moore, T., Smith, M. (eds.) FC 2014 Workshops. LNCS, vol. 8438, pp. 72–86. Springer, Heidelberg (2014)
22. Kar, D., Fang, F., et al.: "A game of thrones": when human behavior models compete in repeated stackelberg security games. In: Proceedings of the AAMAS (2015)
23. Katz, J.: Bridging game theory and cryptography: recent results and future directions. In: Canetti, R. (ed.) TCC 2008. LNCS, vol. 4948, pp. 251–272. Springer, Heidelberg (2008)
24. Kiekintveld, C., Jain, M., et al.: Computing optimal randomized resource allocations for massive security games. In: Proceedings of the AAMAS (2009)
25. Kunreuther, H., Heal, G., et al.: Interdependent security. J. Risk Uncertain. **26**, 231–249 (2003)
26. Laszka, A., Felegyhazi, M., et al.: A survey of interdependent security games. ACM Comput. Surv. CSUR 47 (2014)
27. Laszka, A., Johnson, B., Grossklags, J.: Mitigation of targeted and non-targeted covert attacks as a timing game. In: Das, S.K., Nita-Rotaru, C., Kantarcioglu, M. (eds.) GameSec 2013. LNCS, vol. 8252, pp. 175–191. Springer, Heidelberg (2013)
28. Leeson, P.T., Coyne, C.J., et al.: The Economics of Computer Hacking. SSRN Scholarly Paper, Social Science Research Network (2005)
29. Lou, J., Smith, A.M., et al.: Multidefender Security Games (2015)

30. Maillé, P., Tuffin, B., Reichl, P.: Interplay between security providers, consumers, and attackers: a weighted congestion game approach. In: Altman, E., Katz, J., Baras, J.S. (eds.) GameSec 2011. LNCS, vol. 7037, pp. 67–86. Springer, Heidelberg (2011)

31. Manshaei, M., Zhu, Q., et al.: Game theory meets network security and privacy. ACM Comput. Surv. CSUR 45, 25:1–25:39 (2013)

32. Moore, T., Clayton, R., et al.: The economics of online crime. J. Econ. Perspect. 23, 3–20 (2009)

33. Neumann, J.V., Morgenstern, O., et al.: Theory of Games and Economic Behavior. Princeton University Press, Princeton (1944)

34. Nguyen, T.H., Yang, R., et al.: Analyzing the effectiveness of adversary modeling in security games. In: Proceedings of the AAAI (2013)

35. Nojoumian, M., Stinson, D.R.: Socio-rational secret sharing as a new direction in rational cryptography. In: Walrand, J., Grossklags, J. (eds.) GameSec 2012. LNCS, vol. 7638, pp. 18–37. Springer, Heidelberg (2012)

36. Ogut, H., Menon, N., et al.: Cyber insurance and IT security investment: impact of interdependent risk. In: WEIS (2005)

37. Pita, J., Jain, M., et al.: Deployed ARMOR protection: the application of a game theoretic model for security at the los angeles international airport. In: Proceedings of the AAMAS (2007)

38. Pita, J., Jain, M., et al.: Los angeles airport security. AI Mag. 30, 43–57 (2009)

39. Pita, J., Jain, M., et al.: Robust solutions to stackelberg games: addressing bounded rationality and limited observations in human cognition. Artif. Intell. 174, 1142–1171 (2010)

40. Pita, J., John, R., et al.: A robust approach to addressing human adversaries in security games. In: Proceedings of the AAMAS (2012)

41. Qian, Y., Haskell, W.B., et al.: Robust strategy against unknown risk-averse attackers in security games. In: Proceedings of the AAMAS (2015)

42. Roy, S., Ellis, C., et al.: A survey of game theory as applied to network security. In: Proceedings of the 43rd Hawaii International Conference on System Sciences (2010)

43. Tambe, M.: Security and Game Theory: Algorithms, Deployed Systems, Lessons Learned. Cambridge University Press, New York (2011)

44. Varian, H.: System reliability and free riding. In: Camp, L.J., Lewis, S. (eds.) Economics of Information Security, pp. 1–15. Springer, Heidelberg (2004)

45. Vorobeychik, Y., Letchford, J., et al.: Securing interdependent assets. Auton. Agent. Multi-Agent Syst. 29(2), 305–333 (2014)

46. Yang, R., Ford, B., et al.: Adaptive resource allocation for wildlife protection against illegal poachers. In: Proceedings of the AAMAS (2014)

47. Yang, R., Kiekintveld, C., et al.: Improving resource allocation strategies against human adversaries in security games: an extended study. Artif. Intell. 195, 440–469 (2013)

48. Yang, R., Kiekintveld, C., et al.: Improving resource allocation strategy against human adversaries in security games. In: Proceedings of the IJCAI (2011)

49. Yin, Z., Jain, M., et al.: Risk-averse strategies for security games with execution and observational uncertainty. In: Proceedings of the AAAI, April 2011

Uncertainty in Games:
Using Probability-Distributions as Payoffs

Stefan Rass[1]([✉]), Sandra König[2], and Stefan Schauer[2]

[1] Institute of Applied Informatics, System Security Group,
Universität Klagenfurt, Klagenfurt, Austria
stefan.rass@aau.at
[2] Digital Safety & Security Department,
Austrian Institute of Technology GmbH, Klagenfurt, Austria
{sandra.koenig,stefan.schauer}@ait.ac.at

Abstract. Many decision problems ask for optimal behaviour in (often competitive) situations, where optimality is understood as maximal revenue. The axiomatic approach of von Neumann and Morgenstern establishes the existence of suitable revenue functions, assuming an ordered revenue space. A prominent materialization of this is game theory, where utility functions map actions of several players onto comparable payoffs, typically real numbers. Inspired by an application of that theory to risk management in utility networks, we observed that the usual game-theoretic models are inapplicable due to intrinsic randomness of the effects that an action has. This uncertainty comes from physical and environmental factors that affect the game-play outside of any players influence. To tackle such scenarios, we introduce games in which the payoffs are entire probability distributions (rather than numbers). Towards a sound decision theory, we define a total ordering on a restricted subset of probability distribution functions, and demonstrate how optimal decisions and even basic game theory can be (re)established over abstract revenue spaces of probability distributions. Our results belong to the category of risk control, and are applicable to contemporary security risk management, where decisions must be made under uncertainty and the effects of management actions are almost never deterministic.

1 Introduction

Traditional decision and game theory quantifies decisions in terms of scalar-valued utility functions that admit a sound definition of optimality for a decision. Following the usual axiomatic approach (see [5, Sect. 2.2] for example), one starts from a (total) ordering on the space of rewards and extends this ordering to the space of probability distributions that represent (randomized) actions. Practically, this extension usually boils down to defining a mapping from the action space into a totally ordered set (typically a subset of $\mathbb{R}$), in which different actions can be weighed against each other on grounds of their expected revenue. This approach implicitly assumes that revenues, hereafter called payoffs, are crisp and certain. Reality, however, is neither crisp nor certain in most practically

© Springer International Publishing Switzerland 2015
MHR Khouzani et al. (Eds.): GameSec 2015, LNCS 9406, pp. 346–357, 2015.
DOI: 10.1007/978-3-319-25594-1_20

interesting cases. Uncertainty can occur in various ways, such as in the player's moves or in the revenues themselves. Imprecise playing is captured by concepts like the trembling-hands equilibrium and related. A straightforward mean to handle uncertain payoffs is by taking their average (first moment), taken over the payoff distribution function, which restores things back to the standard setting where everything is crisp. In general, casting uncertainty into crisp values seems unavoidable, since the entire existence of utility functions (upon which games are defined) rests on a totally ordered revenue space, whereas the space of probability distributions cannot be ordered. Thus, the axiomatic approach of von Neumann and Morgenstern [3] (and hence all that builds upon it) would fail when the payoffs are probability distributions.

In this work, we show how we can nevertheless define games in which payoffs *are* probability distribution functions, by imposing some mild restrictions that admit the construction of a total ordering on a (useful) subset of probability distributions. In this way, we can define games that fully account for uncertain payoffs without simplifying random variables into scalars.

In brief, we achieve the following: (1) we construct a total ordering on a subset of the space of payoff probability distributions, (2) we show how this ordering can be decided algorithmically, and (3) we transfer solution concepts like the Nash equilibrium to this new setting, thus making the entirety of game-theory applicable to cases where the revenues are intrinsically random rather than crisp.

This work is inspired by an attempt to apply game-theory to risk management in large utility or SCADA (supervisory control and data acquisition) networks. As an example, take a water supply infrastructure: such an infrastructure is composed from pipes (the utility network), with an additional control layer on top of the utility network (the SCADA network) that manages and monitors the supply. Given the intrinsic complexity and influence on environmental factors, a utility infrastructure will never react entirely deterministic on any change to its configuration parameters. That is, anything that a risk control strategy would prescribe towards optimal utility network provisioning would result in somewhat random effects. Thus, the problem of risk management essentially is a control problem over a system with partially random dynamics (where determinism arises only to the extent where effects obey known laws and models of physics).

Such uncertainty is an intrinsic property of many other kinds of infrastructures too (like power supply that depends on the momentarily consumption by private households, gas supply, etc.). The problem is particularly relevant in matters of social risk response, where risk communication as part of a risk management strategy is done towards "controlling" the public opinion and trust. Here, socio-economic effects that may be much less understood than physical dynamics determine the effect (payoff) of an action, which turns the game-theoretic models inevitably into ones that assign random payoffs to their participants.

Somewhat surprisingly, the existing notions and variations of games, including trembling hands equilibria, disturbed games, stochastic games, etc. seemingly do not consider the full spectrum of possible uncertainty being extended to the payoffs. As a typical example of games with imperfect information, let us look at Bayesian games: these consider a classical notion of uncertainty where the player is unaware about the competitor's decisions and/or the current state of the game. The resolution is based on each player working with beliefs that can be updated (in a Bayesian fashion) over repetitions of the game; and the equilibrium would take the beliefs into account. Although such beliefs are more than common in security modeling, their role is entirely different there in the sense of concerning the effects of an action, rather than being a hypothesis about the payoffs or the opponent's moves. In the context of risk management, the payoff structure is determined by the infrastructure that we seek to protect, and hypotheses about the attacker's actions are exactly the presumed action space of the opponent (whereas the actual move is usually unobservable, as its consequence is mostly noticeably after the damage has happened). Thus, Bayesian games (or more generally, games with imperfect information), do not accurately capture uncertain outcomes of an action as we investigate, as the uncertainty in these known cases comes from different ways of playing the game. Our use-case, however, is actions whose consequences are intrinsically uncertain, and especially not drawn from a number of known possible cases that nature selects at random.

The need to consider actions with random outcomes is further substantiated by an inspection of the foundations of game-theoretic decision-making, which requires a total ordering on the revenues that would no longer be available. Therefore, and to the best of our knowledge, this work appears novel relative to the existing decision-theoretic literature, in the sense of presenting a so far widely unexplored new approach to working with uncertainty. Our vehicle will be nonstandard analysis and the *hyperreal numbers*, which we did not find elsewhere used in the relevant literature. Most closely related to our work is the theory of stochastic orders [7,8], which, however, offers a variety of odering relations on random variables that are somewhat application-specific, and not generally applicable or recommendable for game-theoretic risk management purposes. Our theory partially adds to this area, where we exhibit some explicit relations below.

This work is organized as follows: Sect. 2 introduces the necessary concepts and preliminary results to construct a total ordering on payoffs that are random variables. Since the construction of games is only the last step based on this fundament, we assume the reader to be familiar with the basic concepts of game-theory at this point. Section 3 defines the ordering relation and gives details on how the ordering can be decided efficiently (algorithmically). Section 4 shows how (and why) games can soundly be defined on top of our ordering relation. Section 5 discusses practical aspects of computation, especially aiming at applications on which we elaborate in Sect. 6. Section 7 draws conclusions and discusses open directions for future research.

2   Preliminaries and Notation

We denote sets and random variables by upper-case letters like X, and write $\operatorname{supp}(X)$ to mean the support of X (roughly speaking, the set of events that occur with nonzero probability). Distribution functions are written as upper-case letters, e.g. F, with the respective density being denoted by the according lower-case letter, e.g., f. The symbol $X \sim F$ means that X has distribution function F. Scalar values, e.g., realizations of a random variable, are denoted by lower-case letters like x. Calligraphic letters denote sets (families) of sets. Vectors and sequences are printed as bold characters.

Let $\mathbb{R}^\infty$ be the set of all infinite sequences $a = (a_n)_{n \in \mathbb{N}}$ over $\mathbb{R}$. On $\mathbb{R}^\infty$, we can define per-element additions and multiplications in the sense of $(a_n)_{n \in \mathbb{N}} + (b_n)_{n \in \mathbb{N}} = (a_n + b_n)_{n \in \mathbb{N}}$ and $(a_n)_{n \in \mathbb{N}} \cdot (b_n)_{n \in \mathbb{N}} = (a_n \cdot b_n)_{n \in \mathbb{N}}$. Obviously, the structure $(\mathbb{R}^\infty, +, \cdot)$ is a ring, but not a field, since there is no multiplicative inverse for the nonzero element $(1, 0, 0, \dots) \in \mathbb{R}^\infty$. It can be cast into a *field* that is even totally ordered, by partitioning $\mathbb{R}^\infty$ appropriately and resorting to equivalence classes. To this end, we take a family $\mathcal{U}$ of subsets of $\mathbb{N}$ with the following properties: (i) $\emptyset \notin \mathcal{U}$, (ii) $(A \subset B, A \in \mathcal{U}) \implies B \in \mathcal{U}$ and (iii) $A, B \in \mathcal{U} \implies A \cap B \in \mathcal{U}$. We call such a set an *ultrafilter*.

On $\mathbb{R}^\infty$, we can use $\mathcal{U}$ to define the equivalence relation $a \sim b \iff \{i \in \mathbb{N} : a_i = b_i\} \in \mathcal{U}$ on $\mathbb{R}^\infty$ and define the set of *hyperreal numbers* (w.r.t. $\mathcal{U}$) as ${}^*\mathbb{R} = \{[a]_\mathcal{U} : a \in \mathbb{R}^\infty\}$. Likewise, $\mathcal{U}$ also induces a total ordering on ${}^*\mathbb{R}$ by defining $a \preceq b \iff \{i \in \mathbb{N} : a_i \leq b_i\} \in \mathcal{U}$. On the set ${}^*\mathbb{R}$, the above operations $+$ and $\cdot$ inherit the necessary properties to make $({}^*\mathbb{R}, +, \cdot, \preceq)$ a totally ordered field (for example, the aforementioned element $(1, 0, 0, \dots)$ would under any $\mathcal{U}$ be equivalent to the neutral element).

On the field ${}^*\mathbb{R}$, one can soundly define a full-fledged calculus, which is nowadays known as *nonstandard analysis* [6]. Unfortunately, the existence of $\mathcal{U}$ can be proven only nonconstructively (as a consequence of Zorn's lemma applied to the Fréchet filter $\{U \subseteq \mathbb{N} : U \text{ is cofinite}\}$), but it can be shown that our results are in fact invariant to the particular $\mathcal{U}$. Thus, existence of $\mathcal{U}$ is fully sufficient, even though we lack an explicit representation and cannot practically do arithmetic in lack of $\mathcal{U}$.

3 Optimal Actions with Random Effects

To follow the common construction of games, we first need a sound understanding of optimality when the objects under comparison are random variables. So, let $X \sim F_X, Y \sim F_Y$ be two random variables with respective distribution functions. As a technical condition, suppose that $X, Y \geq 1$ have finite moments of all orders, so that F_X and F_Y are uniquely characterized by their sequence of moments $(\mathsf{E}(X^n))_{n \in \mathbb{N}}$ and $(\mathsf{E}(Y^n))_{n \in \mathbb{N}}$ (by virtue of a Taylor-series expansion of the respective characteristic functions of X and Y). Existence and finiteness of all moments is immediately assured when X and Y have compact supports; an assumption that we will adopt throughout the rest of this work. This is indeed a mild restriction, since any probability distribution with infinite support can be approximated by another distribution with compact support up to arbitrary (fixed) precision.

More importantly, the representation of a random variable by its sequence of moments makes the sequence $x = (\mathsf{E}(X^n))_{n \in \mathbb{N}} \in {}^*\mathbb{R}$ a natural hyperreal representative of the random variable X. Since $^*\mathbb{R}$ is totally ordered, this ordering automatically applies to the random variables themselves, so the following notation is well-defined.

Definition 1. *We prefer a random variable X over a random variable Y if the sequence of its moments diverges slower. Formally:*
$$X \preceq Y : \iff \boldsymbol{x} = (\mathsf{E}(X^n))_{n \in \mathbb{N}} \preceq (\mathsf{E}(Y^n))_{n \in \mathbb{N}} = \boldsymbol{y} \; on \; ^*\mathbb{R}.$$

Assuming that the supports of X and Y are both compact subsets of $[1, \infty)$, it is not hard to show that for any two such random variables with continuous distribution functions, the moment sequences will eventually diverge in the sense that there is a finite index n_0 for which $\mathsf{E}(X^n) \leq \mathsf{E}(Y^n)$ whenever $n \geq n_0$. The same conclusion is analogously obtained for discrete random variables. Indeed, it is the "overlapping" support that determines which moment sequence grows asymptotically faster. This has a twofold positive effect, since (i) every ultrafilter must contain the (cofinite) set $\{n \in \mathbb{N} : n \geq n_0\}$, and (ii) the decision of whether $X \preceq Y$ or $Y \preceq X$ can be made only by comparing the supports of X and Y. The first of these two observations tells us that the particular ultrafilter $\mathcal{U}$ is indeed irrelevant for the ordering relation, and the second observation gives us a handy tool to decide the ordering efficiently.

Let us collect our findings so far as a lemma:

Lemma 1. *Let $\mathcal{F}$ be the set of probability distributions that are compactly supported on $\mathbb{R}^+$, and assume that all measures in $\mathcal{F}$ are absolutely continuous w.r.t. the Lebesgue measure. On $\mathcal{F}$, there exists a total ordering $\preceq$ that can be decided efficiently. The same assertion holds when all measures in $\mathcal{F}$ are absolutely continuous w.r.t. the counting measure.*

Throughout the rest of this work, let thus $\mathcal{F}$ be the set as defined in Lemma 1.

Proof (Sketch; cf. [4]). The proof is merely a compilation of facts stated up to here: let two random variables $X \sim F_X$ and $Y \sim F_Y$ have their distribution functions be represented by the respective sequence of moments. Whichever support "extends" in magnitude over the other has a faster growing sequence of moments (here, we use the hypothesis that all measures admit density functions as implied by their absolute continuity). Without loss of generality, assume that $\max(\mathrm{supp}(X)) \leq \max(\mathrm{supp}(Y))$, and note that this inequality is trivial to check when the supports of X and Y are known. Then, the set $\{n \in \mathbb{N} : \mathsf{E}(X^n) \geq \mathsf{E}(Y^n)\}$ is cofinite and must therefore be contained in every ultrafilter $\mathcal{U}$ on $\mathbb{R}$. Thus, the total ordering induced by $\mathcal{U}$ applies to X and Y in a way that is in fact independent of the particular $\mathcal{U}$.

The $\preceq$-relation also enjoys a physical meaning that will be useful in our application to risk management in Sect. 6. If one of the distributions is strongly skewed but has a fat tail, a crucial issue of the $\prec$-relation is exposed. Figure 1a shows two possible distributions according to two possible actions $i_1, i_2 \in PS_1$, with random effects captured by the distributions F_1, F_2, respectively. Action $i_1 \in PS_1$ gives the right-skewed distribution F_1 that leads to lower damage on average, compared to the more narrow distribution F_2 arising from taking the alternative move $i_2 \in PS_1$ in the game. Here, the overlap of supports would clearly make F_2 preferable over F_1, although F_1 appears to be the better choice for obvious reasons.

This counter-intuitive outcome can be avoided by fixing an acceptance threshold $0 < \alpha < 1$, and cutting off the payoff distributions at the α-quantile. Besides correcting the paradoxical $\preceq$-preference of actions, observe that this also puts infinitely supported distributions directly into our set $\mathcal{F}$. This procedure is not only a handy tool but agrees with common practice in risk management. Furthermore, if the distributions are estimated from observed data, this helps avoiding results influenced by outliers.

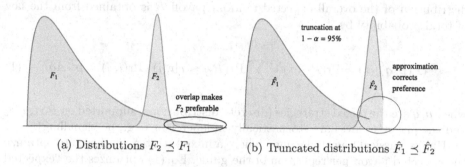

(a) Distributions $F_2 \preceq F_1$ (b) Truncated distributions $\hat{F}_1 \preceq \hat{F}_2$

Fig. 1. Correcting paradoxical comparisons by quantile-based approximations

The practical determination of α is a yet open issue, even the decision about whether or not a tail-cut is advisable. If a practical risk acceptance threshold is available, then one could set the α-quantile to a value so that the expected damage above the α-quantile is less than the acceptable residual risk (which is usually covered by insurances).

Asking for a physical interpretation of the $\preceq$-relation, it is not difficult (cf. [4]) to establish the following result:

Theorem 1. *Let* $X \sim F_1, Y \sim F_2$, *where* $F_1, F_2 \in \mathcal{F}$. *If* $F_1 \preceq F_2$, *then there exists a threshold* $x_0 \in supp(X) \cup supp(Y)$ *so that for every* $x \geq x_0$, *we have* $\Pr(X_1 > x) \leq \Pr(X_2 > x)$.

Proof (Sketch). This easily follows from the observation that $X \preceq Y$ cannot hold unless at some point x_0, we must have $f_X \leq f_Y$. Integrating from x_0 until ∞ then establishes the claim.

From a purely theoretical viewpoint, Theorem 1 exhibits the standard stochastic ordering [7,8] as a subset of $\preceq$. Other orders, such as stochastic integral orders, conversely, include $\preceq$ as a special case.

4 Games with Probability-Distribution-Valued Payoffs

Towards defining game-theory on grounds of the ordering $\preceq$, note that the ordering induces a topology on $^*\mathbb{R}$, so that we can soundly speak about continuous payoff functions. In a matrix game, we would have two players with finite pure strategy spaces PS_1, PS_2, and a payoff structure A that is now a matrix of random variables from $\mathcal{F}$. During the game-play, each player takes its actions at random, which determines a row and column for the payoff distribution $F_{ij} \in \mathcal{F}$. Thus, the payoff matrix is $A \in \mathcal{F}^{n \times m}$, where $n = |PS_1|, m = |PS_2|$. Repeating the game, each round delivers a different random payoff $R_{ij} \sim F_{ij}$ whose distribution is conditional on the chosen scenario $i \in PS_1, j \in PS_2$. Thus, we have the function $F_{ij}(r) = \Pr(R_{ij} \leq r | i, j)$. By playing mixed strategies, the distribution of the overall expected random payoff R is obtained from the law of total probability by

$$(F(\boldsymbol{p}, \boldsymbol{q}))(r) = \Pr(R \leq r) = \sum_{i,j} \Pr(R_{ij} \leq r | i, j) \cdot \Pr(i, j) = \boldsymbol{p}^T A \boldsymbol{q}, \qquad (1)$$

when $\boldsymbol{p}, \boldsymbol{q}$ are the mixed strategies (discrete distributions) supported on PS_1, PS_2 and the player's moves are stochastically independent (e.g., no signalling).

Unlike classical repeated games, where a mixed strategy is chosen to optimize the expected payoff per repetition of the game, Eq. (1) optimizes the "expected payoff distribution" $F(\boldsymbol{p}, \boldsymbol{q})$ for every repetition of the game. It is in that sense static, as it resembles the way in which a normal game rewards the players by numeric payoffs, whereas our form of game rewards the players with a random revenue following an optimal distribution "on average".

Let us illustrate this with a little example: consider a museum director (player 1) intending to prevent his two most expensive paintings from theft by player 2. He thus instructs a guard to check the rooms with these pictures, i.e. $PS_1 = \{1, 2\}$. Similarly the intruder is assumed to make its way towards these rooms and $PS_2 = \{1, 2\}$. Obviously, the damage for the museum is low if they coordinate (depending on how the thief entered, maybe a broken door), so the loss distributions F_{11} and F_{22} could be an exponential distribution with small mean. Otherwise, the loss might be enormous and distributions F_{12} and F_{21} could Gamma distributed around 100,000. For each attack of player 2 they choose a scenario (i, j) and hence a random variable $R_{ij} \sim F_{ij}$.

In standard game theory with real-valued utility functions, the existence of Nash-equilibria is assured in this situation:

Theorem 2 ([1,2]). *If for a game in normal form, the strategy spaces are non-empty compact subsets of a metric space, and the utility-functions are continuous (w.r.t the metric), then at least one Nash-equilibrium in mixed strategies exists.*

Using the properties of hyperreal numbers it can be concluded that $F(\boldsymbol{p}, \boldsymbol{q})$ as a function of the mixed strategies is continuous w.r.t. the ordering topology on $\mathcal{F}$. Thus, Theorem 2 applies [4] and delivers the following result:

Theorem 3. *Every finite game whose payoffs are random variables that are compactly supported on* $\mathbb{R}^+$ *has at least one Nash-equilibrium in mixed strategies.*

We stress that the assumptions on $\mathcal{F}$ made by Theorem 1 imply that in the game, we never combine continuous and discrete probability distributions in the same payoff structure, as a mix of the two would yield singular distributions w.r.t. the Lebesgue measure. This restriction avoids pathological cases in which we would attempt to compare categorial outcomes to continuous outcomes (although a comparison between discrete and continuous random variables is indeed theoretically possible using the moment sequence representation).

As for the meaning of Nash-equilibria under our setting, recall that we assumed the payoffs to be supported on the nonnegative real line $\mathbb{R}^+$. This, in connection with Theorem 1 means that an $\preceq$-optimal behavior will favour strategies whose effects are concentrated closer to zero. Since the payoffs in risk management are considered as damage, optimizing our behavior in terms of $\preceq$ as an equilibrium gives us a loss distribution whose mass accumulates in the closest possible proximity of "zero damage".

A slight catch in our construction is the loss of the straightforward definitions for zero-sum games or similar. For example, $\boldsymbol{x} = (\mathsf{E}(X^n))_{n\in\mathbb{N}}$ may represent a distribution for the random payoff X, but $-\boldsymbol{x} = (-\mathsf{E}(X^n))_{n\in\mathbb{N}}$ certainly does not correspond to any probability distribution any more. Thus, the usual definition of two-person zero-sum games by taking the opponent's revenue negative does not transfer to our generalized setting here (not surprisingly at a second glance, since adding opposite random payoffs will not necessarily add up to zero in every round).

5   Practical and Algorithmic Aspects

So far, everything as been established nonconstructively, so let us turn to questions of how to put things to practice here. For that sake, we go back to the ordering relation and work our way forward up to how Nash-equilibria can be computed. Observe that our setting is generalized in the sense that we deal with distributions rather than real numbers, but also restricted, since the representatives (hyperreal numbers) have an arithmetic in which we cannot practically carry out any operations in lack of an explicit ultrafilter $\mathcal{U}$. We will show how to bypass this difficulty when computing equilibria.

5.1   Modeling Payoffs and Deciding $\preceq$

Since the payoff distribution models have not been restricted beyond assuming compact support, a practical obstacle is first the question where to get the distribution models from. A simple solution is learning the payoff models from available data by kernel density estimation using the Epanechnikov kernel (this one is our choice here as it is compactly supported, thus making the resulting kernel density estimator also compactly supported). More precisely, consider a fixed scenario $(i, j) \in PS_1 \times PS_2$, for which we seek to model the payoff distribution F_{ij} in the game-matrix $\boldsymbol{A}$. Using the Epanechnikov kernel function $K(u) = \frac{3}{4}(1 - u^2)\mathbf{I}_{\{|u|\leq 1\}}(u)$ (where $\mathbf{I}$ is an indicator function), the density function $\hat{f}_{ij}$ for the payoff distribution F_{ij} can be estimated from historial/past recorded payoffs $r_1, r_2, r_3, \ldots, r_k$

$$\hat{f}_{ij}(r) = \frac{1}{k \cdot h} \sum_{i=1}^{k} K\left(\frac{r - x_i}{h}\right), \tag{2}$$

where h is the bandwidth parameter, whose estimation is commonly up to various heuristics known from statistics (e.g., the software suite R (www.r-project.org) has several such methods implemented). A known theoretical necessary constraint on h is given by Nadaraja's theorem [9] that assures convergence of $\hat{f}_{ij}(r)$ in probability towards an (unknown) uniformly continuous limiting distribution, provided that $h(n) = \frac{c}{n^\alpha}$ for (any) two constants $c > 0$ and $0 < \alpha < 1/2$.

If a sufficient lot of data is available on the effects of the actions $(i, j) \in PS_1 \times PS_2$, then (2) supplies us with a compactly supported probability measure that is directly usable to decide the ordering relation $\preceq$ on two such distributions as follows: let $\hat{f}_X$ and $\hat{f}_Y$ (both representing different scenarios $(i_1, j_1) \in PS_1 \times PS_2$ and $(i_2, j_2) \in PS_1 \times PS_2$) be given, and write $X \sim \hat{F}_X, Y \sim \hat{F}_Y$ for the random variables and respective empirical distribution functions. Assume that $\hat{f}_X$ was constructed from n_1 (ordered) data points $x_1 \leq \ldots \leq x_{n_1} \in \mathbb{R}$ and $\hat{f}_Y$ arose from n_2 (ordered) data points $y_1 \leq y_2 \leq \ldots \leq y_{n_2} \in \mathbb{R}$. Equation (2) implies the supports to be $\mathrm{supp}(X) = \bigcup_{i=1}^{n_1}[x_i - h_1, x_i + h_1]$ and $\mathrm{supp}(Y) = \bigcup_{i=1}^{n_1}[y_i - h_2, y_i + h_2]$.

Deciding whether $X \preceq Y$ or $Y \preceq X$ is then a trivial matter of looking which support overlaps the other (in which case the respective moment sequences ultimately exceed one another). The algorithm is simple:

1. $i \leftarrow n_1, j \leftarrow n_2$.
2. if $i = 0$ or $x_i + h_1 < y_j + h_2$ then return "$X \preceq Y$".
3. else if $j = 0$ or $x_i + h_1 > y_j + h_2$ then return "$Y \preceq X$".
4. else, abandon the two points by setting $i \leftarrow i - 1, j \leftarrow j - 1$ and go back to step 2.

The conditions $i = 0$ or $j = 0$ are meant to cover cases in which all observations from one model have been used up, in which case this would be the $\preceq$-optimal one. This means that the decision of $\preceq$ on models of the form (2) is meaningful only between models that are based on equal amounts of observations.

5.2   Computing Nash-Equilibria

One limitation induced by our use of hyperreal calculus is the inability to do arbitrary arithmetic in $\mathcal{F}$ being a subset of $^*\mathbb{R}$, as we lack an explicit model of the necessary ultrafilter $\mathcal{U}$. As a consequence, most of the (more sophisticated) algorithms to compute Nash equilibria are inapplicable, but fictitious play (FP) remains doable. Leaving the full details of FP aside here, the important observation is that the algorithm merely requires selecting a $\preceq$-maximum or -minimum from a finite set, which is easy and efficient by the above procedure to decide $\preceq$. As in ordinary FP players iteratively choose a best response to optimize their payoff (i.e. find the optimal payoff distribution) and estimate their mixture based on these choices.

The remaining details and foundations of calculus that establish the convergence of FP are inherited from the reals to the hyperreals, so that fictitious play is guaranteed to converge under known conditions. However, a subtle issue must be stressed here: the theoretical argument to lift convergence of FP from games over $\mathbb{R}$ to games over $^*\mathbb{R}$ is the transfer principle (more generally, Łos' theorem). This delivers a first-order logic formulation of convergence in terms of stating success of FP along a sequence of hyperreal integers. Unfortunately, as an inspection of the original convergence proof of J. Robinson [6] reveals, convergence sets in after an (in the hyperreal sense) *infinite* (hyper)integer number of iterations. Thus, FP *cannot* be used as usual to solve these games (although in theory, it perfectly works). To bypass this theoretical obstacle, we must approximate a distribution being represented by an infinitude of moments, by a finite vector of moments or other quantities that help us decide preferences on distributions. In any case, the approximation should become more accurate at the tails of the distribution, to retain the ordering as good as possible. A promising candidate for continuous payoff distributions are polynomial approximations, say Taylor-series expansions. The theoretical details of this are subject of ongoing investigations and will be reported in a follow-up article (this will be "part two" of [4]).

In our case, the FP process is applied to the two-person matrix game with payoff structures $(A, -A)$, although $-A$ when defined over the hyperreals *does not* correspond to a proper payoff structure in our setting. Intuitively, this issue can be resolved by switching to a strategically equivalent game in which all payoffs are strictly positive (w.r.t. the ordering on $^*\mathbb{R}$) by shifting all payoffs in the matrix $-A$ by a sufficiently large amount towards becoming positive. In any case, however, since the convergence is established within the set of hyperreal numbers, the arguments that establish the proof of convergence remain intact (although they do not directly deliver a practical algorithm unlike over the reals). Therefore, FP converges towards an equilibrium (p^*, q^*), with the "value" of the game being given by (1) as the distribution $F(p^*, q^*)(r)$, although the correspondence of the relevant mathematical objects to physical entities (payoffs, players, etc.) is lost.

6   Applications to Risk Management

As announced in the introduction, this work has been motivated by an application of game-theory to risk management in utility networks. In these settings, the "payoff" from the game is usually quantified in terms of expected damage, so that we seek to take actions towards minimizing the damage w.r.t. $\preceq$-relation while an attacker tries to maximize it.

The physical interpretation of the $\preceq$-relation given in Theorem 1 in Sect. 3 is particularly relevant for risk management due to its interpretation: *if $F_1 \preceq F_2$, then "extreme problems" are less likely to occur under F_1 than under F_2.* A slight refinement to Theorem 1 applies if the distributions are cut off, in which case the "extreme problems" refer only to events up to a likelihood of at most $1 - \alpha$.

Another way of looking at the meaning of $\preceq$ in risk management can be derived from the moment sequences: for distributions in $\mathcal{F}$, the decision can be made on the average damage (first moment). Upon equal first moments, the $\preceq$-preferred action is the one whose outcome is more certain in the sense of having less variance (second moment). If the first two moments between X and Y agree, then the better action is the one whose effect-distribution is more skewed towards lower damage, etc. (Fig. 1 shows an example of that case). Our discussion following Theorem 3 further substantiates the positive effect for risk management, as equilibria in the $\preceq$-sense leads to random effects with more likely less damage (the probability mass assigned by $F(p^*, q^*)$ under the equilibrium (p^*, q^*) is by the optimization somewhat squeezed towards zero, since the damage is never negative).

Compiling the usual benchmarks of risk management, say the common quantitative formula "risk = damage × likelihood", is a simple matter of computing moments from the payoff distribution as given by (1). Going beyond the above rule of thumb is then a mere matter of computing higher order moments or other quantities of interest from the equilibrium payoff distribution $F(p^*, q^*)$.

7 Conclusions and Outlook

Various directions have been left unexplored in this work, such as details and issues of comparing random variables of different nature (discrete vs. continuous) that live in the same metric space (where a comparison could be meaningful). Furthermore, comparing deterministic to random outcomes is another aspect to receive attention along future research. Further generalizations are possible (and most likely relevant for practical applications) in the area of extreme value modeling. Payoff distributions with fat tails that model extreme, perhaps even catastrophic, effects of certain actions usually violate our assumption on compactness (and hence boundedness) of the support. It is indeed possible to generalize the $\preceq$-relation to such distributions, but this extension comes at the cost of loosing the simple decidability procedure as described in Sect. 5.1. Further practical issues (limitations) arise from the restriction to avoid algebra beyond using the ordering to compute equilibria. Better versions of fictitious play or the exploration of alternative techniques to compute Nash equilibria inside the hyperreals are more intricate issues of future considerations.

Acknowledgment. This work was supported by the European Commission's Project No. 608090, HyRiM (Hybrid Risk Management for Utility Networks) under the 7th Framework Programme (FP7-SEC-2013-1).

References

1. Fudenberg, D., Tirole, J.: Game Theory. MIT Press, London (1991)
2. Glicksberg, I.L.: A further generalization of the Kakutani fixed point theorem, with application to nash equilibrium points. Proc. Am. Math. Soc. **3**, 170–174 (1952)
3. von Neumann, J., Morgenstern, O.: Theory of Games and Economic Behavior. Princeton University Press, Princeton (1944)
4. Rass, S.: On Game-Theoretic Risk Management (Part One) - Towards a Theory of Games with Payoffs that are Probability-Distributions. ArXiv e-prints, June 2015
5. Robert, C.P.: The Bayesian Choice. Springer, New York (2001)
6. Robinson, A.: Nonstandard Analysis. Studies in Logic and the Foundations of Mathematics. North-Holland, Amsterdam (1966)
7. Stoyan, D., Müller, A.: Comparison Methods for Stochastic Models and Risks. Wiley, Chichester (2002)
8. Szekli, R.: Stochastic Ordering and Dependence in Applied Probability. Lecture Notes in Statistics, vol. 97. Springer, Heidelberg (1995)
9. Wand, M.P., Jones, M.C.: Kernel Smoothing. Chapman & Hall/CRC, London (1995)

Incentive Schemes for Privacy-Sensitive Consumers

Chong Huang[1]([⊠]), Lalitha Sankar[1], and Anand D. Sarwate[2]

[1] Arizona State University, Tempe, USA
{chong.huang,lalithasankar}@asu.edu
[2] Rutgers, The State University of New Jersey, New Brunswick, USA
anand.sarwate@rutgers.edu

Abstract. Businesses (*retailers*) often offer personalized advertisements (*coupons*) to individuals (*consumers*). While proving a customized shopping experience, such coupons can provoke strong reactions from consumers who feel their privacy has been violated. Existing models for privacy try to *quantify* privacy risk but do not capture the subjective experience and heterogeneous *expression* of privacy-sensitivity. We use a Markov decision process (MDP) model for this problem. Our model captures different consumer privacy sensitivities via a time-varying state, different coupon types via an action set for the retailer, and a cost for perceived privacy violations that depends on the action and state. The simplest version of our model has two states ("Normal" and "Alerted"), two coupons (targeted and untargeted), and consumer behavior dynamics known to the retailer. We show that the optimal coupon-offering strategy for a retailer that wishes to minimize its expected discounted cost is a stationary threshold-based policy. The threshold is a function of all model parameters: the retailer offers a targeted coupon if their belief that the consumer is in the "Alerted" state is below the threshold. We extend our model and results to consumers with multiple privacy-sensitivity states as well as coupon-dependent state transition probabilities.

Keywords: Privacy · Markov decision processes · Retailer-consumer interaction · Optimal policies

1 Introduction

Programs such as retailer "loyalty cards" allow companies to automatically track a customer's financial transactions, purchasing behavior, and preferences. They can then use this information to offer customized incentives, such as discounts on related goods. Consumers may benefit from retailer's knowledge by using more of these targeted discounts or coupons while shopping. However, the coupon offer may imply that the retailer has learned something sensitive or private about the consumer (for example, a pregnancy [1]) – such violations may make consumers skittish about purchasing from such retailers.

© Springer International Publishing Switzerland 2015
MHR Khouzani et al. (Eds.): GameSec 2015, LNCS 9406, pp. 358–369, 2015.
DOI: 10.1007/978-3-319-25594-1_21

However, modeling the privacy-sensitivity of a consumer is not always straight-forward: widely-studied models for quantifying privacy risk using differential privacy [2] or information theory [3] do not capture the subjective experience and heterogeneous *expression* of consumer privacy. We introduce a framework to model the consumer-retailer interaction problem and better understand how retailers can develop coupon-offering policies that balances their revenue objectives while being sensitive to consumer privacy concerns. The main challenge for the retailer is that the consumer's responses to coupons are not known *a priori*; furthermore, consumers do not "add noise" to their purchasing behavior as a mechanism to stay private. Rather, the offer of a coupon may provoke a reaction from the consumer, ranging from "indifferent" through "partially concerned" to "creeped out." This reaction is mediated by the consumer's sensitivity level to privacy violations, and it is these levels that we seek to model via a Markov decision process. In particular, the sensitivity of the consumers are often revealed indirectly to the retailer through their purchasing patterns. We capture these aspects in our model and summarize our main contributions below.

Main Contributions: We propose a partially-observed Markov decision process (POMDP) model for this problem in which the consumer's state encodes their privacy sensitivity, and the retailer can offer different levels of privacy-violating coupons. The simplest instance of our model is one with two states for the consumer, denoted as "Normal" and "Alerted," and two types of coupons: untargeted *low privacy* (LP) or targeted *high privacy* (HP). At each time, the retailer may offer a coupon and the consumer transitions from one state to another according to a Markov chain that is independent of the offered coupon. The retailer suffers a cost that depends both on the type of coupon offered and the state of the consumer. The costs reflect the advantage of offering targeted HP coupons relative to untargeted LP ones while simultaneously capturing the risk of doing so when the consumer is already "Alerted".

Under the assumption that the retailer (via surveys or prior knowledge) knows the statistics of the consumer Markov process, i.e., the likelihoods of becoming "Alerted" and staying "Alerted", and a belief about the initial consumer state, we study the problem of determining the optimal coupon-offering policy that the retailer should adopt to minimize the long-term discounted costs of offering coupons. We show that the optimal stationary policy exists and it is a threshold on the probability of the consumer being alerted; this threshold is a function of all the model parameters. The simple model above is extended to multiple consumer states and coupon-dependent transitions. We model the latter via two Markov processes for the consumer, one for each type (HP or LP) of coupon such that a persnickety consumer who is easily "Alerted" will be more likely to do so when offered an HP (relative to LP) coupon. Our structural result (a stationary optimal policy) holds for multiple states and coupon-dependent transitions. While the MDP model used in this paper is simple, its application to the problem of privacy cost minimization with privacy-sensitive consumers is novel. In the conclusion we describe several other interesting avenues for future work. Our results use many fundamental tools and techniques from the theory of

MDPs through appropriate and meaningful problem modeling. We briefly review the related literature in consumer privacy studies as well as MDPs.

Related Work: Several economic studies have examined consumer's attitudes towards privacy via surveys and data analysis including studies on the benefits and costs of using private data (e.g., Aquisti and Grossklags in [4]). On the other hand, formal methods such as differential privacy are finding use in modeling the value of private data for market design [5] and for the problem of partitioning goods with private valuation function amongst the agents [6]. In these models the goal is to elicit private information from individuals. Venkitasubramaniam [7] recently used an MDP model to study data sharing in control systems with time-varying state. He explicitly quantifies privacy risk in terms of equivocation, an information-theoretic measure, and his objective is to minimize the weighted sum of the utility (benefit) that the system achieves by sharing data (e.g., with a data collector) and the privacy risk. In our work we do not quantify *privacy risk* directly; instead the retailer learns about the *privacy-sensitivity* of the consumer indirectly through the cost feedback. Our MDP's state space is the privacy sensitivity of the consumer. To the best of our knowledge, models capturing this aspect of consumer-retailer interactions and the related privacy issues have not been studied before; in particular, our work focuses on explicitly considering the consequence to the retailer of the consumers' awareness of privacy violations.

Markov decision processes (MDPs) have been widely used for decades across many fields [8]; in particular, our formal model is related to problems in control with communication constraints [9,10] where state estimation has a cost. However, our costs are action and state dependent and we consider a different optimization problem. Classical state-search problems [11,12] also have optimal threshold policies; however the retailer's objective in our model is to minimize cost, and not necessarily estimate the consumer state. Our model is most similar to Ross's model of product quality control with deterioration [13], which was more recently used by Laourine and Tong to study the Gilbert-Elliot channel in wireless communications [14], in which the channel has two states and the transmitter has two actions (to transmit or not). We cannot apply their results directly due to our different cost structure, but use ideas from their proofs. Furthermore, we go beyond these works to study privacy-utility tradeoffs in consumer-retailer interactions with more than two states and action-dependent transition probabilities. We apply more general MDP analysis tools to address our formal behavioral model for privacy-sensitive consumers.

2   System Model

We model interactions between a retailer and a consumer via a discrete-time system (see Fig. 1). At each time t, the consumer has a discrete-valued state and the retailer may offer one of two coupons: high privacy risk (HP) or low privacy risk (LP). The consumer responds by imposing a cost on the retailer that depends on the coupon offered and its own state. For example, a consumer who is "alerted" (privacy-aware) may respond to an HP coupon by refusing to shop at the retailer. The retailer's goal is to decide which type of coupon to offer at each time t to minimize its cost.

2.1   Consumer Model

Modeling Assumption 1 *(Consumer's State). We assume the consumer is in one of a finite set of states that determine their response to coupons – each state corresponds to a type of consumer behavior in terms of purchasing. The consumer's state evolves according to a Markov process.*

For this paper, we primarily focus on the two-state case; the consumer may be Normal or Alerted. Later we will extend this model to multiple consumer states. The consumer state at time t is denoted by $G_t \in \{\text{Normal}, \text{Alerted}\}$. If a consumer is in Normal state, the consumer is very likely to use coupons to make purchases. However, in the Alerted state, the consumer is less likely to use coupons, since it is more cautious about revealing information to the retailer. The evolution of the consumer state is modeled as an infinite-horizon discrete time Markov chain (Fig. 1). The consumer starts out in a random initial state unknown to the retailer and the transition of the consumer state is independent of the action of the retailer. A *belief state* is a probability distribution over possible states in which the consumer could be. The belief of the consumer being in Alerted state at time t is denoted by p_t. We define $\lambda_{N,A} = Pr[G_t = \text{Alerted}|G_{t-1} = \text{Normal}]$ to be the transition probability from Normal state to Alerted state and $\lambda_{A,A} = Pr[G_t = \text{Alerted}|G_{t-1} = \text{Alerted}]$ to be the probability of staying in Alerted state when the previous state is also Alerted. The transition matrix Λ of the Markov chain can be written as

$$\Lambda = \begin{pmatrix} 1 - \lambda_{N,A} & \lambda_{N,A} \\ 1 - \lambda_{A,A} & \lambda_{A,A} \end{pmatrix}. \tag{1}$$

We assume the transition probabilities are known to the retailer; this may come from statistical analysis such as a survey of consumer attitudes. The one step transition function, defined by $T(p_t) = (1 - p_t)\lambda_{N,A} + p_t\lambda_{A,A}$, represents the belief that the consumer is in Alerted state at time $t + 1$ given p_t, the Alerted state belief at time t.

Modeling Assumption 2 *(State Transitions). Consumers have an inertia in that they tend to stay in the same state. Moreover, once consumers feel their privacy is violated, it will take some time for them to come back to Normal state.*

To guarantee Assumption 2 we consider transition matrices in (1) satisfying $\lambda_{A,A} \geq 1 - \lambda_{A,A}$, $1 - \lambda_{N,A} \geq \lambda_{N,A}$, and $\lambda_{N,A} \geq 1 - \lambda_{A,A}$. Thus, by combining the above three inequalities, we have $\lambda_{A,A} \geq \lambda_{N,A}$.

2.2   Retailer Model

At each time t, the retailer can take an *action* by offering a coupon to the consumer. We define the action at time t to be $u_t \in \{\text{HP}, \text{LP}\}$, where HP denotes

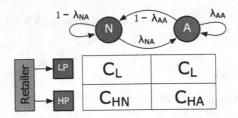

Fig. 1. Markov state transition model for a two-state consumer.

offering a high privacy risk coupon (e.g. a targeted coupon) and LP denotes offering a low privacy risk coupon (e.g. a generic coupon). The retailer's utility is modeled by a *cost* (negative revenue) which depends on both the consumer's state and the type of coupon being offered. If the retailer offers an LP coupon, it suffers a cost C_L independent of the consumer's state: offering LP coupons does not reveal anything about the state. However, if the retailer offers an HP coupon, then the cost is C_{HN} or C_{HA} depending on whether the consumer's state is Normal or Alerted. Offering an HP (high privacy risk, targeted) coupon to a Normal consumer should incur a low cost (high reward), but offering an HP coupon to an Alerted consumer should incur a high cost (low reward) since an Alerted consumer is privacy-sensitive. Thus, we assume $C_{HN} \leq C_L \leq C_{HA}$.

Under these conditions, the retailer's objective is to choose u_t at each time t to minimize the total cost incurred over the entire time horizon. The HP coupon reveals information about the state through the cost, but is risky if the consumer is alerted, creating a tension between cost minimization and acquiring state information.

2.3   The Minimum Cost Function

We define $C(p_t, u_t)$ to be the expected cost acquired from an individual consumer at time t where p_t is the probability that the consumer is in Alerted state and u_t is the retailer's action:

$$C(p_t, u_t) = \begin{cases} C_L & \text{if } u_t = \text{LP} \\ (1 - p_t)C_{HN} + p_t C_{HA} & \text{if } u_t = \text{HP} \end{cases} . \qquad (2)$$

Since the retailer knows the consumer state from the incurred cost only when an HP coupon is offered, the state of the consumer may not be directly observable to the retailer. Therefore, the problem is actually a Partially Observable Markov Decision Process (POMDP) [15].

We model the cost of violating a consumer's privacy as a short term effect. We adopt a discounted cost model with discount factor $\beta \in (0, 1)$. At each time t, the retailer has to choose which action u_t to take in order to minimize the expected discounted cost over infinite time horizon. A policy π for the retailer is a rule that selects a coupon to offer at each time. Given that the belief of the consumer being in Alerted state at time t is p_t and the policy is π, the infinite-horizon discounted cost starting from t is

$$V_\beta^{\pi,t}(p_t) = \mathbb{E}_\pi \left[\sum_{i=t}^\infty \beta^i C(p_i, A_i) | p_t \right], \tag{3}$$

where $\mathbb{E}_\pi$ indicates the expectation over the policy π. The objective of the retailer is equivalent to minimizing the discounted cost over all possible policies. We define the minimum cost function starting from time t over all policies to be

$$V_\beta^t(p_t) = \min_\pi V_\beta^{\pi,t}(p_t) \text{ for all } p_t \in [0,1]. \tag{4}$$

We define p_{t+1} to be the belief of the consumer being in Alerted state at time $t+1$. The minimum cost function $V_\beta^t(p_t)$ satisfies the Bellman equation [15]:

$$V_\beta^t(p_t) = \min_{u_t \in \{\mathsf{HP},\mathsf{LP}\}} \{ V_{\beta,u_t}^t(p_t) \} \tag{5}$$

$$V_{\beta,u_t}^t(p_t) = \beta^t C(p_t, u_t) + V_\beta^{t+1}(p_{t+1}|p_t, u_t). \tag{6}$$

An optimal policy is *stationary* if it is a deterministic function of states, i.e., the optimal action at a particular state is the optimal action in this state at all times. We define $\mathcal{P} = \{[0,1]\}$ to be the belief space and $\mathcal{U} = \{\mathsf{LP}, \mathsf{HP}\}$ to be the action space. In the context of our model, the optimal stationary policy is a deterministic function mapping $\mathcal{P}$ into $\mathcal{U}$. Since the problem is an infinite-horizon, finite state, and finite action MDP with discounted cost, there exists an optimal stationary policy [16] π^* such that starting from time t,

$$V_\beta^t(p_t) = V_\beta^{\pi^*,t}(p_t). \tag{7}$$

We only consider the optimal stationary policy because it is tractable and achieves the same minimum cost as any optimal non-stationary policy.

By (5) and (6), the minimum cost function evolves as follows: if an HP coupon is offered at time t, the retailer can perfectly infer the consumer state based on the incurred cost. Therefore,

$$V_{\beta,\mathsf{HP}}^t(p_t) = \beta^t C(p_t, \mathsf{HP}) + (1 - p_t)V_\beta^{t+1}(\lambda_{N,A}) + p_t V_\beta^{t+1}(\lambda_{A,A}). \tag{8}$$

If an LP coupon is offered at time t, the retailer cannot infer the consumer state from the cost since both Normal and Alerted consumer impose the same cost C_L. Hence, the discounted cost function can be written as

$$V_{\beta,\mathsf{LP}}^t(p_t) = \beta^t C(p_t, \mathsf{LP}) + V_\beta^{t+1}(p_{t+1}) = \beta^t C_L + V_\beta^{t+1}(T(p_t)). \tag{9}$$

Correspondingly, the minimum cost function is given by

$$V_\beta^t(p_t) = \min\{ V_{\beta,\mathsf{LP}}^t(p_t), V_{\beta,\mathsf{HP}}^t(p_t) \}. \tag{10}$$

3   Optimal Stationary Policies

The first main result is a theorem providing the optimal stationary policy for the two-state basic model in Sect. 2.

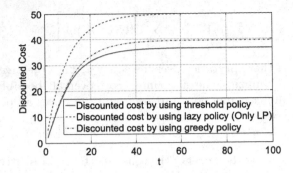

Fig. 2. Discounted cost from by using different decision policies

Theorem 1. *There exists a threshold* $\tau \in [0,1]$ *such that the following policy is optimal:*

$$\pi^*(p_t) = \begin{cases} \text{LP} & \text{if } \tau \leq p_t \leq 1 \\ \text{HP} & \text{if } 0 \leq p_t \leq \tau \end{cases}. \tag{11}$$

More precisely, assume that $\delta = C_{HA} - C_{HN} + \beta(V_\beta(\lambda_{A,A}) - V(\lambda_{N,A}))$,

$$\tau = \begin{cases} \frac{C_L - (1-\beta)(C_{HN} + \beta V_\beta(\lambda_{N,A}))}{(1-\beta)\delta} & T(\tau) \geq \tau \\ \frac{C_L + \beta\lambda_{N,A}(C_{HA} + \beta V_\beta(\lambda_{A,A})) - (1-\beta(1-\lambda_{N,A}))(C_{HN} + \beta V_\beta(\lambda_{N,A}))}{(1-(\lambda_{A,A} - \lambda_{N,A})\beta)\delta} & T(\tau) < \tau \end{cases}, \tag{12}$$

where for $\lambda_{N,A} \geq \tau$,

$$V_\beta(\lambda_{N,A}) = V_\beta(\lambda_{A,A}) = C_L/(1-\beta) \tag{13}$$

and for $\lambda_{N,A} < \tau$,

$$V_\beta(\lambda_{N,A}) = (1 - \lambda_{N,A})[C_{HN} + \beta V_\beta(\lambda_{N,A})] + \lambda_{N,A}[C_{HA} + \beta V_\beta(\lambda_{A,A})], \tag{14}$$
$$V_\beta(\lambda_{A,A}) = \min_{n \geq 0}\{G(n)\}, \tag{15}$$

where

$$G(n) = \frac{C_L\frac{1-\beta^n}{1-\beta} + \beta^n[\bar{T}^n(\lambda_{A,A})(C_{HN} + C(\lambda_{N,A})) + T^n(\lambda_{A,A})C_{HA}]}{1 - \beta^{n+1}[\bar{T}^n(\lambda_{A,A})\frac{\lambda_{N,A}\beta}{1-(1-\lambda_{N,A})\beta} + T^n(\lambda_{A,A})]}, \tag{16}$$

$$T^n(\lambda_{A,A}) = \frac{(\lambda_{A,A} - \lambda_{N,A})^{n+1}(1 - \lambda_{A,A}) + \lambda_{N,A}}{1 - (\lambda_{A,A} - \lambda_{N,A})} \tag{17}$$

$$\bar{T}^n(\lambda_{A,A}) = 1 - T^n(\lambda_{A,A}) \tag{18}$$

$$C(\lambda_{N,A}) = \beta\frac{(1 - \lambda_{N,A})C_{HN} + \lambda_{N,A}C_{HA}}{1 - (1 - \lambda_{N,A})\beta}. \tag{19}$$

The full proof of Theorem 1 is in the extended version of this paper [17]. We illustrate our policy's performance by comparing its discounted cost to two other

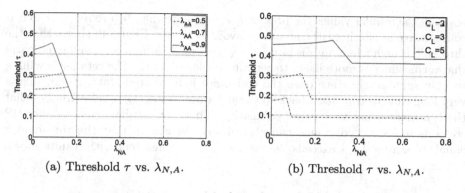

(a) Threshold τ vs. $\lambda_{N,A}$.

(b) Threshold τ vs. $\lambda_{N,A}$.

Fig. 3. Threshold τ vs. β for different values of $\lambda_{A,A}$ and $\lambda_{N,A}$

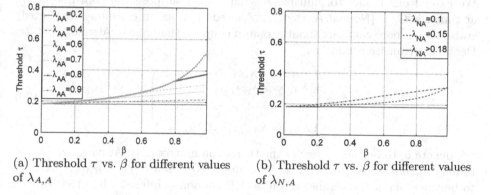

(a) Threshold τ vs. β for different values of $\lambda_{A,A}$

(b) Threshold τ vs. β for different values of $\lambda_{N,A}$

Fig. 4. Threshold τ vs. β for different values of $\lambda_{A,A}$ and $\lambda_{N,A}$

policies: a greedy policy which minimize the instantaneous cost at each decision epoch and a lazy policy which the retailer only offers LP coupons. Figure 2 shows the discounted cost averaged over 1000 independent MDPs versus the time t for these different decision policies. The illustration demonstrates that the proposed threshold policy performs better than the greedy policy and the lazy policy.

Figure 3a shows the optimal threshold τ as a function of $\lambda_{N,A}$ for three fixed choices of $\lambda_{A,A}$. The threshold increases when $\lambda_{N,A}$ is small because the consumer is less likely to transition from Normal to Alerted so the retailer can more safely offer an HP coupon. When $\lambda_{N,A}$ gets larger, the consumer is more likely to transition from Normal to Alerted, so the retailer is more conservative and decreases the threshold for offering an LP coupon. When $\lambda_{N,A} \geq \kappa$, the retailer uses κ as the threshold for offering an HP coupon. With increasing $\lambda_{A,A}$, the threshold τ decreases. On the other hand, for fixed C_{HN} and C_{HA}, Fig. 3b shows that the threshold τ increases as the cost of offering an LP coupon increases, making it more desirable to take a risk and offer an HP coupon. Figure 4 shows the relationship between the discount factor β and the threshold τ as functions of transition probabilities. Figure 4a shows that τ increases as β increases. When β

is small, the retailer values the present rewards more than future rewards so it is conservative in offering HP coupons to avoid low costs. Figure 4b shows that the threshold is high when $\lambda_{A,A}$ is large or $\lambda_{N,A}$ is small. A high $\lambda_{A,A}$ value indicates that a consumer is more likely to remain in Alerted state. The retailer is willing to play aggressively since once the consumer is in alerted state, it can take a very long time to transition back to Normal state. A low $\lambda_{N,A}$ value implies that the consumer is not very privacy sensitive. Thus, the retailer tends to offer HP coupons to reduce cost. One can also observe in Fig. 4b that the threshold τ equals to κ after $\lambda_{N,A}$ exceeds the ratio κ. This is consistent with results shown in Fig. 3.

4 Consumer with Multi-level Alerted States

We extend our model to multiple Alerted states: suppose the consumer state at time t is $G_t \in \{\text{Normal}, \text{Alerted}_1, \dots \text{Alerted}_K\}$, where a consumer in Alerted_k state is even more cautious about targeted coupons than one in Alerted_{k-1} state. Define the transition matrix

$$\Lambda = \begin{pmatrix} \lambda_{N,N} & \lambda_{N,A_1} & \cdots & \lambda_{N,A_K} \\ \lambda_{A_1,N} & \lambda_{A_1,A_1} & \cdots & \lambda_{A_1,A_K} \\ \vdots & \vdots & \ddots & \vdots \\ \lambda_{A_K,N} & \lambda_{A_K,A_1} & \cdots & \lambda_{A_K,A_K} \end{pmatrix}. \tag{20}$$

We denote $\bar{\mathbf{e}}_i$ to be the ith row of the transition matrix (20). At each time t, the retailer can offer either an HP or an LP coupon. We define $C_{HN}, C_{HA_1}, \dots, C_{HA_K}$ to be the costs of the retailer when an HP coupon is offered while the state of the consumer is Normal, $\text{Alerted}_1, \dots, \text{Alerted}_K$, respectively. If an LP coupon is offered, no matter in which state, the retailer gets a cost of C_L. We assume that $C_{HA_K} \geq \cdots \geq C_{HA_1} \geq C_L \geq C_{HN}$. The belief of the consumer being in Normal, $\text{Alerted}_1, \dots, \text{Alerted}_K$ state at time t is defined by $p_{N,t}, p_{A_1,t}, \dots, p_{A_K,t}$, respectively. The expected cost at time t has the following expression:

$$C(\bar{\mathbf{p}}_t, u_t) = \begin{cases} C_L & \text{if } u_t = \text{LP} \\ \bar{\mathbf{p}}_t^T \bar{\mathbf{C}} & \text{if } u_t = \text{HP} \end{cases}, \tag{21}$$

where $\bar{\mathbf{p}}_t = (p_{N,t}, p_{A_1,t}, \dots, p_{A_K,t})^T$ and $\bar{\mathbf{C}} = (C_{HN}, C_{HA_1}, \dots, C_{HA_K})^T$. Assume that the retailer has perfect information about the belief of the consumer state, the cost function evolves as follows: by using an LP coupon at time t,

$$V_{\beta,\text{LP}}^t(\bar{\mathbf{p}}_t) = \beta^t C_L + V_\beta^{t+1}(\bar{\mathbf{p}}_{t+1}) = \beta^t C_L + V_\beta^{t+1}(T(\bar{\mathbf{p}}_t)), \tag{22}$$

where $T(\bar{\mathbf{p}}_t) = \bar{\mathbf{p}}_t^T \Lambda$ is the one step Markov transition function. By using an HP coupon at time t,

$$V_{\beta,\text{HP}}^t(\bar{\mathbf{p}}_t) = \beta^t \bar{\mathbf{p}}_t^T \bar{\mathbf{C}} + \bar{\mathbf{p}}_t^T \begin{pmatrix} V_\beta^{t+1}(\bar{\mathbf{e}}_1) \\ V_\beta^{t+1}(\bar{\mathbf{e}}_2) \\ \vdots \\ V_\beta^{t+1}(\bar{\mathbf{e}}_{K+1}) \end{pmatrix}. \tag{23}$$

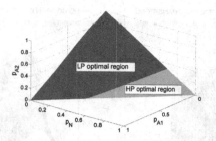

Fig. 5. Optimal policy region for three-state consumer.

Therefore, the minimum cost function is given by (10). In this problem, since the instantaneous costs are nondecreasing with states when the action is fixed and the evolution of belief state is the same for both LP and HP, the existence of an optimal stationary policy with threshold property for finite many states is guaranteed by Proposition 2 in [18]. The optimal stationary policy for a three-state consumer model is illustrated in Fig. 5. For fixed costs, the plot shows the partition of the belief space based on the optimal actions and reveals that offering an HP coupon is optimal when $p_{N,t}$ is high.

5   Consumers with Coupon-Dependent Transition

Generally, consumers' reactions to HP and LP coupons are different. To be more specific, a consumer is likely to feel less comfortable when being offered a coupon on medication (HP) than food (LP). Thus, we assume that the Markov transition probabilities are dependent on the coupon offered. If an LP\HP coupon is offered, the state transition follows the Markov chain

$$\Lambda_{\mathsf{LP}} = \begin{pmatrix} 1 - \lambda_{N,A} & \lambda_{N,A} \\ 1 - \lambda_{A,A} & \lambda_{A,A} \end{pmatrix}, \ \Lambda_{\mathsf{HP}} = \begin{pmatrix} 1 - \lambda'_{N,A} & \lambda'_{N,A} \\ 1 - \lambda'_{A,A} & \lambda'_{A,A} \end{pmatrix}, \tag{24}$$

respectively. According to the model in Sect. 2, $\lambda_{A,A} > \lambda_{N,A}, \lambda'_{A,A} > \lambda'_{N,A}$. Moreover, we assume that offering an HP coupon will increase the probability of transition to or staying at Alerted state. Therefore, $\lambda'_{A,A} > \lambda_{A,A}$ and $\lambda'_{N,A} > \lambda_{N,A}$. The minimum cost function evolves as follows:

$$V^t_{\beta,\mathsf{HP}}(p_t) = \beta^t C(p_t, \mathsf{HP}) + (1 - p_t)V^{t+1}_\beta(\lambda'_{N,A}) + p_t V^{t+1}_\beta(\lambda'_{A,A})$$
$$V^t_{\beta,\mathsf{LP}}(p_t) = \beta^t C_L + V^{t+1}_\beta(p_{t+1}) = \beta^t C_L + V^{t+1}_\beta(T(p_t)),$$

where $T(p_t) = \lambda_{N,A}(1 - p_t) + \lambda_{A,A}p_t$ is the one step transition defined in Sect. 2.

Theorem 2. *Given action dependent transition matrices* Λ_{LP} *and* Λ_{HP}, *the optimal stationary policy has threshold structure.*

A full proof of Theorem 2 is in the extended version of this paper [17]. Figure 6 shows the effect of costs on the threshold τ. The threshold for offering an HP

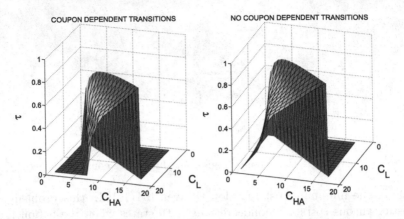

Fig. 6. Optimal τ with/without coupon dependent transition probabilities.

coupon to a consumer with coupon dependent transition probabilities is lower than our original model without coupon-dependent transition probabilities. The retailer can only offer an LP coupon with certain combination of costs; we call this the LP-only region. It can be seen that the LP-only region for the coupon-independent transition case is smaller than that for the coupon-dependent transition case since for the latter, the likelihood of being in an Alerted state is higher for the same costs.

6   Conclusion

We proposed a POMDP model to capture the interactions between a retailer and a privacy-sensitive consumer in the context of personalized shopping. The retailer seeks to minimize the expected discounted cost of violating the consumer's privacy. We showed that the optimal coupon-offering policy is a stationary policy that takes the form of an explicit threshold that depends on the model parameters. In summary, the retailer offers an HP coupon when the Normal to Alerted transition probability is low or the probability of staying in Alerted state is high. Furthermore, the threshold optimal policy also holds for consumers whose privacy sensitivity can be captured via multiple alerted states as well as for the case in which consumers exhibit coupon-dependent transition. Our work suggests several interesting directions for future work: cases where retailer has additional uncertainty about the state, for example due to randomness in the received costs, game theoretic models to study the interaction between the retailer and strategic consumers, and more generally, understanding the tension between acquiring information about the consumers and maximizing revenue.

References

1. Hill, K.: How target figured out a teen girl was pregnant before her father did (2012). http://www.forbes.com/sites/kashmirhill/2012/02/16/how-target-figured-outa-teen-girl-was-pregnant-before-her-father-did/

2. Dwork, C.: Differential privacy. In: van Tilborg, H.C.A., Jajodia, S. (eds.) Encyclopedia of Cryptography and Security, pp. 338–340. Springer, New York (2011)
3. Sankar, L., Kar, S., Tandon, R., Poor, H.V.: Competitive privacy in the smart grid: an information-theoretic approach. In: 2011 IEEE International Conference on Smart Grid Communications (SmartGridComm), pp. 220–225. IEEE (2011)
4. Acquisti, A.: The economics of personal data and the economics of privacy. Background Paper for OECD Joint WPISP-WPIE Roundtable, vol. 1 (2010)
5. Ghosh, A., Roth, A.: Selling privacy at auction. Games Econ. Behav. (2013). Elsevier
6. Hsu, J., Huang, Z., Roth, A., Roughgarden, T., Wu, Z.S.: Private matchings and allocations. arXiv preprint arXiv:1311.2828 (2013)
7. Venkitasubramaniam, P.: Privacy in stochastic control: a markov decision process perspective. In: Proceedings of Allerton Conference, pp. 381–388 (2013)
8. Feinberg, E.A., Shwartz, A., Altman, E.: Handbook of Markov Decision Processes: Methods and Applications. Kluwer Academic Publishers, Boston (2002)
9. Lipsa, G.M., Martins, N.C.: Remote state estimation with communication costs for first-order LTI systems. IEEE Trans. Autom. Control **56**(9), 2013–2025 (2011)
10. Nayyar, A., Başar, T., Teneketzis, D., Veeravalli, V.V.: Optimal strategies for communication and remote estimation with an energy harvesting sensor. IEEE Trans. Autom. Control **58**(9), 2246–2260 (2013)
11. MacPhee, I., Jordan, B.: Optimal search for a moving target. Probab. Eng. Informational Sci. **9**(02), 159–182 (1995)
12. Mansourifard, P., Javidi, T., Krishnamachariy, B.: Tracking of real-valued continuous markovian random processes with asymmetric cost and observation. In: American Control Conference (2015)
13. Ross, S.M.: Quality control under markovian deterioration. Manag. Sci. **17**(9), 587–596 (1971)
14. Laourine, A., Tong, L.: Betting on gilbert-elliot channels. IEEE Trans. Wirel. Commun. **9**(2), 723–733 (2010)
15. Bertsekas, D.P.: Dynamic Programming and Optimal Control, vol. 1, 2, issue 2. Athena Scientific, Belmont (1995)
16. Ross, S.M.: Applied Probability Models with Optimization Applications. Courier Dover Publications, New York (2013)
17. Huang, C., Sankar, L., Sarwate, A.D.: Designing incentive schemes for privacy-sensitive users. ArXiV, Technical report arXiv:1508.01818 [cs.GT], August 2015
18. Lovejoy, W.S.: Some monotonicity results for partially observed markov decision processes. Oper. Res. **35**(5), 736–743 (1987)

Author Index

Printed in the United States
By Bookmasters